ERGONOMICS AT WORK

ERGONOMICS AT WORK

DAVID J. OBORNE

Department of Psychology
University College of Swansea

175 YEARS OF PUBLISHING
1807 1982

JOHN WILEY & SONS LTD

Chichester · New York · Brisbane · Toronto · Singapore

Library of Congress Cataloging in Publication Data:
Oborne, David J.
  Ergonomics at work.
  Bibliography: p.
  Includes index.
  1. Human engineering.   I. Title.
TA166.025   620.8′2   81-14642
ISBN 0-471-10030-7   AACR2

British Library Cataloguing in Publication Data:
Oborne, David J.
  Ergonomics at work.
  1. Human engineering
  1. Title
  620.8′2   T59.7

  ISBN 0 471 10030 7

Photosetting by Thomson Press (India) Ltd., New Delhi
and printed at Page Bros. (Norwich) Limited

*D
620.82
OBO*

Contents

Preface

Ergonomics experienced its birth during the Second World War and it is perhaps fitting that during this period the cry was continuously heard 'Give us the tools, and we will finish the job'. The 'job' then was particularly straightforward: victory over a fearsome aggressor. The present day 'job' is perhaps more diverse: victory over the oppressive forces which continue to make work less productive, less pleasant, less comfortable and less safe.

Today, Churchill's plea is as real as it was during the dark days of the Second World War. Modern work would be almost impossible without adequate tools: a screwdriver to turn a screw, or a hammer to force home a nail. Microscopes allow otherwise invisible objects to become visible; and computers supplement people's cognitive capacities to enable them to perform operations unthinkable a decade or so ago.

Unfortunately, however, machines suffer at least one major drawback, they have to be operated: the screw driver must be turned, the hammer wielded, the microscope adjusted, and the computer programmed. The tools are useless without the energy and decision-making abilities of the human operator. Until the era arrives of machines which are able to carry out intelligent thought, make and take decisions and perform complex actions, the working system needs the help of 'man'. This symbiotic relationship between the human operator and the environment must remain.

This book is concerned with the ways in which the working relationship can be adapted to perform at its most efficient. It will consider how the behaviour of both the operator in the system and the system itself can either enhance or reduce the system's effectiveness. In short it will examine the various ways in which the two major components in the system—'man' and 'environment'—interact and communicate with each other.

In some senses the book is intended to be evangelical in its approach. It does not set out to provide the reader with a comprehensive databank with which he can design an environment. Rather it intends to present the fundamental aspects of man–machine interactions, and to explain ergonomics (and its American sister discipline human factors) and its role within the context of modern day work. It will emphasize aspects of the working situation which need to be considered, and explain the reasons why various solutions are proposed. It then hopes to motivate and to point the reader in the right direction to discover more about the various facets of the man–environment interaction within his own working system.

'Information' and 'communication' are the two keywords to this interactive process; one is transferred by the other. As the reader progresses through this book he should become aware of their importance. Without them no interaction can take place. More importantly if they are of poor quality, then very poor interaction will occur. Either of these lead to inefficient, uncomfortable or unsafe situations which are good for neither the operator nor the employer.

After the first chapter which considers ergonomics in its widest context (its basis and history, its position amongst the disciplines, and its financial value), the following two chapters examine the way in which the human body allows the communicative processes to occur. Chapter 2 considers the communication channels from the environment into the operator, through the senses, while Chapter 3 examines the flow in the opposite direction, through the operator's limbs to the environment.

In the following three chapters, the behavioural aspects of these communication channels will be discussed. First communication between operators (primarily sensory); second communication between the environment and the operators (again primarily sensory) and finally communication between the operators and the environment (mainly using the limbs, but also sensory aspects).

Having considered, in fine detail, these communication requirements, Chapters 7, 8 and 9 discuss the ways of arranging the environment to the operator's benefit—again to enhance the communication channels. In Chapter 7 the operator's immediate work area—perhaps the machine, the console or even the writing pad—is investigated. Chapter 8 then discusses the most appropriate ways of building up the workplace—which consists of many machines and men—again to enhance the communication. Finally, and possibly as an extension of Chapter 8, Chapter 9 considers the function and value of appropriate seating and posture at work.

At this point, then, the communication channels are open and are efficient. However they do not operate *in vacuo*. Their efficiency can be affected, sometimes radically, by the environment in which the operator is expected to perform. Four environmental aspects are considered in Chapters 10 and 11: the vibration, the noise, the temperature, and the illumination. Although it is stressed that these are by no means the only environments at work, they are, perhaps, the most relevant for most operators.

The purpose of Chapters 12 and 13 is to demonstrate the importance of applying the ergonomic principles discussed and described earlier. Chapter 12 deals specifically with safety, and illustrates the cost to the individual, to the organization, and to the nation, of unsafe behaviour and environments. It demonstrates how the disruption of efficient communication channels and the establishment of alternative and less optimal channels can reduce safety and lead to accidents.

Death, injury and substandard work, however, are not the only result of a lack of ergonomic considerations, as Chapter 13 illustrates. In many situations it can lead to reduced efficiency caused by poor feedback to the operator. In this way the maintenance of system effectiveness and the inspection process can suffer.

Having roused the reader to consider his system more fully, Chapter 14 illustrates many of the pitfalls ready to trap the unwary investigator who has to deal with the behaviour of human beings as subject-matter in an experimental design. Armed with such information, however, it is hoped that the reader will be in a fit position to progress further in implementing ergonomic solutions to the man–environment interaction problems.

To paraphrase Churchill, after having read this book it is hoped that the reader will cry 'Give us the ergonomic tools, tools that have been designed so that we can interact efficiently with them, and we will finish the job more effectively'.

D. J. OBORNE

CHAPTER 1

Ergonomics Past and Present

THE EXORCISM OF PROCRUSTES

Once upon a time there lived a Greek robber called Procrustes who had devised a cunning way of extorting money from weary travellers unfortunate enough to pass his door. He simply offered them hospitality on the strict understanding that they would either sleep in one of his two spare beds or pay for the food and drink that they had consumed. If the traveller opted to take the bed, as most did, Procrustes added one further stipulation: that he should fit one of the beds exactly. After being wined and dined the unsuspecting victim would be shown to his bedchamber in which there were two beds; one was very long, the other very short, but neither would fit him exactly. At this point Procrustes' trick soon became apparent even to the least intelligent wayfarer. Unless he paid the exorbitant fees demanded, Procrustes would threaten to make his victim fit one or other of the beds—either by putting him on the short bed and cutting off his legs, or by stretching his body enough to fit the long bed. Needless to say, most tired travellers took the easy way out and paid up.

Ever since man first began to interact with his environment in any complex way, this form of Procrustean approach has been widespread. Industrial man in particular has constantly been 'tailored' to 'fit' the demands of his physical world, with most victims normally accepting a fair degree of discomfort and disability without too much fuss. In metaphorical terms, arms have been elongated to reach inaccessible controls and perceptual abilities stretched to be able to hear or to see virtually inaudible or invisible signals. At the other end of the Procrustean scale, legs have been cut to fit cramped workplaces and cognitive capacities shrunk to fit boring tasks.

The problem has become increasingly important since the Industrial Revolution, particularly with the expansion in the complexity of both work and machines. Because of a poor 'fit' between the human operator and his environment, lives have been lost, productivity reduced, and errors have been incurred in countless thousands of cases. Until relatively recently the demands of the environment have been paramount, with the needs and abilities of the man in the environment having to take secondary importance.

Gradually, however, the ghost of Procrustes is being exorcised. Indeed, one of the aims of this book is to illustrate the areas in which diehard Procrustean conditions may be exposed and righted. However before the new order can be fully appreciated, it is useful to consider in more detail some of the limitations of

1

the Procrustean approach in which man is adapted to suit the requirements of his environment. (This approach is sometimes embodied in the concept of training.) These limitations can be considered in three areas: first, the cost of making the operator fit his environment; second, the effectiveness of this approach; and third, the possible disruption of the man's performance which can occur when he is placed under stress.

Training the operator in tasks which are difficult to carry out is a costly procedure. It is, of course, accepted that training schedules cannot be dispensed with altogether but, as will be shown throughout this book, in the majority of cases training and production times can be reduced considerably if the machine is designed to reflect the operator's abilities.

Although it is possible to train operators to a high level of competence, training alone will not solve their problems and increase their output. Given the increasing complexity of industrial plant and machinery, if the operator's environment is also not also suitably designed his behaviour may still not be effective enough. He needs further help in the form of appropriately designed controls, information displays and other aspects of his environment before the effects of any training can be maximized. An example of this problem can be seen in a study by Chaney and Teel (1967), who compared the detection efficiency of experienced machine parts inspectors either after a four-hour training programme or after being given a set of specially designed visual aids and displays to help them to detect defects. Their results showed that whereas the training programme resulted in a 32 per cent increase in detected defects, the use of the appropriate visual aids resulted in a 42 per cent increase. Although training was useful, therefore, an appropriately designed environment was even more so.

The effectiveness of training can also be measured in terms of the degree to which the trained behaviour is maintained after long periods with little or no practice. In a laboratory study, for example, Ellis and Hill (1978) demonstrated that numbers formed from the now common seven-segment liquid crystal displays (see Chapter 5) are more difficult to read under short viewing times (that is, they lead to more reading errors), than are numbers from conventional displays. Although these difficulties could be overcome after appropriate training, the skills which had been acquired decreased significantly over the period of a month without the opportunity for practice.

The final limitation of the Procrustean approach, suggested by Taylor and Garvey (1966) is, perhaps, one of the most important. No matter how well the operator has been trained, his behaviour can break down under stress and this can result in his making inappropriate responses to a situation. For example, Murrell (1971) cites the case of a hydraulic press which was wrecked during an emergency action (that is, when the operator was under stress). For normal operation, a lever needed to be pushed down to raise the press and the operator had been trained to carry out this action very efficiently. However, because the normal expectation would be to lift a lever to raise some part of a machine (see Chapter 7), when the emergency occurred the operator, wishing to raise the press, forgot his training and pulled the lever up. This caused the platen to move down and wreck the press. In the majority of cases no amount of training will overcome the tendency of an operator to 'do what comes naturally' when he is placed under stress. (It might, of course, be argued that some types of training regimes (for example, in the military) are designed to overcome this breakdown in behaviour which occurs under stressful conditions. This may be so, but in most working environments the cost of such training would possibly be prohibitive and the type of training would probably not be tolerated.)

It is clear from these examples that training alone will not utilize the full potential of the human operator. Only when training schedules are linked to a full understanding of the task which is being carried out, so that the work is designed to be in harmony with the operator's physical, cognitive and emotional capacities, will performance be optimal under a range of conditions. It is the role of ergonomics to attempt to highlight this concordance between the environment and the man.

Ergonomics, then, is a discipline which attempts to redress the balance which has previously been biased towards the Procrustean approach. Its basic aim is to measure the capabilities of the man and then to arrange the environment to fit such abilities. As Rodger and Cavanagh (1962) describe it, ergonomics attempts to 'fit the job to the man' rather than to 'fit the man to the job'.

THE RISE OF ERGONOMICS

The birth date of ergonomics can be pinpointed fairly accurately to 12 July 1949. A meeting was held at the Admiralty at which an interdisciplinary group was formed for those interested in human work problems (Edholm and Murrell, 1973). Later at a meeting on 16 February 1950, the term *ergonomics* was adopted and the discipline was born. (The word ergonomics was coined from the Greek: *ergon*—work; *nomos*—natural laws.)

Although the birth of ergonomics can be fairly well defined, the gestation period of this new discipline was long and tortuous, and certainly no such precise date can be given for its conception. However, the initial rise of interest in the relationship beween man and his working environment could be said to have commenced at about the time of the First World War. Workers in munition factories were essential in maintaining the war effort, but with the drive for a higher output of arms a number of unforeseen complications arose. The attempt to resolve some of these problems led, in 1915, to the establishment of the Health of Munitions Workers' Committee which included some individuals trained in physiology and psychology among its investigators. At the end of the war, this Committee was reconstituted as the Industrial Fatigue Research Board (IFRB), chiefly to carry out research into fatigue problems in industry.

In 1929 the IFRB was renamed the Industrial Health Research Board and its scope broadened to investigate general conditions of industrial employment, 'particularly with regard to the preservation of health among the workers and to industrial efficiency'. It had investigators who were trained as psychologists, physiologists, physicians and engineers and who worked, both separately and together, on problems covering a wide area. These included posture, carrying loads, the physique of working men and women, rest pauses, inspection, lighting, heating, ventilation, 'music-while-you-work', selection, and training. As Murrell (1967) points out, two features of the work carried out beween the wars are important. First, it was at times interdisciplinary, and second, it was largely 'exploratory in character being in the nature of a probing of the "natural history" of industry'.

With the outbreak of the Second World War there occurred a rapid development in military field. However, as if the stresses of battle were not enough, the military equipment became so complex and the operating speeds so high that the additional stresses which resulted caused men either to fail to get the best out of their equipment or to suffer operational breakdown. It became essential, therefore, for more to be known about man's performance capabilities and limitations. Naturally this produced extensive research programmes in many diverse fields, and it was as reaction to the desire to draw together the new-found knowledge that the Admiralty meeting took place and the discipline *ergonomics* was finally born.

THE SCOPE OF ERGONOMICS

Ergonomics, therefore, developed via the interests of a number of different professions, and it still remains a multidisciplinary field of study. It crosses the boundaries between many scientific and professional disciplines and draws on the data, findings and principles of each. Present-day ergonomics is an amalgam of physiology, anatomy, and medicine as one branch; physiological and experimental psychology as another; and physics and engineering as a third. The biological sciences provide information about the structure of the body: the operator's physical capabilities and limitations; the dimensions of his body; how much he can lift; the physical pressures he can endure, etc. Physiological psychology deals with the functioning of the brain and the nervous system as they determine behaviour; while experimental psychologists attempt to understand the basic ways in which the individual uses his body to behave, to perceive, to learn, to remember, to control his motor processes, etc. Finally, physics and engineering provide similar information about the machine and the environment with which the operator has to contend.

From these areas an ergonomist takes and integrates data to maximize the operator's safety, efficiency, and reliability of performance, to make his task easier to learn, and to increase his feelings of comfort.

These criteria, however, are by no means independent. For example, an operator's efficiency is highly dependent on his accuracy, but accuracy is not the only component of efficiency—others include reliability, speed and the reduction of effort and fatigue. Arguing in the same vein, ergonomics seeks to increase safety. In turn this should result in a reduction of time lost through illness and (perhaps) a corresponding increase in (worker) efficiency. By the same token, however, safety will itself depend on efficiency. Indeed, throughout this book many examples will be advanced to illustrate the fact that the margin of safety which is left in an operation is largely a function of the operator's speed or reliability.

A further aim of ergonomics is to attempt to reduce the unpredictability of operator performance, in other words, to increase his (or her) reliability. Thus the human operator should not only be fast and efficient, he should be reliably so. Again, although reliability is related to accuracy, they may also be independent. An operator may perform his task accurately most of the time but, because of some intermittent action of his work situation, he may be unreliable in his accuracy.

The question of ease of learning has already been discussed. Thus a system which has been designed to produce a series of tasks which are easier to learn will reduce training time and costs, and may produce less errors under stress.

The final aspect, comfort, is a subjective criterion that is becoming increasingly important in present-day situations and refers to a sense of wellbeing and ease induced by the system. The concept of comfort, and the controversies surrounding its definition will be discussed in later chapters, but it is sufficient to point out here that an uncomfortable operator is prone to errors and is likely to perform less efficiently.

In summary, therefore, the task of the ergonomist is first to determine the capabilities of the operator, and then to attempt to build a work system around these capabilities. In this respect ergonomics is often referred to as the science of 'fitting the environment to the man'. Only when this approach has been fully accepted can the spirit of Procrustes finally be said to be exorcised.

ERGONOMICS AND RELATED DISCIPLINES

It is pertinent at this point to question where ergonomics stands in relation to seemingly related disciplines such as operations research, work study, and time-and-motion study. Each tries to maximize the worker's effectiveness, and certain areas of overlap are bound to exist. Despite this similarity of objectives, it is possible also to perceive differences between the disciplines.

As its title suggests, *time-and-motion study* is concerned primarily with increasing performance by measuring and then minimizing the times taken to perform various operations (motions). The fundamental philosophy of this discipline suggests that: (a) although there are usually numerous ways to perform any task, one method will be superior to others, and (b) the superior method can be determined by observing and analysing the time taken to carry out parts of the activity.

Thus Barnes (1963) defines time and motion study as

the systematic study of work systems with the purpose of (i) developing the preferred system and method, usually the one with the lowest cost; (ii) standardising this system and method; (iii) determining the time required by a qualified and properly trained person working at normal pace to do a specific task or operation; and (iv) assisting in training the worker in the preferred method.

Proponents of the discipline argue that information from time-and-motion analyses should enable the operator's activities to be rearranged and be carried out within 'standard times'. These can then be used to set production schedules, to determine supervisory objectives and operating effectiveness, to set work-time standards, to determine the number of machines a person may run, to coordinate workers for increased effectiveness, to determine costs, and to provide a basis for the setting of incentive wages (Mundel, 1950).

Because of its image, essentially of being the panacea of all productivity ills (but in many cases failing to live up to expectations), because of its high reliance on work speed as the main criterion, and because of its potential misuse by some managements when setting production goals, over the years time-and-motion study has become mistrusted by management and workers alike. This is unfortunate for ergonomics since, as will become apparent, the discipline also relies to some extent on analysing the times taken to perform various actions. In the eyes of many individuals, therefore, ergonomics is tarred with the same brush. The major difference between the two disciplines, however, lies both in the use to which the data obtained are put, and in the fact that time-and-motion analyses are not the sole sources of information available to the ergonomist. As will become evident in future chapters, criteria for ergonomics also involve operator accuracy, comfort and satisfaction, in addition to considering the effects of aspects of his environment such as the noise, illumination and temperature. Thus the aim of the ergonomist is to consider and to optimize the total work system rather than merely to manipulate the human link in the chain.

In many respects *work study* evolved from time-and-motion study, but places less emphasis on the derivation of time standards (as well as on the use of such standards towards the financing of incentive plans). de Jong (1967) suggests that work study includes such considerations as the total work system and its technology; the work environment; the tasks needed to be carried out; the instructions, the methods of working, and the training in the preferred methods; standards of performance, including time standards; and job evaluation and wage payment plans.

From this list it is apparent that some overlap occurs between ergonomics and work study. Both consider the man at work, both attempt to analyse the work process to try to optmize performance, and both place less reliance on time and more emphasis on the total process and worker wellbeing. However, differences between their aims and objectives may still be observed. Work study still emphasizes such managerial problems as job evaluation and wage incentive plans; its techniques still revolve, primarily, around time-and-motion analysis and omit many of the other methods employed by ergonomists to gather information; and its aims appear to conclude at the level of simply identifying and analysing the work situation and problems, rather than attempting, in any systematic way, to match the requirements of the situation to the capabilities of the operator. As Moores (1972) concludes, in work study 'tasks are examined with little reference to the individual and this is often reflected in the job being tailored to the lowest common denominator in the catalogue of ability. It could be said, therefore, that work contains some ergonomics philosophy, but not enough to make the two disciplines identical.

The scope of *operations research* (OR) is at a more molar level than that of either time-and-motion or work study, or even ergonomics. OR strives to produce an optimum total work system by forecasting the future requirements of the system and then planning the work load and system to meet these requirements. In this respect it is similar to its sister discipline *organization development* (OD).

OD involves major changes in organizations—changes in the way work is done, changes in the structure of the system, and perhaps changes in the physical plant. The aim is to improve the functioning of the organization, to make it more flexible and adaptable to change, to make it better equipped to solve problems and to handle conflicts. The central concept of OD is that these major changes will not be realized successfully without focusing a great deal of attention on people as elements in the system. They must be involved in implementing and planning the change; they must be educated about the change; they must look at their feelings about the change and about other people in the organization and they need to understand about their organization in relation to the environment. Without such detail being observed, proponents of OD argue that the necessary changes are unlikely to be successful.

It is apparent, therefore, that each of the above disciplines involves some ergonomics and ergonomics may, in turn, have borrowed some of its philosophy, methods and techniques from each. However, it is also apparent that the subject matter and emphasis of each do differ, sometimes significantly so. Ergonomics takes as its central concern the human operator, his performance, safety and comfort. OR and OD are also interested in performance and, to a certain extent wellbeing, but they are interested in the human operator only as he is valuable to the system or organization and then at a more qualitative level. Time-and-motion study, although concerned with the human operator, works at a far more molecular level. Its main objective is to analyse the operation into its various motions and then to measure the time taken to carry out these motions. The primary emphasis is on man as a worker, that is as a source of mechanical power, so that time and energy expenditure standards for various activities can be produced. Time-and-motion study does not consider, in any concerted way, the effects of the working environment on such additional aspects as the worker's safety and comfort.

THE MAN–MACHINE SYSTEM

Ergonomics seeks to maximize safety, efficiency and comfort by matching the requirements of the operator's 'machine' (or indeed any aspect of his workplace which he has to use) to his capabilities. By linking the man to his machine in this way a relationship is established between these two components, so that the machine presents information to the man via his sensory apparatus to which he may respond in some way—perhaps to alter the machine's state via various controls. For example to be able to drive a car along a road safely and efficiently, a relationship must be built up between the driver and the vehicle such that any deviation of the car from a prescribed path (which is determined by both the driver and the shape of the road), will be displayed back to the driver by his visual (and perhaps auditory) senses. These deviations can then be corrected by his limb movements via the steering-wheel and perhaps the brake. In their turn the corrections will be perceived as displayed information and the sequence continued until the journey's end. In this way information passes from the machine to the man and back to the machine in a closed, information-control, loop (see Figure 1.1). It is the ergonomist's task to preserve and to enhance the operation of loops of this nature. For example the speed of information transmission may be increased (perhaps by a cleaner windscreen or by more understandable road signs) or the operation of a control may be made more efficient (perhaps by servoassisted steering or by altering the control position or dimensions).

Many examples of such single man-machine loops can be seen in different work situations. However in modern-day working environments these single loops are often combined to produce more complex systems which, being composed of groups of different components (both men and machines), have to be designed to work together.

From an ergonomics standpoint the combination of different, single, man–machine loops into a complex work system creates problems. Two loops may act efficiently when considered separately, but when combined into a simple

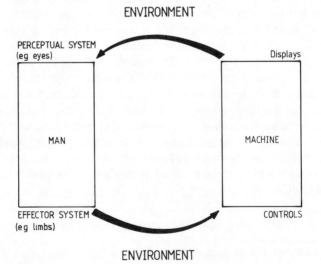

Figure 1.1 The 'man–machine' loop. The machine displays information to the human operator who operates his controls to affect the machine. The environment can interfere with the efficiency of this loop

system they might act antagonistically owing to unexpected interactions. For example, one loop may involve an operator pressing a lever in response to a deflection on a display. A second loop might involve him in pushing a button in response to a warning sound. Taken separately the two loops may act extremely efficiently; the display is well designed, the lever compatible with the operator's force capabilities, the warning sound audible and attention getting and the push button easy to operate. When operating together, however, the two loops might interact to produce a breakdown in behaviour. For example, if put in the 'wrong' places, the operator's natural response may be to (incorrectly) operate the lever in response to the warning sound, and (unsuccessfully) attempt to alter the machine state by pushing the button. It is not too difficult to conceive how problems of this nature can be increased in proportion to the number of extra loops which are added to the system.

For this reason, the emphasis of modern ergonomics has been to investigate the man and his environment within the system, rather than to examine minutely details of the components which constitute any one man–machine loop. As Singleton (1967) has pointed out,

It is no use concentrating on control design, for instance, such that a lever is just at the right height, with a handle nicely moulded to the grasp and forces appropriate to the limb flexion if, when the control is moved, quite uncotrollable things start to happen in the machine. Similarly it is no use having easily legible scales and pointers with a well-designed lighting system if the data, now so easily visible, are either useless or need further computation before a decision can be made.

Ergonomics, therefore, seeks to consider aspects of the job beyond a 'knobs and dials' approach, 'beyond the panel' (Murrell, 1969), to the man–machine system—to the total interaction between the man and his environment.

THE ALLOCATION OF FUNCTIONS BETWEEN
MAN AND MACHINE

It has been argued that one of the first and most important problems in man–machine system design concerns the allocation of functions between men and machines. What functions of the system should be assigned to men and what to machines? Or, what kind of things can and should human operators be doing in a man–machine system?

In response to questions such as these, many authors have attempted to compile lists of operations which are carried out most efficiently by men and by machines (for example, Chapanis, 1960; Murrell, 1971). Taken together these lists suggest that man is a better decision-maker, particularly in cases in which unexpected events may occur; is able to improvise; has a fund of past experience; and is able to perceive and to interpret complex forms involving depth, space and pattern. Machines, on the other hand, are highly efficient computing, integration and differentiation devices; are able to deal with predictable events in a reliable fashion; and are useful in hazardous environments.

Although such statements are helpful in directing an ergonomist's thinking towards man–machine problems, and reminding him of some of the characteristics that man and machines have as systems components, they do have their limitations. For example, to decide on the respective advantages and disadvantages of man and machine, one also needs to know something about the operator's preferences. If a man does not accept willingly some of the functions which he is asked to carry out then undesirable byproducts, such as absenteeism, high turnover and low productivity, may result.

In addition, Chapanis (1965a) suggests three further problems: First, general man–machine comparisons can be wrong or misleading. In many cases the system itself determines the adequacy of its individual components. For example, the general statement that man is superior to a machine for decision-making tasks is just that—a general statement. It cannot be true for all men or for all machines.

The second argument against such lists is that it is not always important to decide which component can do a particular job better. In many cases the question should be 'is the component good enough for the job?' Fitts (1962) rephrases this question to ask 'which component will do an adequate job for less money, weight or power, or with a smaller probability of failure and less need for maintenance?'

This leads to Chapanis's final criticism of such lists—that general comparisons between men and machines give no consideration to tradeoffs. When choosing between a man and a machine to perform a particular function, many additional considerations need to be taken into account, for example weight, cost, size or availability. Variables such as these need to be traded off, one against the other, before the ideal system can be designed. The type of question which may need to be asked, therefore, could be 'is it better to design a system which, because of size limitations, includes a human operator and less equipment, or would it be better to reject the human operator and so lose some flexibility but include more equipment'. Simple lists which compare the values of men and machines separately do not help to solve this type of tradeoff problem.

ERGONOMICS COSTS AND REWARDS

Deciding on the relative values of men and machines, therefore, is a difficult task which is made more complex when the question of their respective costs is included. It is germane at this point, therefore, to consider how ergonomics fares when it is subjected to some sort of cost-benefit analysis. Any manager who contemplates either implementing an ergonomics investigation of part of his plant or introducing a system designed to ergonomic principles must be able to justify the cost in relation to the rewards.

Chapanis (1976) points out that a full cost-benefit equation is extremely difficult to devise since many factors, some of them 'invisible', are involved when assessing the value of a system. Those which he considers important include, on the benefit side, the value of all goods and services produced by the system, and the values which accrue from any incidental or 'spinoff' products. On the negative side of the equation Chapanis includes equipment costs, costs of replacement or the maintenance of parts, operating costs, the costs of job aids, auxiliary equipment and manuals, of personnel selection, of training, of salaries and wages, of accidents, errors and breakages or wastage, and the social costs of implementing the system (for example, the long-term costs of pollution).

Many of these factors can be expressed in tangilbe, monetary terms. Others, however (for example, the cost of pollution, selection, accidents, etc.), are less quantifiable (see Chapter 12). Nevertheless they make important contributions to reducing the efficiency and productiveness of a system and must be taken into account.

In addition to these factors Corlett and Parsons (1978) describe other criteria which ought to be included in the assessment. These include reduced stress, turnover costs and increased work interest and satisfaction.

A further problem when considering how ergonomics relates to the cost-benefit equation is that few well-controlled studies have been carried out for an answer to be given. From an economic viewpoint this is understandable—a redesigned piece of equipment, for example, which is thought to be better than its existing counterpart will naturally be used throughout the plant as soon as possible. However, it does mean that comparable measurements of the system effectiveness with and without the new equipment and under otherwise similar conditions (pay, time of year, type of operator, etc.) are extremely difficult to obtain.

Despite these problems, however, some studies are available which illustrate the cost effectiveness of ergonomics.

Sell (1977) describes detailed work which was carried out at the British Iron and Steel Association to design overhead crane cabs. One problem which had occurred involved damage to railway wagons caused by the drivers of some cranes swinging the magnetic hooks of the cranes against the side of the railway truck. It was estimated that this damage was costing the Company around £60 per week in wagon repairs in 1959. Simple observations showed that a factor which contributed to the lack of operator's control over the hook was that he was unable to reach all of his controls, and at the same time look out of the crane to see the position of the magnet. By moving both the magnet and the other controls so that the operators could see the hook and reach his controls, the design fault was eliminated. Beevis and Slade (1970) have suggested that the modifications cost £270, which meant that the costs were repaid within about six weeks.

Along the same theme, Teel (1971) describes the costs and rewards accruing from two separate ergonomics studies. The first was conducted to evaluate the effectiveness of specially prepared visual aids in improving the performance of machine parts inspectors. (These inspectors examined precision parts for defects which might make them unsuitable for use.) The cost of the study, in other words, of preparing such aids for the inspectors along ergonomics principles, was calculated to be $7200. The savings in the first year alone, however, were over $1000. Since the production run was to continue for three years, Teel argues that the total savings accounted for more than four times the cost of the study.

In his second study, the workplace of electronic assemblers was considered according to ergonomic principles. Using a redesigned console, for example, the time taken by each operator to complete a task was reduced by 64 per cent with 75 per cent fewer errors. The cost of the study amounted to $4200, whilst the savings per year were calculated to be in excess of $28 000.

These are merely two examples of the benefits to be obtained from considering the work from the point of view of the operator. However as Chapanis's (1965a) list discussed earlier indicates, productivity benefits to a company may not always be measurable simply in terms of the number of the units produced. For example, in some cases productivity may rise simply because all of the workers remain at their jobs. In these terms sickness and accidents may diminish the workforce. Even if a worker is only temporarily hurt by an accident, and he takes perhaps only half of an hour off work to visit the sick bay or doctor, his productive times is reduced by at least that hour and probably more.

As will be seen in Chapter 12, in many cases accidents arise through poor ergonomic design, that is, by not designing the machine to fit the capacities and expectations of the operator. Indeed Powell *et al.* (1971), in a survey of 2000 machine-shop floor accidents in four different industries, were able to attribute to poor machine and environment design 43 per cent of lathe accidents, 27 per cent of press section accidents, and 80 per cent of accidents in the gear section. They describe the genesis of one simple accident involving a forklift truck and discuss two ways which it may have been avoided. The important point to remember is that the accident, in addition to being painful for the operator, reduced the effective workforce for a few days:

On three of our sites, forklift trucks were used extensively. In one department each driver normally worked on a particular truck. On the day of this accident, the driver was not operating his normal truck because this was being overhauled. After he had been using the truck allocated to him for a while, he had to adjust its forks to increase the distance between them for a new load. Both his usual truck and the allocated one had forks which ran along a horizontal slide.

On the driver's normal truck, the forks did not run very easily on the slide and so some force was needed to move them. He applied this same force to one of the forks on the strange truck. He pulled the fork right off the slide and it landed on his feet giving him a severe injury which necessitated several days' absence.

This mishap could have been 'designed out':

(a) The best arrangement would probably be some form of screw-controlled slider, so that the driver could easily adjust the width between the forks by turning a handle. (It might even be a remote and power operated control so that the driver need not leave his seat.)
(b) Given that a driver does not have to heave on a fork to shift it, some way of preventing the fork falling free must be provided. An end-stop on the slide is one way

of doing this but some end-stops could make a nip for the fingers of the driver moving the fork. A chain of appropriate length between each fork and the centre of the slide could be used (p. 101).

SUMMARY

This chapter has considered how ergonomics, and its American sister discipline human factors, arose as a response to the need to consider how the human operator manages to cope with his environment. The important concept is that discrepancies between what the environment requires and what the operator is able to give should be resolved by adapting the environment to the operator, rather than the normal approach of making the operator suit the environment. If such a basic change is made, it is argued, then the efficiency, safety, comfort and productivity of the total man–machine system will be enhanced. The remainder of this book considers the various aspects of the system.

CHAPTER 2

The Structure of the Body:
I   The Sensory Nervous System

At a structural level the human operator can be said to be little more than a complex system of bones, joints, muscles, tissues, nerves and fluids. Such a statement, however, paints an extremely simplistic picture of the human being as we know him. The rich complexity of human life is made up of the infinite variations which can occur in the arrangement of the body structures and in the individual ways in which they are used. However, since it is impossible to understand fully how a proper relationship between the human operator and his mechanical environment can be created without some idea of his body structure, the purpose of this and the following chapter is to use this simplistic approach to look at the manner in which the bones, joints, tissues and nerves are arranged to make up the human body. The limitations of these structures define the most basic concept of ergonomics: literally that of fitting the environment to the man.

As should have become apparent already, the relationship between man and his environment hinges, primarily, on a complex closed-loop system. At its simplest level, this involves firstly the display of information from the machine to the operator: a light flashes, a colour appears, a sound is made, a lever changes position, etc. On the basis of this information, and perhaps more importantly on his interpretation of the meaning of the information, the operator may be required to carry out some action—pull a lever, push a switch, adjust a rudder, etc. This transmits commands to the machine which alters the display and, with the display of information back to the man, the loop is completed.

The efficient operation of this closed-loop system requires a number of body structures to be brought into play. First, there are the receptor agencies of the human body, the sense organs; through these the information is initially passed to the operator and they represent the first possible area in which errors can occur. Second, the nerves carry information from the sense organs to the interpretation and decision-making areas of the brain, and from the brain to the muscles. Although their speed of conduction is very fast, they have their limitations. Third are the body structures which carry out various actions—the effector processes. Once a decision to act has been made, the information is transmitted to the muscles in the body which control the action of the bones, joints and tendons. In addition to decision-making capacities, these effector processes represent possibly the greatest limitation to the operator's mechanical efficiency.

14

This chapter considers the structure and functions of the nerves and the sense organs. Chapter 3 deals with the structure of the bones, joints and muscles and the limitations which they place on the body's movements.

THE SENSORY MECHANISMS

The basic unit which transmits all sensory information is the single nerve cell or neurone, and this comes in many shapes and sizes depending on its position in the body. It conducts the information in terms of very small potential differences (in the region of 70 mV) from one part of the body to another, and this is accomplished through a system commonly known as the neural net.

Perhaps the simplest way to conceive of the net is to use the analogy of a telephone system. The German physicist and biologist Helmholtz expressed it in 1868 in the following way (Helmholtz, 1889):

The nerve fibres have often been compared with telegraphic wires traversing the country, and the comparison is well fitted to illustrate the striking and important peculiarity of their mode of action. In the network of telegraphs we find everywhere the same copper or iron wires carrying the same kind of movement, a stream of electricity, but producing the most different results in the various stations according to the auxiliary apparatus with which they are connected. At one station the effect is the ringing of a bell, at another a signal is moved, at a third a recording instrument is set to work.... In short, every one of the hundred different actions which electricity is capable of producing may be called forth by a telegraphic wire laid to whatever spot we please and it is always the same process in the wire itself which leads to these diverse consequences.... All the difference which is seen in the excitation of different nerves depends only upon the difference of the organs to which the nerve is united and to which it transmits the state of excitation.

Each neurone in the body essentially consists of three parts. First, the cell body which contains the nucleus; second, a mass of hair-like protrusions extending from the cell body, known as dendrites; and third, a long, single, thin extension of the cell body, the axon. This may be from less than a millimetre up to more than a metre long, and ends as dendrites. In many neurones, the axon is covered by a fatty insulating sheath known as the myelin sheath.

The chain of command, therefore, is for the cell body dendrites to convey information to the cell body itself. This is transmitted along the axon to other dendrites at the far end which are closely entangled with the cell body dendrites of the next neurones. The information is transmitted across the dendrite gaps (synapses) and so on throughout the body.

The rate at which impulses travel along a nerve fibre depends on the thickness of the myelin sheath surrounding the axon and also, to some extent, on the diameter of the axon itself. When the sheath is thick, the rate of travel may be as high as 120 m/sec, while if the sheath is very thin or non-existent the rate of travel could be as low as 0.6 m/sec.

The speed of transmission through the body also depends on the number of synaptic gaps between neurones, since the synapses create a slight delay in the transmission process. Although the synaptic gap is of the order of 100 angstroms (0.000 001 mm), the delay between the arrival of an impulse at the axon terminal and the initiation of an impulse the other side of the synapse is in the order of 0.5–1.0 m sec. During that time, another impulse could be travelling a metre along the fibre.

The Senses

Various types of information are fed into the sensory system through receptors, which are classified into three groups. The exteroceptors receive information about the state of the world outside the body and so include the eyes, ears, and touch receptors in the skin. Using the same terminology, the interoceptors inform the individual about the internal state of his body, for example its state of hunger or fullness of the bladder. Finally, the proprioceptors are concerned with motor functions and give information about the position of the body or parts of the body in space. They comprise two groups of receptors: the kinaesthetic and vestibular systems.

The receptors in which ergonomists are mainly interested are the exteroceptors and proprioceptors. The exteroceptors are important because they allow information to be transmitted from the environment to the operator, while the proprioceptors tell the operator what his body is doing and his position relative to the environment and to his machine. The importance of the efficient functioning of these two groups of receptors, therefore, should be clear: in addition to influencing his decision-making capacities they effectively control the operator's behaviour in carrying out his part in the closed-loop system.

Before describing the senses in more detail, it is useful to consider the difference between sensation and perception. Whichever sensory system is considered, sight, hearing, taste, touch or smell, the particular organ receives energy from the outside world, converts it into small potential differences and passes these along the system to the brain. This is the process of sensation (or reception). It is determined entirely by the quality of the stimulus and of the particular organ and nervous system in operation, and it is an objective process. On entering the brain, however, the nerve impulses are interpreted to produce a recognizable pattern of sight, smell, sound, etc., and this process is influenced greatly by the individual's past experience, his expectations, his feelings, and his wishes. This process of perception is entirely subjective in nature.

Thus the distinction between sensation and perception is very important: although two people may receive and sense the same object, they may not perceive it identically, and the implications of this for aspects of ergonomics such as designing appropriate displays should be clear. This distinction will be emphasized again in Chapter 4 when discussing in more detail the channels of communication which can be set up between two or more operators.

THE VISUAL SYSTEM

Of all of the senses, vision has been the most thoroughly studied. It is also, perhaps, the system which is most overloaded at work. In essence the system consists of two eyes, each connected to the the visual cortex of the brain by an optic nerve (see Figure 2.1). The two nerves meet at the optic chiasma at the base of the brain, where parts of each nerve cross over to terminate in the visual cortex on the opposite side of the brain to the eye from which it originated. In fact fibres from the left-hand side of each eye terminate in the left visual cortex, and fibres from the right-hand side of each eye terminate in the right visual cortex.

The effects of this crossing of fibres may prove to be important when information is presented in very short periods of time or when extremely fast responses are required. For example, it is now fairly well understood that the two halves of the brain do not 'perform' equally well for all types of material. For instance speech appears to be analysed better in the left half (hemisphere) whereas the right hemisphere is dominant with respect to spatial ability. In terms of material normally presented to the visual system, the left hemisphere is better than the right at analysing words, whereas there is some evidence which suggests

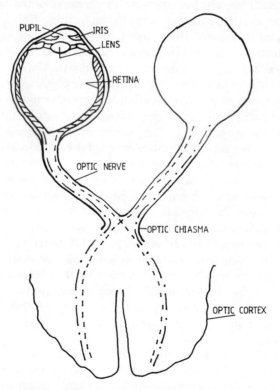

Figure 2.1   The eye and visual system, showing
the pathways of portions of the optic nerve

that numbers are recognized faster in the right hemisphere (Kimura and Durnford, 1974).

Depending on the nature of the visual material, therefore, it might be acted on more speadily or efficiently if presented to one half of the visual field than the other. This right or left-hand advantage, however, is only likely to be noticed when very fast responses are required and so it is unlikely to be important for most work situations.

With regard to the eye itself perhaps the easiest way to understand its structure is to compare it with a camera. A controlled amount of light enters the eye through the pupil (aperture), the diameter of which is controlled by the coloured portion of the iris (stop). It is then bent and focused by the lens to fall on to the retina which acts as a photosensitive layer.

Although the analogy with the camera is suitable at a descriptive level, differences do exist between the structure of the eye and that of a camera which give the eye more flexibility. First, the lens is in a fixed position relative to that of the photosensitive area—it is not moved to and fro to focus on objects at different distances as would be the case in a camera. Focusing is accomplished by the muscles which surround the eye changing its curvature. Second, the photosensitive layer is curved, which compensates for the curvature of the lens. Both of these differences allow quicker and more efficient focusing. Third, the size of the pupil is regulated by the iris. This allows the eye to operate over a far greater range of illumination intensities than a camera, which has to use a fixed type of film sensitivity at any one time. Fourth, the retina does not have the same level of sensitivity to light over its surface; it is highest at the centre and falls off quickly towards the periphery. A photographic film of this type would produce a picture with very good detail at the centre only, with reduced quality the further away from the centre the image falls. To compensate for this the eye constantly scans the visual field, using a set of six muscles attached to each side, above and below the eyeball. In this way different images can be pointed at the part of the retina with the highest sensitivity in rapid succession. Finally, by using both eyes together it is possible to obtain binocular vision. This helps us to perceive spatial relationships between objects and the environment, so that the eyes will record in three dimensions, whereas the camera can only 'see' in two.

Perhaps the most important part of the eye itself is the retina. This is the layer at the back which acts in the fashion of a complex photodiode to convert luminous energy to electrical energy. It is made up of three layers of neurones, the most important being on the outside surface which contains two different neurones called the rods and cones (so called because of their shape). These appear to be the main receptor cells and, as shown in Table 2.1, they perform different functions.

A number of important points emerge from Table 2.1. First because the rods, which function at low illumination levels, are primarily towards the periphery of the retina, we can see dim objects much more effectively if we look at them off-centre. This means staring a little to the right or left of the object to focus it towards the periphery of the retina where the density of the rods is high. Off-

Table 2.1 The different properties of rods and cones (adapted from Morgan, 1965, and reproduced by permission of McGraw Hill)

Rods	Cones
Function at low levels of illumination (such as at night)	Function at higher illumination levels (such as during daylight)
Differentiate between shades of black and white	Differentiate between colours
Most sensitive in the greenish part of the spectrum	Most sensitive in the yellowish part of the spectrum
More numerous in the periphery of the eye	More numerous in the central part of the eye
Sensitive to very weak stimuli	Mainly involved in space perception and visual acuity

centre viewing is a trick used frequently by amateur astronomers, men on night lookout duty, and others who have to work in dimly lit environments. Second, our sensitivity to light changes with changes in the ambient illumination level. At some point, for example at dusk, the responsibility for the conversion of light energy to electrical impulses switches from the cones to the rods. Third, the rods and cones are sensitive to different wavelengths of light. These two latter points are important when considering the way in which we adapt to light and dark.

Light and Dark Adaptation

Because of the two types of receptors in the retina (the rods and the cones), the human eye is able to function over an extremely wide range of illumination levels. Cone vision provides acute vision at daytime (photopic) levels of illumination, whereas rod vision allows for the high degree of light sensitivity that is essential for seeing at night when light levels are low (scotopic). With increasing or decreasing luminance levels, however, there will always be a point at which one set of photoreceptors ceases to operate as the other takes over. If the increase or decrease in illumination is relatively slow this adaptation to dark or light conditions is fairly smooth, but with fast illumination changes the well-known experience of temporary blindness results. For example moving from pitch darkness to bright sunlight sharply increases the level of illumination falling on the retina to such levels that the previously functioning rods are unable to cope and the cones have not had a chance to operate fully. In this case the eyes normally have to be closed, or dark glasses worn, to allow them to become gradually adapted to the change in the light level, after which time the eyes can again function efficiently within a new dynamic range of intensities.

Because the cones are relatively fast-reacting, their light adaptation is often complete within a minute or two. Rods, however, are much slower in their action, and dark adaptation may take half an hour or even longer, depending on the previous illumination levels. For this reason, coloured goggles are often worn by people having to work in dark environments (for example radar operators or maintenance men) for some time before entering the dark room. Red is the colour normally used because it is one which does not greatly affect the visual pigment in the rods. Thus the cones may operate fairly normally during daylight while the rods are able to become and to remain adapted for the dark (see, for example, Cushman, 1980).

The Visual Perception of Movement

There are many situations in which a human operator might wish to perceive motion accurately. For example, the movement of a pointer across a dial; the movement of a vehicle in which he is seated; or the movement of an object falling towards him.

In general, movement can be perceived in two ways. In the first an object is kept in view by moving the eyes, so the observer receives information about the speed and direction of the object from the contraction of the muscles which surround and position the eye. The second case occurs when the eye is stationary and the object's image moves across the retina; the moving object is then perceived by the stimulation of different retinal cells. Under these circumstances the minimum velocity which can normally be detected is about 1–2 min of arc/sec. If the object moves faster than this the normal reaction will be for the moving object passing in the peripheral area of the visual field to be detected by the high density of rods in that part of the retina, and for the eye to move across and track the object to maintain a clearer image.

The lowest level of movement that can be detected (the movement threshold) is reduced considerably (by an order of approximately 10) if another (stationary) figure is also present in the visual field. This provides a reference point against which the moving image can be compared, and it is likely that other visual processes are also involved, perhaps concerning the ways in which we perceive spatial relationships.

The Visual Perception of Space

The perception of the spatial relationships between a series of objects in the visual field is normally accomplished by one or both of two processes—the use of cues obtained from different objects in the visual field, and the accommodation and convergence of the eyes.

Two types of visual cues are normally used, binocular and monocular. In the former the images received by the two eyes are compared by the brain and the disparity between them is used to indicate the relative positions of objects in space. For example, if a near and far object are both straight in front of the observer and he fixates on the near object, on closing his right eye the left eye sees the far object as having moved to the left. Similarly, looking throughout the right eye only, the far object is observed to the right. These two different images, which are received by the two visual cortices, tell the observer something about the relative position of the two objects.

The observer also uses a number of monocular cues to perceive spatial relationships, and each relates to his past visual experience. These cues include the relative sizes of objects (if two objects are the same size the object further away appears to be smaller); covering and shadow (if one object is in front of another it may partially obscure it, or the degree of shadow may provide information relating to the distance between the two objects); and texture. For example Gibson (1950) has argued that if there is any regular marking or visible texture in the object, for example on a floor, this texture undergoes a transformation in perspective so that in the retinal image there is a gradient of texture density. This gradient, then, suggests space.

In addition to binocular and monocular cues obtained from the images of different objects in the visual field, the fact that the eyes have had to converge to fixate on the various objects also tells the observer about their relative position in space. Fixating a far object, for example, causes double images of the near object to be observed and this disparity can be reduced by converging the eyes. This convergence may be used as a cue that the second (nearer) object is nearer than the initial fixation point.

Accommodation and convergence, therefore, concern the action of focusing the image on each retina so that a single 'picture' is produced. Thus the two eyes converge on the object, while the shapes of the lenses are altered to accommodate the two slightly different images. The degree to which these two processes take place provides valuable information about the relative positions of objects in the visual field, and this information is obtained from the position (proprioceptive) receptors in the eye and lens muscle. (The value of the proprioceptive system is discussed later in this chapter.)

Visual Acuity

Having considered some of the principles underlying our ability to register spatial aspects of the visual scene, it is appropriate to turn to the topic of visual acuity. This refers to the process by which we are able to see fine details. Again many aspects of work require this ability—to register the fact that a very slowly moving pointer has moved; to detect differences in the position of two controls; to recognize the presence of an object in the visual field; to localize and to distinguish between two close objects in space, etc. The three types of acuity most commonly recognized are line acuity (the ability to see very fine lines of known thickness), space acuity (the ability to see two spots or lines as being separate or, in other words, the ability to see a space between the lines), and vernier acuity (the ability to detect a discontinuity in a line when one part of the line is slightly displaced) (Murrell, 1971).

In essence three factors determine the degree of acuity under any given condition. First, the size of the pupil—acuity is fairly linearly related to pupillary diameter down to a value of about 1 mm. High ambient illumination levels and some drugs, however, may cause the pupil to be constricted, and this could be an important factor for the safe and efficient operation of machines which require high levels of acuity. A second factor is the light intensity being reflected from the object (its luminance). An object which is perhaps too fine to be seen at all in low illumination may become clearly visible when illumination is increased. (As was mentioned above, however, too high a level of illumination is likely to cause the pupil to become constricted so reducing acuity.) Over the range of illuminations normally found, acuity varies linearly with a logarithmic increase in illumination between a visual angle (that is, the angle subtended by the object at the eye) of about 0.2 to 1.5 min of arc. Finally, and within limits, acuity is related to the time allowed to view the object so that a reduced exposure time reduces acuity. The exposure times experienced in normal work, however, are usually above those needed for this to be a problem (above 200 m sec at normal daylight levels of illumination).

Visual Flicker

In many cases the human operator may be called on to respond not to a visual stimulus which has a relatively steady illumination level, but to one which flickers, for instance one which alternates between 'on' and 'off'. Flashing warning or indicator lights are, perhaps, the most obvious examples of such displays.

Flicker is a temporal phenomenon, its perception depending heavily on the ability of the visual system to react to the fast changes in light intensity. In many cases, therefore, it may be considered as a type of acuity task, since the observer must be able to 'perceive' the period of darkness between two flashes.

This task is carried out fairly well for flashes of light occurring relatively infrequently, when a flickering sensation will be obtained with separate flashes being perceived. As the flash frequency is increased, however, the flashes appear to merge to produce light which is indistinguishable from a steady state illumination. The frequency at which this merging occurs is normally called the critical fusion frequency (c.f.f.), and depends on many factors. For instance c.f.f. has been shown to range from about two to three flashes per second at very low illumination levels (that is, scotopic vision using the rods) to around 60 flashes per second with extremely high levels (that is, photopic vision). At the changeover point between scotopic and photopic vision, at illumination levels normally experienced at dusk, c.f.f. is around 15 flashes per second.

Colour Vision

A quick glance around a modern workplace will illustrate immediately the importance of colour. As will be discussed in later chapters, colour is used to help the operator distinguish between different parts of his working area, his controls, his displays and parts of a display. It creates mood (it is generally accepted that reds and yellows are 'warm' colours whereas blues are 'cool') and, using contrasts, colour helps to improve visibility.

Different colours are perceived as a result of the eyes receiving different wavelengths of light, which can be reflected from a coloured surface or might emanate from a coloured light source. The normal eye is able to sense light in the spectrum with a wavelength of about 400 and 700 nanometres ($1 \text{ nm} = 10^{-9} \text{ m}$), but the retina is not equally sensitive to wavelengths of light. This means that different colours of the same intensity will appear to be either brighter or less bright according to their wavelength. When the eye is light-adapted, the brightest spectral colour is at about 550 nm which gives an impression of yellowish-green, and the brightness progressively diminishes as the wavelength approaches either the 400 nm (red) or the 700 nm (violet) ends of the spectrum.

The process by which all the colours and hues normally experienced are decoded are too complex to be explained here, and indeed large gaps exist in our knowledge of this area. Of more interest to the practising ergonomist, however, is the proportion of the population who are deficient in their colour discrimination. These people, who represent about 6 per cent of the male and 0.5 per cent of the female populations (Morgan, 1965) may experience difficulty in work, particularly when colour is used to code various aspects of their machines.

Colour-blindness may be classified in a number of ways. The most common is on the basis of the ability to discriminate between the colours of red, green and blue. Normal people are able to discriminate all three colours, hence they are called trichromats. The most common type of colourblind individual is the dichromat—he might confuse red with green or yellow with blue (red–green blind individuals are considerably more common than yellow–blue individuals). The relatively rare person who is totally colourblind (0.003 per cent) sees only white, black and shades of grey. He is described as a monochromat.

The existence of colourblindness may be fairly simply determined using cards which make up the Ishihara colour test. Each card, which should be presented under standardized conditions, contains a number of different coloured dots, some of which form a pattern—either a number or a wavy line. Because of his inability to discriminate particular colours the colourblind individual has difficulty in perceiving these patterns and his deficiency may be demonstrated by an inability either to name the number or (for illiterate subjects) to trace the wavy line. Although the test appears simple to administer, it must be emphasized that it only produces vaild and reliable results in the hands of a skilled tester and under controlled lighting conditions.

THE AUDITORY SYSTEM

If the eye can be likened to a camera, the ear can be thought to perform like a microphone. The main job of both is to convert the noises which they receive in the form of sound pressure waves into electrical patterns, which are then

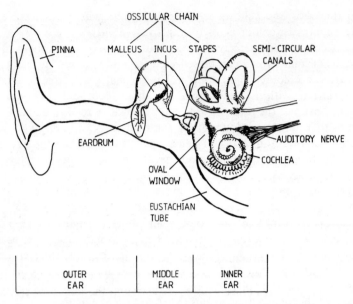

Figure 2.2   The structure of the ear

recognized by a decoding apparatus. However, unlike the eye which functions in a superior way to the camera, in many respects the human ear can be considered as being inferior to the highly sophisticated modern microphones which are available today.

The ear itself is composed of three recognizable sections: the outer ear, the middle ear, and the inner ear, and these are illustrated in Figure 2.2. What most of us commonly call the ear is to the anatomist the pinna of the outer ear, which in animals like the dog or the cat can be moved in different directions to help 'collect' sound waves. In human beings, however, the pinna is a less effective sound trapper. The outer ear also consists of a tube which runs inwards from the pinna and is terminated at the eardrum (the tympanic membrane). It is through these parts of the external ear that sounds are conducted to the middle ear and then to the inner ear.

The middle ear performs two main functions: to transmit sound waves and to protect the inner ear. Sound wave transmission is caried out by three small bones, the malleus, the incus and the stapes, which are collectively called the ossicular chain. They are so arranged that they span across the middle ear, and connect the eardrum to a thin 'oval window' on the other side. For this part of the ear to function properly, it is important that the air pressure remains the same as that in the environment and this is achieved by a tube (the eustachian tube) which connects the inner ear to the back of the throat. However, sudden changes in the air pressure can close the eustachian tube thus creating pressure differences between the middle ear and the outside atmosphere. The effect of this may be to cause the excruciating pain experienced by some air travellers, or even permanent damage to the ear.

The outer and middle ear appear to have the function not only of transmitting sound to the inner ear, but also of protecting it from having to operate on sound pressure levels which are outside its capacity. Kryter (1970) points to three ways in which this can occur. First, the action of the eustachian tube may prevent pressure waves which have fast (less than 200 m sec) rise times (the time taken for the noise sound pressure level to reach its maximum intensity) being transmitted to the inner ear. (Events such as explosions can easily produce sounds with such fast rise times.) Second, if high intensity pressure waves are experienced, small muscles in the middle ear can contract to stiffen the ossicular chain and attenuate the sound. Third, the mass and stiffness of the ossicular chain are such as to prevent the transmission of a pressure wave with extremely fast rise times (of less than $50\,\mu$ sec).

The inner ear performs two separate functions. The first concerns the process of hearing and the second (which is discussed in the next section) the maintenance of posture. The primary receptor organ for hearing is the cochlea, which derives its name from its coiled structure, similar to the shape of a snail shell. It tapers slightly along its length, with the broader end incorporating the 'oval window' of the inner ear, and it is filled with fluid. Running the length of the cochlea is a membrane (the basilar membrane) which acts in a similar way to the ribbon in an old carbon ribbon microphone.

Opinions differ as to the precise mechanism of hearing. However, a simple model would suggest that the sound pressure waves are transmitted as vibrations across the inner ear from the eardrum to the oval window by the action of the ossicular chain. These in turn set up hydrostatic travelling waves along the cochlea and basilar membrane, which cause the membrane to vibrate and its covering of hairs to be compressed. The sensation of pitch is produced because different parts of the membrane are sensitive to different frequencies. High-frequency sounds are perceived at the base of the basilar membrane, whereas the apex is sensitive to low-frequency sounds, while loudness is discriminated by the extent to which the hairs are compressed.

Sound Localization

Although the action of the ear may be considered to be purely mechanical, our interpretation of what we hear is certainly not. For example, just as our two eyes enable us to perceive depth, so the provision of an ear on each side of the head enables us to tell from which direction a sound is coming. This information can be extremely beneficial in, for example, a dangerous situation when, in addition to registering the presence of a warning sound, being able to locate the sound could provide the 'split second' needed to take appropriate avoiding action.

As with visual depth perception, our ability to localized sound sources also utilizes cues. Instead of being related to the size of the object, however, these cues are in terms of the time different taken by the sound to reach each ear.

It is possible to understand how sound localization occurs by imagining a sound source positioned to the right (north east) of a fixed head (which is facing north). Because the distance is slightly further to the left ear than to the right, it will take correspondingly longer for the sound pressure waves emanating from the sound source to reach the left ear. In fact if the sound source is positioned directly to the right side of the listener it will take 0.029 m sec longer to reach the left ear for every centimetre difference between the two ears. This time difference is enough to give a cue as to the left–right location of the sound source. Furthermore, in addition, to the time cues there is also likely to be an intensity difference at the two ears because of the shadowing effect of the head. The action of the head as a barrier to sounds reaching one ear rather than another, therefore, helps to provide another cue as to the direction of the sound source.

These cues can perhaps best be demonstrated if the operator is wearing earphones. If pulses of equal intensity are led to each ear, and if each pulse arrives at the same time, the listener will perceive the sound as being situated in the middle of his head. If the pulse to the right ear is made to arrive slightly earlier, however, the composite sound which is perceived will appear to originate from a direction more to the right. Similar effects can be obtained with intensity cues. Thus if the intensity of the stimulus to the right ear is increased, even when the pulses arrive simultaneously at each ear, the composite sound will again appear to be coming from a direction to the right. Combinations of time and intensity cues, therefore, are likely to produce the ability which we have to localize sounds accurately.

Accurate localization, however, only occurs when the sound source is situated to the right or to the left of the listener. If the head is fixed, front–back and up–down discrimination is very poor. For this reason it is very important that the head be free to move to allow us to place the sound source along a left–right line relative to the ears. When this occurs, of cource, we begin to integrate information derived from the neck and shoulder muscles (proprioceptive information) with that obtained from the time and intensity cues to localize the sound source more efficiently.

Tone and Loudness Discrimination

Frequency and intensity (and their corresponding subjective attributes, tone and loudness) are the two defining characteristics of any auditory stimulus. However our ability to discriminate between (and thus act upon) different sounds is limited.

In this respect McCormick (1976) differentiates between two types of discrimination process. First, a relative type of judgement in which there is an opportunity to compare two or more stimuli, for example, comparing two sound in terms of their loudness, or two colours in terms of their hue. Second, there is an absolute type of judgement in which no such obvious opportunity exists to be able to compare the stimuli. In this case the task becomes more one of identification or labelling.

As might be expected, people are able to make more accurate discriminations on a relative judgement basis. Using colour as an example McCormick (1976) suggests that most people can differentiate as many as 100 000 to 300 000 different colours on a relative basis when comparing two at a time. The number of colours that can be identified on an absolute basis, however, is limited to no more than a dozen or so.

In most work settings, discrimination is likely to be based primarily on absolute judgements. The operator may have to decide to press the 'red' button, to react to the 'soft' noise, or he may need to operate the 'small' control. Little opportunity for immediate comparison between stimuli exists in each situation. Unfortunately, however, the ability of people to make absolute discriminations between individual stimuli of most types is not very large. In this connection Miller (1956) refers to the 'magical number seven plus or minus two', implying that the range of such discriminations is somewhere around 7 ± 2 (or 5 to 9). Thus we are able to discriminate four to five sounds of different loudnesses and about the same number of different tones.

If different dimensions are combined, however, the number of absolute judgements which can be made is increased. For example Pollack and Ficks (1954) asked subjects to discriminate between over 15 000 tones, each having different frequencies, intensities, durations, rates of interruption, ontime fractions and locations in space. Under these conditions their listeners were able to discriminate about 150 separate tones without error, although their accuracy diminished as more tones were added. Miller (1956) sums up these and similar results by saying that 'as we add more variables (dimensions). . . we increase the total capacity but we decrease the accuracy for any particular variable. In other words we can make relatively crude judgements of several things simultaneously'.

In terms of auditory discrimination, therefore, the message is clear: to increase the number of dimensions carried in the meaningful sound.

The Limitations of the Ear

From an ergonomics standpoint, the main imperfection of the auditory system as an analyser lies in its intermittent failure to be able to detect a signal in noise. When this happens the signal is said to be *masked* by the noise. It is an important problem in many working situations, for example, when listening to speech in 'static' or perceiving a warning sound in a noisy environment.

Masking can be considered to be both a central and a peripheral phenomenon. It occurs at the periphery (in the ear) due to the direct interaction of the two sounds (signal and noise) on the eardrum and on the ossicular chain. It also occurs as an interaction of the two signals in the central processing system (the brain), but the reason for a masking effect at the central level is far from clear. Masking, and possible ways of overcoming it, are discussed in more detail in Chapter 10.

THE PROPRIOCEPTIVE SENSES

The proprioceptive senses are those concerned with the perception of the body's own movement and with informing the individual of his position, and the position of his limbs, in space. In essence this sensory system is composed of two separate systems: the vestibular system in the ear which is mainly concerned with maintaining the body's posture and equilibrium, and the kinaesthetic system which consists of sensors in the muscles and tendons to indicate the relative positions of the limbs and of different parts of the body.

The Vestibular System

Situated within the inner ear and just above the oval window of the cochlea are sense organs which are collectively known as the vestibular apparatus. When describing this system it is usual to distinguish between two types of vestibular organs: the semicircular canals and the utricles.

The semicircular canals comprise three, almost circular, tube-like structures which are attached to one another in the vertical and two horizontal planes. Each canal is filled with a fluid which, by virtue of its inertia, flows through the canal whenever the head moves in the plane of that canal. The semicricular canals thus act as angular speedometers which are capable of discriminating rotational movement in any direction as the head is moved (Reason and Brand, 1975). In addition to conveying information about the rate at which the head is being turned, the signals produced by the fluid movements also generate reflex eye movements, termed nystagmus, which help to keep the visual world stable. Since each canal is part of a three-part system, when the sensations from all canals are integrated they will inform the person about the direction of his rotary head movements as well as its speed of movement.

The second organ, the utricle, is found at the point where the three canals meet. As with the semicrircular canals the utricle is filled with a fluid, but it also contains a flattish blob of jelly called the otolith which is covered with dense crystals. When the otolith moves in the fluid it probably stimulates hair cells around the cavity, and this gives information about the head's orientation with respect to gravity. Thus the urticles inform the individual about the tilt of his head, and also its linear displacement.

The vestibular receptors, therefore, enable man to maintain his upright posture and to control his position in space. The utricle informs the body about the static position of the head—whether the individual is upright, standing on his head or leaning over. The semicircular canals provide similar information about the head's rotation, its speed and direction and thus its equilibrium.

The Kinaesthetic System

Because a man at work needs to know what each part of his body is doing during any operation, the kinaesthetic senses form an extremely important system. This system is operated by receptors which are situated in the muscles and tendons and convey information to the brain concerning the extent to which these structures are deing stretched.

Three types of kinaesthetic receptors can be found in the body tissues. The first, which are spindle-shaped, are located in the muscles and provide information concerning both the extent to which the muscle is being stretched and to the rate of stretching. The second type of receptor is located at different positions in the tendons. These give information relating to the extent to which a joint is moved so that, again, speed and direction of movement will be indicated. The third type of kinaesthetic receptor (Pacinian corpuscle) is located in the deeper tissues. These receptors furnish information regarding deep pressure and are sensitive to any deformation in the tissue within which they are embedded. In addition they can frequently be stimulated by a squeezing action whenever the body or the limb changes position, and are thought also to be stimulated by vibration.

Each of these receptors will give the operator some idea of where his body or his limbs are positioned in space, without him having to use his eyes. For example by integrating the information obtained from the biceps and triceps in his arm an operator can tell by how much his arm is extended. With further information from the biceps and triceps tendons and from his shoulder muscles, he should be able to tell by how much the arm is having to be supported, in other words, its position with respect to the horizontal.

The proprioceptive system is one which is frequently overlooked when considering an operator's behaviour in a working environment. This is possibly because no one visible organ is responsible as, for example, the eye is the organ for vision or the ear for hearing. In many respects it provides 'unconscious' information. A lack of consciousness, however, does not imply a lack of importance, for the proprioceptive system is absolutely necessary for notifying the operator about the activities of his body without him having to monitor every part of it, for example his feet and toes while walking or his hands while operating a control above his head.

As an example of the importance of this system Fleishman (1966) has listed eleven important ability traits, found by himself and his co-workers, which underlie a large range of physical skills. The names provided are arbitrary but, as Dickinson (1974) points out, many of the abilities either include proprioception to a greater or lesser extent or are totally measures of different kinds of proprioceptive sensitivity. The eleven factors are:

1. Control precision: this factor is common to tasks which require fine, highly controlled muscular adjustments, primarily where larger muscle groups are involved.
2. Multilimb coordination: this is the ability to coordinate the movements of a number of limbs simultaneously.
3. Response orientation: this ability has generally been found in tasks which involve rapidly discriminating direction and orientating movement.

4. Reaction time: this represents the speed with which the individual is able to respond to a stimulus when it appears.
5. Speed of arm movement: this is similar to reaction time but represents the speed with which an individual can make gross, discrete arm movements where accuracy is not required.
6. Rate control: this ability involves making continuous anticipatory motor adjustments relative to changes in the speed and direction of a continuously moving target or object.
7. Manual dexterity: this ability involves skilful, well-directed arm–hand movements and is involved in manipulating fairly large objects under speed conditions.
8. Finger dexterity: this is the ability to make skill-controlled manipulations of tiny objects using the fingers.
9. Arm–hand steadiness: this is the ability to make precise arm–hand positioning movements; the critical feature, as the name implies, is the steadiness with which such movements can be made.
10. Wrist and finger speed: this ability could be called 'tapping', and relates to the ability to move the wrist and finger quickly and in time with some external stimulus.
11. Aiming: speed and accuracy of placement are critical features of this ability.

The proprioceptive system, therefore, plays a major role in training skilled behaviour, since much of the development of complex motor skills depends on the efficient feedback mechanisms. For example, in a skill such as typing, the feedback obtained from the kinaesthetic receptors in the fingers, arms, shoulder muscles and joints allows the operator to be able to sense where his fingers ought to be placed without any conscious placement on his part. In addition the efficient use of the kinaesthetic system (in other words when he is fully skilled) enables the operator to be able to sense when a limb is in an incorrect position and rapidly move it to the correct location. Car drivers, for example, have learned that the right foot controls the accelerator and brake, and the left foot the clutch. It takes a conscious decision to operate the brake with the left foot and if he did so, the driver would feel 'uncomfortable' in his lower leg and ankle.

As Dickinson's list suggests, a further use of the proprioceptive system in skilled behaviour lies in the information which it provides with respect to the timing of motor responses. In a large proportion of physical skills it is not enough merely to be able to predict appropriate responses; the timing with which a response is initiated is also important for smooth performance. For instance, when hitting a moving ball the subject needs to anticipate the arrival of the ball and to time his response so that the ball is struck when it is in a specific position. In addition to helping him to place his arm and hand at the correct place, the proprioceptors provide information regarding the speed and direction of his arm movement so that he can reach the correct place at the correct time.

Finally it should be emphasized that muscles operate many parts of the body in addition to the skeletal system. For example the position of the eyes is maintained by the presence of muscles which attach the eye to the socket. Proprioceptive receptors are also present in these eye muscles and provide information about the degree and direction of the eye's movement. Furthermore, if the eye is fixated at a particular point, the proprioceptive feedback from these muscles informs the individual of his body's orientation around the eye. The role of the muscle's proprioceptive systems, therefore, should never be overlooked.

SUMMARY

This chapter considered the roles of various aspects of the sensory nervous system in providing the channel for information to flow from the outside world to the human operator. Since they represent the first link in the man–machine system, the importance and modes of function of the visual, auditory and proprioceptive systems need to be understood before the efficiency of these channels can be increased. The next chapter will perform the same service for the body's motor processes—the communication channel in the other direction.

CHAPTER 3

The Structure of the Body:
II Body Size and Movement

The sensory physiology and psychology discussed in the previous chapter essentially describe the ways in which information is received and decoded by the operator. The limitations of the sensory and decision-making apparatus perhaps represent the first set of restrictions to the efficient working of the closed-loop, man–machine–man, system.

The efficient reception and interpretation of information by the man, however, is only part of the problem. For the loop to be continued the information normally needs to be responded to in some way, requiring an appropriate movement to be made by the man to transmit his information to the machine. In terms of the structure of the body, therefore, the second possible level of limitations for the closed-loop system lies in the operator's ability to use his bones, joints and muscles to move his body, or parts of it, in the desired way.

One of the more obvious restrictions of movement which the operator could experience is likely to be his own physical size. A tall man in a small room, a large hand operating a small control, a small pair of legs trying to cover a large distance each illustrates how the body dimensions themselves restrict the ability to move. The study of body dimensions, often referred to as *anthropometry*, represents an essential aspect of any ergonomics investigation.

Whereas the sizes of the bones and tissue may restrict movement in the initial stages, complete mobility can be brought about only by using the system of joints and muscles which connect the limbs. When considering such aspects as, for example, the action of the back muscles and joints in taking strains during lifting or sitting, or the range of arm movements during lifting and of leg movements during walking, the ergonomist enters the territory of the physicist and applied mathematician. Because the actions of the bones and joints are analysed and interpreted in terms of a complex system of levers, this aspect of body mobility is known as *biomechanics*.

The purpose of this chapter is to consider the ways in which man carries out and controls his motor behaviour and the factors which limit his peformance.

BODY MOVEMENT: BONES, JOINTS AND MUSCLES

The 206 bones which make up the human skeleton perform one or both of two functions. A few, such as those which comprise the skull or the breastbone, protect the vital body organs from mechanical damage. The majority of bones, however, do not fulfil a protectionist role but give the body rigidity and enable it to perform its required tasks. In this respect, and from an ergonomics standpoint, the bones most directly concerned with doing work are the long bones of the arms and legs and small long bones of the toes and fingers. These are characterized by having a shaft with two enlarged ends which form suitable surfaces for the joints with other bones. A third group of bones can be added to that of the protectors and the workers, of which the ribs are an example. These perform both a protectionist (of the lungs) and a working (aid to breathing) function.

In essence the bony skeleton of the body consists of two lever systems, the arms and the legs, joined together by an articulated column, the spine. Since most bones in the body are designed to aid movement of the body parts, they are connected to each other at joints and are held together by the ligaments and muscles. The prime function of the ligaments, therefore, is to hold the joint tightly together and to resist any sideways movement which may damage them, but by doing so they also tend to limit movement when a muscle is fully stretched.

In addition to being determined by the ways in which the muscles are distributed around the joints, the direction and degree of movement of parts of the body also depend largely on the shapes of the joint surfaces. For example, the joints in the fingers and in the elbow and knee have a simple hinge action which permits movement in one plane only. Some joints, however, such as those in the wrist and ankle, can allow movement in two planes owing to their surfaces being less flat. Finally, the hip and the shoulder are ball and socket joints which permit a wide range of movement. The ball of the hip joint is almost completely enclosed in a deep socket which gives great mechanical strength but limits the range of movement. In the shoulder, however, where a much wider range of movement is needed, the socket is more shallow, but a dislocated shoulder is quite common as a result.

The joints in the spinal column are of a special type. Instead of the two smoothed surfaces of the joint moving against each other, spinal joints have between them a fibrocartilage disc whose function is to act as a shock absorber and to allow a wide range of gross body movement. The individual vertebrae in the spine are joined to each other by a series of elastic ligaments which help to maintain the normal curvature of the spine. As Murrell (1971) points out, they are probably the only ligaments which maintain a steady strain. The effect of the whole of this assembly (vertebrae, discs, ligaments) is that the body can bend forwards readily by about 180 degrees, but not very far backwards.

In addition to bending, the spine also allows the body to rotate—the degree of movement varying between 90 degrees in the vertebrae of the neck to about 30 degrees in the lumbar region. Murrell (1971) emphasizes the importance of such rotational movements in being able to scan the area surrounding the body. Thus the combined movement of the neck, lumber region and the eyes enable the full horizon of 360 degrees to be scanned with the pelvis remaining in the sitting position.

The importance of the spine in maintaining posture will be considered later (Chapter 9) but it is useful to note here that should undue strain be put on the mechanism, the result may be an aching back. A more severe strain can cause damage to the muscles, to the elastic ligaments and, if the strain is sudden, to the intevertebral discs. The result of this may be to cause part of the disc to be squeezed into the spinal canal producing the extremely painful back condition known as a 'slipped disc'.

Despite the perfect mechanical engineering of the body's bones and joints, no work could ever be done without the adequate functioning of the muscular system. Three types of muscles are present in the human body. The first controls the action of the main working bones and is known as striated (or skeletal) muscle. They are composed of long, thin, cylindrical fibres which are connected in bundles, and are attached to the bones via tendons. These may be so short as to be invisible or quite long as in the tendons which operate the fingers. Approximately 40 per cent of the total muscles of the body are striated and, except in some clinical conditions, their operation is under the control of the operator. For this reason it is the striated muscle and its functioning which is most interesting to ergonomists. The second type of muscle, whose action is not under voluntary control, has a smooth appearance and maintains the functioning of vital body organs such as the stomach and intestines. Finally the heart is made from a muscle system unique to itself, cardiac muscle, which is similar to a mixture of both striated and smooth muscles.

Muscles are able to contract only in one direction, and when they do so they become about one-half of their original length. This limits the amount of limb movement to a function of the length of its individual fibres. On the other hand the force which can be exerted by a fibre is independent of its length, the maximum strength of the muscle being determined by the number of fibres which it contains. Because of the way in which nerve impulses act, individual muscle fibres are only able to be in a state of contraction or non-contraction (a so-called 'all-or-none' state). Controlled, gradual movement, therefore, is brought about by extra fibres successively being brought into operation. Thus the greater the force which is required, the greater are the number of fibres which are brought into play. As was discussed in Chapter 2, such actions are monitored by the kinaesthetic receptors in the muscles, and this provides the feedback loop needed to tell the operator the extent to which his action has been carried out.

MUSCLE STRENGTH, ENDURANCE AND FATIGUE

The work levels which the human operator is called on to perform must clearly be within his physical as well as his cognitive capabilities. This is the reason for discussing such aspects as anthropometrics and biomechanics. However it is often forgotten that the muscles themselves are contstrained in their ability to carry out work. This is due in the first place to limits on their strength and secondly on their ability to maintain that strength (in other words, their endurance and their resistance to fatigue).

When discussing any of these factors a clear distinction must be made between the type of work which the muscle is called on to do, that is between static and dynamic work. This distinction is normally made in terms of whether or not motion accompanies the muscular tension. The work is said to be static if no motion occurs, for example when holding a weight in the palm of the hand with the arm outstretched but not moving. If the arm moves up and down, however, then the upper arm and shoulder muscles are said to be doing dynamic work.

Strength

Kroemer (1970) has defined strength as 'the maximal force muscles can exert isometrically in a single, voluntary effort'. He accepts, however, that such a definition places the concept of strength strictly in the realms of static muscular load. Very little work has been carried out to determine levels of dynamic muscular strength.

From an ergonomics standpoint, it is necessary to understand muscular strength because of the levels and types of resistance which are built into all machines. This concept of resistance will be discussed in more detail in Chapter 6 when considering controls, but in essence it refers to the fact that machine controls, for example, need to work actively against the operator's movement to be effective. This allows for more precise movement and guards against accidental operation. At a relatively low level, therefore, some strength is required to operate a control. Other situations require higher levels of strength, for example, in the case of mechanical foot controls which operate heavy machinery. In these cases the resistance occurs naturally as a result of mechanical linkages and is not purposefully built into the machine. Finally, it is relatively easy to understand the concept of resistance to change in some situations, for example when heavy loads need to be moved. Here the resistance is caused by vertical or lateral forces, and they need high levels of muscular strength to be overcome.

Ergonomists need information on muscular strength, therefore, to be able to suggest appropriate control and movement systems; to determine maximum and optimum control resistances; to define the forces required in various manual tasks and to ensure the adequate arrangements for safe, efficient lifting or carrying. Levels of human strength are also relevant to the design of equipment which is used under abnormal or special conditions such as space-travel— because of restrictions on area and weight, conventional sources of power may be logistically impractical or expensive. Finally, it is important to remember that most muscular actions which interest ergonomists commonly require the integrated exertion of many muscle groups. For example thrusting on a pedal requires turning the ankle, extending the knee and the hip and stabilizing the pelvis and trunk on the seat. In these cases the maximum force which can be exerted in a complex action will be determined by the weakest link in the muscular chain concerned.

Many factors are related to (and perhaps influence) muscular strength, perhaps the most obvious being age; tasks requiring strength which are easily exerted by young workers may exceed the capacity of other, older groups. In terms of the many individual factors Damon, Stoudt, and McFarland (1971) provide a list which is by no means exhaustive, but which includes age (strength increases rapidly in the teens, reaches a maximum in the mid to late twenties and remains at this level for five to ten years), sex (women are generally about 30 per cent weaker than men of the same age). Although this broad generalization implies that the differences in muscle strength between the sexes are the same for all muscle groups, and that men and women use their limbs in the same way, Redgrove (1979) argues that no evidence exists to support this suggestion. Indeed, Hettinger (1961) produces evidence to suggest that, for different muscle groups, women's strength varies on average between 55 per cent and 80 per cent that of men); body position; fatigue; exercise; health; diet (hunger or inadequate diets decrease body strength); drugs; diurnal variation; environmental factors; motivation and occupation (although this may be related to the different amounts of exercise taken by people in various occupations). Additional factors include weight (Rasch and Pierson (1963) have demonstrated that body weight is an important determinant of arm strength) and height (Caldwell (1963) obtained a correlation of 0.76 between pulling strength and height).

Not withstanding such variables, Davis and Stubbs (1977a, b; 1978) have produced safe levels of the single loads or forces which can be lifted safely by fit young males, standing, sitting, kneeling and squatting with the back erect. Their data, however, were obtained from a military population and care must be taken if they are to be applied to civilians. However Davis, Ridd, and Stubbs (1980) have reported that a male industrial population produced very similar data up to the age of about 40 years, but the maximum loads which they could lift safely decreased to about 75 per cent of the military data after about age 50.

Musclar Endurance

The term endurance relates to man's ability to continue to work or, in the static case, to continue to exert force. All of the data which have been obtained in this area lead to the same conclusions: that the length of time for which a force can be maintained depends on the proportion of the available strength which is being exerted. Thus the smaller the force which is required, the longer it can be exerted. This relationship between required strength and endurance has been shown to be non-linear (for example Caldwell, 1963, 1964; Carlson, 1969) as may be seen in Figure 3.1 (from Kroemer, 1970). This implies that while total strength may be maintained for only a few minutes (that is how strength is defined), if the contraction is to be maintained for a very long time only about 15 to 20 per cent efficiency will be possible.

PERCENT OF STRENGTH

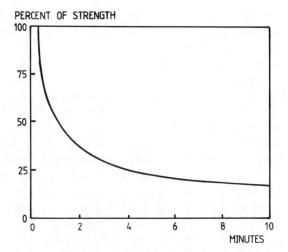

Figure 3.1 The relationship between strength and endurance limits (from Kroemer, 1970, reproduced by permission of The Human Factors Society, Inc.)

At this point a distinction needs to be made between individual and group strength. Although the function shown in Figure 3.1 indicates the relationship between endurance and the percentage of each person's individual strength, the curve was obtained by averaging across many different strengths. Carlson (1969) and others, however, have demonstrated that whereas the relationship between strength and endurance shown in Figure 3.1 remains the same over different groups, weaker people are able to maintain proportionately greater endurances than stronger people. In other words, the curve is shifted slightly up or down according to the individual's maximum strength. Carlson suggests that this paradox is due to the fact that the stronger people, by definition of the term 'proportion of maximum strength', have to exert higher forces. This means that the muscles are working harder and requiring more oxygen-carrying blood. However because the muscles are more contracted, they tend to occlude the blood vessels which limits the length of time that contractions can take place. (The importance of oxygen in reducing muscular fatigue is discussed below.)

Muscular Fatigue

Fatigue is an important aspect of any situation in which work is done, whether it be dynamic or static work. Depending on the degree of fatigue experienced it may cause discomfort, distraction, and possibly a reduction in satisfaction and performance. In many cases such factors rapidly lead to accidents. Muscular fatigue can often be avoided, however, if the manner in which it arises is understood, so that the work is designed to avoid the factors which induce it.

For a muscle to contract (that is, to do work) an extremely complicated chemical reaction is set up in the muscle itself. Described in its simplest form,

38

the energy for the contraction is supplied by the breakdown of a chemical in the muscles called adenosine triphosphate (ATP) to adenosine diphosphate (ADP). However the ADP must be regenerated to ATP before another contraction can take place, and the energy for this reversing action is providing by the breakdown of glycogen.

Unfortunately, a byproduct of the glycogen breakdown is a poisonous substance called lactic acid which quickly accumulates in the muscles causing the muscular pain so well associated with fatigue. This is removed by a reaction with oxygen and converted into carbon dioxide and water. The function of oxygen, therefore, is to convert the byproducts of the energy-producing reaction and this may continue for some time after muscular activity has taken place. The role of the blood is to transport oxygen to the muscles and to remove the carbon dioxide and water. The simplified reaction is shown in Figure 3.2.

It is interesting to note that the energy for muscular activity comes from a reaction which does not depend primarily on the presence of oxygen. This means that work can be done even if the immediate supply of oxygen is insufficient, and this allows the body to make a sudden, extreme effort which would be quite impossible if the energy had to be obtained from the oxidation of some substance in the muscle fibre. So the importance of understanding the mechanisms which cause fatigue lies in the fact that the oxygen supplied by the blood, and the blood itself, are the sole agents for either reducing the level of fatigue or for increasing the length of time before fatigue sets in. Conditions need to be designed, therefore, in which the flow of blood to the muscles is maximal.

Blood flow is increased during dynamic work simply as a result of the pumping action set up by the muscles doing the work. Blood is pumped through the blood vessels which supply the muscles, so helping the breakdown of lactic acid and removing the carbon dioxide and water. As long as the supplies of blood and oxygen can be maintained in sufficient quantities, therefore, and are not exceeded by the production of lactic acid, musclar fatigue is likely to be

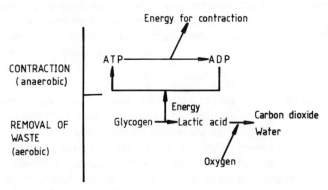

Figure 3.2 A simplified diagram of the process by which the energy needed for muscle contraction is obtained

kept at bay. Under conditions of static load, however, no such pumping action occurs and the muscles soon become starved of the oxygen which they require. When this occurs they are said to be in 'oxygen debt'. (Oxygen debt can also occur, of course, in dynamic work if the supply of blood and oxygen is not sufficient.) This effect can be demonstrated well if work is carried out with arms above the head, in which case the effect of gravity and the lack of muscle contraction serve to restrict blood flow to the muscles, so causing an oxygen debt.

As Murrell (1971) has pointed out, the moral of the tale is that work design which calls for the application of a force over a long period, whether operating a foot pedal, pulling a lever, working above the head or carrying a suitcase, should be avoided. As far as possible all muscular activity should be intermittent so as to allow the blood to flow through the muscle to reduce the possibility of any oxygen debt building up, or to facilitate the paying back of a debt which has been incurred.

When discussing fatigue, it is not commonly realized that the main muscles which do static work continuously are those in the back and shoulders, since they are involved in maintaining posture. Although they have more red muscle fibres than muscles which do dynamic work, they often run into an oxygen debt if a rigid posture is maintained for too long. This is particularly important to the seated operator. For example a typist may develop back and shoulder pain unless adequate opportunity is given to relax and contract the muscles, perhaps by walking about or by doing other work. In addition the design of the seat itself can help to reduce the degree of static load on the postural muscles, and this will be considered in Chapter 9.

The same sensations of pain and fatigue may, in some cases, occur after long periods of continuous work even if an oxygen debt has not built up. This is caused by the fluid produced after the breakdown of lactic acid, building up between the muscle bundles and being unable to be carried away by the blood. This fluid causes the muscles to swell, which may then press on the intramuscular blood vessels further reducing the flow of blood to the muscles. In most cases this fluid will eventually be dissipated by rest after the activity has ended but, if insufficient rest is allowed, then any more activity is likely to cause an even further increase in fluid level. Over time a muscle which is continuously used may become distended by intramuscular fluid and fibrous material can build up. This could ultimately interfere with the normal muscular contraction and cause permanent damage.

Heavy or repeated muscular exertion may have other consequences. For example it is fairly common for frequently used muscles to become hypersensitive, when they are more liable to contract than a muscle which has not been exercised. Localized contractions or muscular spasm may result from such hypersensitivity, and perhaps a buildup of intramuscular fluid could cause local tender areas. In addition, whole groups of muscles may go into contraction spontaneously to give what is commonly called 'cramp'. This can occur when some muscles are used continuously to exert a static force or on repeated movements of a comparatively short range. 'Writer's cramp', for example, is a common affliction amongst most authors!

TRAINING THE MUSCLES

Being able to train the muscles is an important facility for anyone who depends on the work done by muscular activity. This is particularly so in sport but is also true in the case of heavy industries which include a great deal of lifting, pulling or pushing.

The two aspects of muscular work important in this context are those of strength and endurance. With regard to muscle strength, Hettinger (1961) produced the greatest improvements in the strength of different muscle groups by single, maximum, isometric contractions of the muscles for 1–2 sec/day. No further increase in strength was obtained with more than a single maximum contraction per day, which implies that long, continuous periods of exercise are unnecessary for training. With one maximum muscle contraction every other day, the increase in strength fell to 80 per cent of that obtainable with daily sessions; with two sessions per week the comparable figure was 60 per cent; and with a single session weekly only 40 per cent.

Muller (1965), too, has demonstrated that standard isometric training over eight weeks increased the average static muscular strength by about 30 per cent of the original lifting strength, but it did not have similar consequences for dynamic strength or endurance. Dynamic training, however, did increase dynamic endurance, particularly at the higher levels of work output. However it still did not affect dynamic strength.

Muller explains these results by pointing to the fact that endurance depends on how much oxygen the cardiovascular system is able to pump to the working muscles in a given time. Since muscular training also helps to train the cardiovascular supply by making it have to deliver more oxygen carrying blood to the periphery, this helps to lower the load on the heart and produce a higher output per heart beat.

By relating the success of training methods to the amount of blood flow in the periphery, Muller suggests three further techniques to increase endurance. First, massage during rest pauses: he showed that by manually supplying the contractions needed to pump the blood through the muscles without the muscles having to work (and thus produce lactic acid), quicker recovery from work fatigue was obtained. Thus the work which was possible in eight hours could be concentrated in two or three hours using massage. Massage has its limitations, however, since exhaustion was far greater after 2 to 3 hours work with massage than after the eight hours without massage. For this reason Muller suggests that massage has a potential for sports, but not for industrial work.

A second technique which he suggests it to temporarily restrict the blood supply to the appropriate muscles prior to work. He illustrates his argument by the observation that if one presses the skin with a finger hard enough to stop the blood supply, the pressed spot will become red (that is, gorged with blood) when the finger is removed. Using a cuff placed around the thigh for up to ten minutes, Muller was able to demonstrate a 600 per cent increase in endurance on a pedalling task. Again, however, the cost of such increases in endurance is likely to be in terms of more long-lasting fatigue. In addition prolonged occlusion of the blood flow can be dangerous and should not be attempted by untrained personnel.

The final technique proposed by Muller does not increase fatigue. This involves cooling parts of the skin, so increasing the flow of blood to the periphery. He demonstrated that repeated five-minute periods of heavy pedalling work, interspersed with ten-minute rest pauses, could be continued for longer if the legs were immersed in water at 15 degrees C during the pauses.

In summary, therefore, understanding the physiological and biochemical mechanisms of human movement is useful for two reasons. First appropriate work and equipment design may be developed to reduce the number of situations in which restricted blood flow (and thus fatigue) can occur. Secondly the physiological principles may be used to increase human physical work capacity through training and other techniques.

BODY SIZE: ANTHROPOMETRY

For many hundreds of years man has realized the importance of knowing something about his own body sizes. Indeed measuring units of the 'foot' and the 'hand' are still used today, having been derived from the dimensions of standard body parts. The idea that the physical size of the person is somehow related to his ability to function in the world is so old that it is surprising how often the concept is neglected in everyday thought and design.

The term anthropometry is derived from two Greek words: *anthropo(s)*–human, and *metricos*—of, or pertaining to, measurement. Thus the subdiscipline is concerned with 'The application of scientific physical methods to human subjects for the development of design standards and specific requirements and for the evaluation of engineering drawings, mock-ups, and manufactured products for the purpose of assuring the suitability of these products for the intended user population' (Roebuck, Kroemer, and Thomson, 1975).

The ergonomist, therefore, will use anthropometric data to ensure, quite literally, that the machine or the environment fits the man. Whenever the human operator has to interact with his environment it is important to have details of the dimensions of the appropriate part of the body. So, overall stature is an important determinant of, for example, room size, door height, or cockpit dimensions; the dimensions of the pelvis and buttocks limit the size of hatch openings or seats; the size of the hand determines the demensions of controls and supportive stanchions; and it is necessary to have details of arm reach to be able to position control consoles at appropriate distances. The list of possible examples is virtually endless.

Until recent years the main source of anthropometric data has been obtained primarily from military settings. This is possibly due to two reasons. First, large numbers of subjects need to be measured so that representative dimensions of a population can be obtained. The military scientists have at their disposal many thousands of available men and women, particularly during wartime. Second, with the development of faster fighting machines (including tanks, boats, planes and submarines) space has been put increasingly at a premium. The need, therefore, to build effectively a machine around a single operator, but at the same time to build for a number of different operators, has required the appropriate anthropometric data. In addition anthropometric data is needed to design efficient clothing for use in a number of different theatres of war.

Unfortunately, however, data obtained from military personnel tends to be slightly misleading if civilian applications are envisaged. For example, conscripts measured during or just after the war may not have had a full diet or may have been working in difficult surroundings. Over time, factors such as these may affect the development of some body dimensions. In addition, as will be discussed later, the problem is further increased since anthropometric data obtained from members of one country, for example the United States, may not apply to those of another country, for example Great Britain.

The type of anthropometric data which mainly interests ergonomists can be divided into two categories: first, structural anthropometry (often called static anthropometry) which deals with the simple dimensions of the stationary human being—for example, weight, stature, and the lengths, breadths, depths and circumferences of the body structures. The second category is called functional anthropometry (or dynamic anthropometry) which deals with compound measurements of the moving human being—for example, reach, and the angular ranges of various joints.

Table 3.1 Details of some anthropometric surveys of civilian populations

Reference	Date	Sample size	Sex	Age
Kemsley	1950	27 515 33 562	Male ⎱ Female ⎰	14–75
Roberts	1960	78	Female	Average 71.65
Damon and Stoudt	1963	133	Male	Average 81.62
Ward and Fleming	1964	70	Male	?
Ward and Kirk	1967	100	Female	? (elderly)
Lewin	1969	87 77	Male ⎱ Female ⎰	25–49
Garrett (hand anthropometry)	1971	26 23	Male ⎱ Female ⎰	? (students)
Andrews and Manoy	1972	323	Male	24–65
McClelland and Ward	1976	140	Male & ⎱ Female ⎰	18–30; 60+
Haslegrave	1979	1584 416	Male ⎱ Female ⎰	17–64
Guillien and Rebiffé	1980	8005	Male	20–60
Diebschlag and Muller-Limroth	1980	50 50	Male Female	?

For both types of anthropometric measurements, comparative data are slowly becoming available on selected civilian populations. Some of the more easily obtainable reports are listed in Table 3.1. In addition, Barkala (1961) provides details of some earlier surveys undertaken to estimate body measurements in relation to seat design; and the results of the majority of studies until 1971 have been collated and tabulated by Damon, Stoudt, and Mc Farland (1971). These last two references, however, deal mainly with data from military populations.

Regarding functional anthropometry, the simplest forms of such data are tables which indicate the ranges of motion of individual body articulations. In this respect the definitive study of joint ranges remains that of Dempster (1955).

Although some situations may exist in which precise details are required, for some practical purposes simple approximations of the different body dimensions are both suitable and available. For example, a set of cards with rotating dials may be purchased which roughly indicate average dimensions in a number of different situations (Diffrient, Tilley, and Bardagy, 1974). However it must be

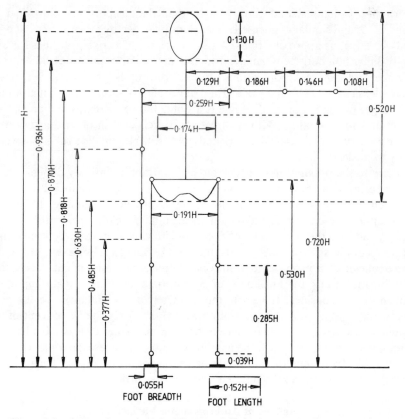

Figure 3.3 Approximate lengths of parts of the body expressed as proportions of the total height (H) (after Drillis and Contini, 1966, reproduced by permission of the U.S. Government Department of Health, Education, and Welfare)

remembered that these simple approximations are just that, and cannot take the place of data from fully controlled anthropometric surveys.

Humanscale contains six sets of dimensions (normal body dimensions, the lengths of different limbs or parts of limbs for normal seating, for sitting at tables, for wheelchair users, and for the elderly and handicapped). The authors accept the caution given above by emphasizing that the cards should not be taken as a panacea for all design problems. They provide no more than a starting point for building models and mockups before the final workplace or object is designed. This point cannot be emphasized too strongly, particularly since the data sets are all based on American samples.

Similar aids have been provided by a number of investigators (see for example, Roebuck, Kroemer, and Thomson, 1975, Chapter 8), and a useful approximation of the relationship between body parts is shown in Figure 3.3 (after Drillis and Conti, 1966).

Variability in Anthropometric Data

It cannot be emphasized too strongly that the proportions illustrated in Figure 3.3, and similar condensed dimensions provided by other sources, should only be taken as approximations. A certain degree of variability exists for any body dimension both between members of a particular population and between members of different populations. In this respect height provides a ready example, since a quick survey of a group of people will soon reveal that although the height of the majority of people is around 160 to 170 cm, some people are taller and some smaller. With such variability reflected in other dimensions it is easy to see how simplified compound graphs, tables or slide-rules may serve only to mislead.

Because the population does exhibit such variability in body dimensions, it is the normal custom when reporting anthropometric data to indicate the extent of such variability. It has become common practice, therefore, to specify anthropometric data in terms of statistical numbers called percentiles, which simply indicate the percentage of the population who have body dimensions up to a certain size (in other words, that size or smaller). For example, taking the diameter of a proposed entrance hatch, an ergonomist might decide that hip breadth (plus the thickness of appropriate clothing) is an important dimension to consider. If he sets the diameter of the hatchway to the 50th percentile (that is, to the average hip breadth) then only the 50 per cent of the potential users who have a smaller hip breadth may enter or leave. Under such circumstances, particularly if the hatchway represents an escape route, it would be sensible to design to the 100th percentile or even larger. Then all the population (100 per cent) would be able to pass through.

Sources of Anthropometic Variability

The wide distribution of body dimensions and shapes which may be encountered in any one population can often be due to slight genetic differences. However other, more readily observable, variables can also affect body dimensions and their variability, and these include age, sex, culture, occupation and even historical trends.

Age

The change in body dimensions from birth to maturity is well known and, indeed, the increases occur consistently although sometimes irregularly. For height, as for most other body lengths, full growth is attained for all practical purposes by the age of 20 in males and 17 in females (Damon, Stoudt, and McFarland, 1971). Some 'shrinkage' of older people has also been noted (Stoudt et al., 1965), but this apparent change could be related to some of the historical trends discussed below. It may also be due to slight degeneration of the joints in old age.

Sex

With the increasing observance of sexual equality in the workplace, designing for differences in body dimensions between the sexes will become an important aspect of the ergonomist's task. In this respect males are generally larger than females for most body dimensions, and the extent of the difference varies from one dimension to another. Examples of this may be seen in the various hand dimensions provided by Garrett (1971). In each of the 34 different dimensions (hand length, thickness, and depth; hand, fist and wrist circumference; digit length and thickness, etc.), male sizes were larger than female sizes with the greatest differences occurring in the thickness dimensions (the dimensions of the males being approximately 20 per cent larger than those of the females). For the length dimensions (hand, finger) males were only approximately 10 per cent larger than females.

Women, however, are consistently larger than men in the four dimensions of chest depth, hip breadth, hip circumference and thigh circumference. Pregnancy markedly affects certain female dimensions, mainly in the abdomen and pelvic regions but also in the breasts. Such changes begin to reach anthropometric significance about the fourth month of pregnancy (Damon, Stoudt, and McFarland, 1971), and Diffrient, Tilley, and Bardagy (1974) suggest that the average female abdominal depth increases from 164 to 290 mm throughout pregnancy.

Culture

The importance of national and cultural differences in anthropometrics has been realized for sometime, but there has been little concerted effort to implement the relevant data in the production of new plant and machinery until relatively recently. As Chapanis (1974) suggests, up to the present time ergonomics has been largely an American and Western European discipline. With more efficient communication and travel facilities, however, and with the emergence of newly industrialized countries, the need to design different environments to fit all populations is becoming felt. He further points out:

Today Americans buy automobiles, typewriters, taperecorders and calculating machines made in France, Germany, Britain and Japan. We in turn sell our computers, aircraft and farm machinery to Europe, Africa, South America, Western Europe and perhaps soon to the Peoples Republic of China. Similarly the European Common Market has resulted in a free interchange not only of products, but of workers among the several countries belonging to that confederation. These are only a few of the internationalis-ations of business and commerce that have become an important fact of modern life. (Reprinted form *Ergonomics* with permission from the publishers.)

The implication of this statement should be clear: in addition to leading to poor worker performance, inappropriate anthropometric design may result in the loss of orders and exports to overseas countries.

The variability in anthropometric dimensions due to national and cultural differences may not all be as dramatic as the difference between the central African pygmy tribes (average male height, 144 cm) and the Northern Nitoles of Southern Sudan (average male height, 183 cm) (Roberts, 1975). However, for design purposes the differences are important. Kennedy (1975) illustrates this point with data relating to the range of heights of seven military populations: United States, France, Italy, Japan, Thailand and Vietnam. Relating these statures to cockpit design, he points out that it is customary in the US Air Force to design for the central 90 per cent of the population (that is, from the 5th to the 95th percentile). Whereas this range would accommodate essentially the same percentage of Germans, as far as height is concerned it would fit only the upper 80 per cent of the French, 69 per cent of Italians, 43 per cent of Japanese, 24 per cent of Thai and 14 per cent of Vietnamese users.

Variations also appear to exist between people in different areas within the same country. For example, Guillien and Rebiffé (1980) analysed the statures of a large number of bus drivers from east and west France. Their data, shown in Figure 3.4, illustrates a large difference between the statures of the subsamples, although the decrease in stature with age is fairly similar.

For reference purposes White (1975) has collated anthropometric data for the weights, heights, sitting heights and chest circumferences of male, military samples from 19 countries.

Occupation

Differences in body size and proportions among occupational groups are common and are fairly well known. For example, many body dimensions of a manual labourer are on average larger than those of an academic. However, such differences may also be related to age, diet, exercise or many other factors, in addition to some degree of self-selection. For example, only males taller than 172 cm or females taller than 162 cm are accepted for recruitment for the British police force. Whatever the reasons for the differences, however, anthropometric

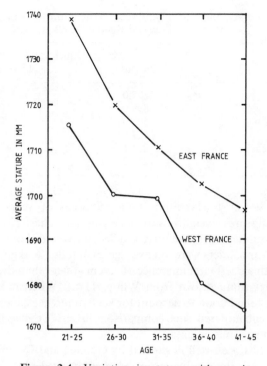

Figure 3.4 Variation in stature with age in people from different areas of France (reproduced with permission from Guillien and Rebiffé, 1980, Anthropometric models of a population of bus drivers. In D. J. Oborne and J. A. Levis (eds.) *Human Factors in Transport Research, Vol. 1.,* copyright by Academic Press Inc. (London) Ltd.)

variability in different occupations must be realized (a) for designing environments for the particular occupation, and (b) before using anthropometric data obtained from members of one occupation to design the environments of another.

Historical trends

Many people have observed that the equipment used in earlier years would be too small for efficient use today. Suits of armour, the heights of doors and the lengths of graves provide indications of the stature of people in former times, and each is smaller than would be required in modern times. This has led to the suggestion that the average size of the population increases over time, perhaps due to better diet and living conditions. Unfortunately no detailed evidence exists either to support or refute this position.

Some evidence exists, however, to suggest that similar but smaller increases

Table 3.2 Estimated increases in anthropometric dimensions over 15 years
(from Guillien and Rebiffé, 1980)

	Average Dimension in mm.	
	Present	15 years time
Stature	1697.0	1720.0
Sitting height	899.0	909.0
Buttock-knee length	593.0	601.0
Knee height	527.0	535.0
Maximum hand reach	861.1	870.0
Shoe length	280.3	282.4

may occur in shorter time periods than hundreds of years. For example comparing the statures, weights, sitting heights and chest circumferences of demobilised US soldiers from the First and Second World Wars, it appears that Second World War soldiers were, on average, 3cm taller, 6.4kg heavier, and sat 0.9cm higher with a chest circumference of 3.9cm greater than their First World War counterparts (data drawn from White, 1975). Whereas many extrinsic factors could be called upon to account for such differences, including diet and types of fighting encountered, such comparisons illustrate the need for continued updating of anthropometric data.

This problem has been well illustrated by Guillien and Rebiffé (1980). They were concerned with the production of adequately sized cabs for bus drivers, particularly since the vehicles are designed to remain in service for two or three decades. From an anthropometric survey of over 8000 male bus drivers they were able to predict that the height of their drivers would increase by the order of 3 cm in ten years although, with the reduction of height with age, this figure would reduce to an increase in height of 2.3 cm over 15 years. From these data they were able to predict other dimensions for their drivers over 15 years, as shown in Table 3.2.

BODY MOVEMENT: BIOMECHANICS

The human body has been built to move by the action of its bones, joints and muscles, and the movement may take many varied and complicated forms. Because of this a separate subdiscipline, biomechanics, has been developed to consider the mechanics and range of human movement. Roebuck, Kroemer and Thomson (1975) define biomechanics as the

Interdisciplinary science (comprising mainly anthropometry, mechanics, physiology and engineering) of the mechanical structure and behaviour of biological materials. It concerns primarily the dimensions, composition and mass properties of body segments, the joints linking the body segments together, the mobility in the joints, the mechanical relations of the body to force fields, vibrations and impacts, the voluntary actions of the body in bringing about controlled movements in applying forces, torques, energy and power to external objects like controls, tools and other equipment.

It is beyond the capacity of this section to consider the total range of interest of biomechanics, but the important actions of the man at work, namely those of walking and lifting will be discussed. The types of movements of the body segments and their respective angular limits may be obtained from reference texts (for example Damon, Stoudt, and McFarland, 1971). However, it should be remembered when reading such figures that the ranges of joint motions vary from person to person due to anatomical differences, and also as a result of other factors such as age, sex, race, body build, exercise, occupation, fatigue, disease, body position and the presence or absence of clothing (Damon, Stoudt, and McFarland, 1971).

The Mechanics of Locomotion

From an ergonomics, standpoint, the mechanics of locomotion are important for a number of reasons. First, locomotion can cause fatigue; second, understanding the way in which people walk may help in the design of suitable footware; third, many accidents are caused by slipping (or inadequate locomotion); and finally understanding how the normal legs work may help in the design of suitable prosthetic devices for the handicapped.

Walking from one place to another on two legs may seem to be a simple act. Actually it is the product of many complex interactions between the forces generated within the body and several external forces acting on it, coordinating in such a way as to produce not any 'step' but rather a particular pattern of movement known as 'normal step'. This is accomplished by means of a control system which automatically integrates changing conditions in the force–motion–position picture at scores of joints within the body and the changing pattern of external gravitational forces. As discussed in the last chapter, the vestibular apparatus in the ear plays a major part in this control system, which also corrects for changes in temperature, pressure, friction, loads being carried, obstacles, height, weight, and orientation in space.

In considering the way in which we walk, it is possible to make two separate levels of observations. The first is to observe the normal gait or movement of the legs and leg segments, and the second is to consider the more objective, but less easily observed, forces acting at the major joints.

Gait

In describing the biomechanics of gait, Peizer and Wright (1974) divide the cycle into two phases: stance and swing (see Figure 3.5). The stance phase begins when the heel of one leg strikes the ground and ends when the toe of the same leg lifts off. The swing phase represents the period between toe off on one foot and heel contact on the same foot. As we alternate from swing to stance on each leg, there is a period when both feet are in contact with the ground simultaneously. This is called the period of 'double support' and occurs between push off and toe off on one foot and heel strike and 'foot flat' on the other

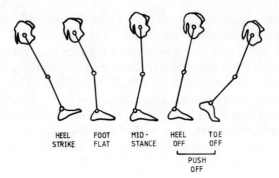

HEEL   FOOT   MID-   HEEL   TOE
STRIKE   FLAT   STANCE   OFF   OFF

PUSH
OFF

STANCE PHASE

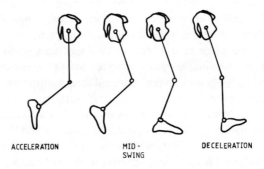

ACCELERATION   MID-   DECELERATION
                 SWING

SWING PHASE

Figure 3.5 Diagrammatic representation of the stance and swing phases of human gait (after Peizer and Wright, 1974, and reprinted by permission of the Council of the Institution of Mechanical Engineers from *Human Locomotion Engineering*)

(the absence of double support indicates that a person is running rather than walking).

At ordinary speeds, a single leg is in its stance phase for approximately 65 per cent of the cycle and in swing for approximately 35 per cent. The period of double suport occupies between 25 and 30 per cent of the gait cycle time.

Motions and forces

The forces which cause locomotion result from those created by the muscles and from external forces, principally the influence of gravity on the body.

When standing erect, the body's centre of gravity is in front of the hip, knee and ankle joints. This force tends to bend (flex) the hip, straighten (extend) the knee and bend (dorsiflex) the ankle. To begin walking we normally relax the muscles in one leg, allowing the body to fall forward to that side, and swing the leg forward at the hip.

The importance of analysing the normal gait and the motions which produce it may be seen in Figure 3.4. The section of gait producing the most instability is at the push off of one leg; at this point minimum contact is made with the ground, with what contact there is occurring solely through the toes of one foot. In addition the pelvis is well forward of the contact point and stability is further reduced by the other leg being swung forward. It is at this point that most slips occur.

Slipping is one of the most common causes of accidents on the job, and depends largely on the static friction which exists between the foot and floor prior to the slip. In this respect Kroemer (1974) presents a table which indicates the coefficients of friction set up between eight different shoe materials and 12 floor materials. Considering the floor material alone, he showed that rubber pads, concrete, and soft wood have the highest coefficients of friction (that is, are most resistant to slipping), and they were not substantially affected by soiling. Of the shoe materials, the standard rubber sole used by the US Air Force and the US Army proved superior to other sole materials such as neoprene. Although a flat rubber and cork sole had a higher overall coefficient of friction under dry conditions, it did not fare too well in 'soiled' conditions.

Carlsoo (1972) suggests that there is not just one but two particularly critical phases of the normal gait at which slipping is likely to occur. The first occurs at heel strike at the beginning of the stance phase. The weight of the body is then lying behind the contact point of the heel and the ground while the movement of the body's centre of gravity has just started the swing phase. The second critical instant is at the actual pushoff when the body's centre of gravity lies in front of the pushing foot. It could well be said that the first of these two instances presents the greatest danger, since if a person slips when his centre of gravity is behind his foot he is liable to fall backwards with little chance of using his hands to catch himself. At least if he slips during the second instant the body is bent forwards and he is likely to fall forwards.

Carlsoo also suggests that slipping is highly likely to occur if the coefficient of friction between the foot is less than that required for normal walking (about 0.4). However, we are able to walk on what would otherwise be slippery surfaces because the coefficient of friction depends on two forces which act normally and tangentially on the foot. Thus for any given combination of floor surface and footware the resultant coefficient of friction can be increased by using the muscles of the ankle, leg and hip to alter the forces which act on the foot. However, if the muscles are used too frequently to overcome the tendency to slip, they will soon become fatigued. Once again, therefore, it is clear that the environment should be adapted to suit the operator, otherwise its poor design will soon become apparent.

In an attempt to aid the designer in his quest for a suitable environment, particularly in terms of slipping, Grieve (1979a and b) has developed a technique which he describes as PSD (Postural Stability Diagram) analysis. His technique is too detailed to be described here but his basic argument is that a work situation will be subjected to two types of constraints: personal (to do with the operator, his strength and the position at any time of his body and body parts) and environmental (to do with the friction of the interface and to the demands of the job and equipment). By measuring levels of these constraints in the push, pull, press and lift dimensions, Grieve describes how the resultant data may be combined graphically to determine contours of safe operations.

The Mechanics of Lifting

Lifting is an action which is frequently required in work. However it is one which, if carried out incorrectly, may result at the very least in back pain and discomfort or at the worst in permanent disability. Indeed Russek (1955) established the high incidence of back injuries in tasks involving weight lifting, and indicated the lumbar area of the spine to be the back region most susceptible. As an example of the problem, Whitney (1958) reports a survey carried out in 1953 which indicated that nearly 30 per cent of all accidents in factories occurred during the manual handling of goods. Furthermore, Benn and Wood (1975) have estimated that over 13 million days a year are lost in industry due to painful back conditions. As they point out, this constitutes a greater loss of time than even strikes (between 6.8 and 11 million days in 1969 and 1970 respectively).

Nevertheless, some caution should be urged before accepting too strong a causal relationship between lifting and back injuries, simply because the epidemiology of these injuries is hard to establish. For example Magora (1972) conducted a survey of over 3000 workers in Israel to determine the incidence of low back pain, and considered their occupational requirements in terms of sitting, standing and lifting. As he points out, the results as shown in Table 3.3 are extremely interesting from a number of points of view. First, more workers experienced pain symptoms who rarely or never sat, stood or lifted in their jobs than those who did so often. Second, the distributions of back pain for weight-lifting jobs were not significantly different from the sitters or standers. Third, they would suggest that the way to reduce the possibility of back pain is to introduce variability in the work.

Many workers, for example Davis and Troup (1964), have shown that manual activity is regularly accompanied by considerable increases in intra-abdominal pressure. This is due to the action which all of the muscles around the abdomen (the diaphragm, pelvic floor and anterior abdominal wall) have on the fluid contents of the abdomen. These pressure increases are related to the amount of force which acts on the lower spine, and so Davis and Stubbs (1977a) have reported that occupations in which peak intratruncal pressures of 100mmHg or more are induced, have an increased liability to report back injuries. (This was

Table 3.3 Proportions of workers complaining of lower back pain (LBP) having different occupational requirements for sitting, standing and lifting (from Magora, 1972)

Sitting	Proportion complaining of LBP (%)
Often	12.6
Sometimes	1.5
Rarely	25.9
Standing	
More than 4 hours per day	13.8
Variable	2.5
Less than 4 hours per day	24.9
Lifting	
Often	13.9
Variable	2.5
Rarely	28.1

particularly so for the mining and construction industries.) Such pressures can be monitored by swallowing a small pressure-sensitive radio pill and it was on this basis that Davis and Stubbs produced their safe levels of manual forces reported earlier.

The biomechanics involved in lifting depend primarily on the body posture and on the techniques employed, of which there are essentially two. The first, commonly known as the derrick action, derives its name from its general similarity to the action of a derrick crane. Throughout the lifting operation the knees are kept fully extended and the back and arms flexed forwards to grasp the object. The lifting action is brought about by extending, or attempting to extend, the lumbar region of the spine and hip joints. As Whitney (1958) has observed, this appears to be the natural method of lifting.

In the second method, known as the knee action, the grasp is made by folding the legs as in squatting. In this technique the trunk is maintained quite erect, and the lifting action occurs primarily as a result of extending the knee joint which, in turn, extends the hip joint.

Some authors have compared the two types of action solely in terms of the forces produced downwards through the feet. Whitney (1958), for example, has demonstrated no difference between the forces which can be exerted using the two techniques, implying that both are equally effective in lifting weights. In general these conclusions have been supported elsewhere. If this is the case, and since the knee action probably requires more initial energy to be consumed in establishing the posture in the first place, this probably exaplains the common observation of the derrick action being the more natural action.

Unfortunately, however, these conclusions are based solely on the forces able to be exerted using the two techniques, and takes no account of other problems associated with the derrick action. Thus as Davies (1972) emphasizes, bending and twisting during a lift causes spinal injury and this is more likely to occur during a derrick lift. Furthermore, in addition to potentially damaging the spine,

54

the increased pressures in the truncal region predisposes the operator to a hernia.

It is for these reasons that the knee action is the type of lifting action most commonly supported these days. In this action there exist four links in the lifting chain (as opposed to three in the derrick action) namely the lower legs, the upper legs, the back and the arms. When the back is kept in its natural, curved position the forces on the intervertebral surfaces and discs will be evenly distributed, and it will be the muscles rather than the ligaments and bony structures that counteract the pull of gravity. These suggestions are supported by Tichauer (1971) who has shown that as the lifting action nears that of the knee action, the load on the back muscles is reduced.

A pamphlet *Lifting in Industry* (Anon, 1966) has been prepared to advance the correct lifting technique. In essence it describes the knee action in terms of: (1) the feet should be far enough apart for balanced distribution of the weight; (2) the knees and hips should be bent and the back kept as straight as possible with the chin tucked in; (3) the arms should be held as near to the body as possible; (4) wherever possible, the whole of the hand should be used to grasp; and (5) lifts should be made smoothly without jerks or snatches.

The following formulae (from Poulsen and Jorgensen, 1971) may be used to predict the maximum lifting load using this type of action:

males: maximum load = 1.1 × isometric back muscle strength
females: maximum load = 0.95 × isometric back muscle strength — 8 kg

These figures, however, refer to a single lift. Jorgensen and Poulsen (1974) have demonstrated that lifting ability falls dramatically when repeated lifts need to be made. The relationship follows closely that seen between strength and endurance. Thus the authors suggest that only between two to three lifts per minute should take place if the load to be lifted is 75 per cent of the maximum load possible. For a load of only 10 per cent of maximum, six to nine lifts per minute can be tolerated. Ronnholm (1962), however, has demonstrated that if such lifts are able to be rhythmically interspersed with short rest periods, efficiency and output may be increased and the work load may be reduced. The sex, age height and body weight of the lifter, however, may affect these conclusions, as Snook (1978) has demonstrated.

Before concluding the discussion of lifting, it should be noted that Procrustes is starting to rear his head again. One might well question whether, by instructing industrial man to adopt the knee action, one is attempting once again to fit man to his environment rather than make the environment fit the man. If the derrick action is the natural lifting posture, then the mere fact that heavier loads could possibly be lifted more safely by some other action should simply mean that reductions should be made in the maximum loads needed to be lifted—not that man should be trained to adopt an unnatural posture. As was illustrated in Chapter 1, training may become ineffective under conditions of stress.

Unfortunately no survey data are available to support or to refute the contention that the derrick action is the natural action; nevertheless, this action does allow more freedom for the operator. For example, restrictive clothing such as skirts, tight trousers and protective aprons, coupled with the extent to which the knees obtrude into the load space, are all factors which may reduce the possibility of the operator adopting the crouch position.

Finally it is still the case that the incidence of reportable back injuries has not lessened since the Second World War despite intensive efforts to introduce the 'correct' methods. It may well be, therefore, that the human operator is still trying to exert his 'authority' over his environment, suggesting that more data is required concerning the safe limits for the derrick action.

SUMMARY

This chapter considered some of the structures in the final link in the man–machine–man system—the bones, joints and muscles. No matter how efficiently a mechanical system has been designed—how fast it works, how reliable it is, how aesthetically pleasing it appears—if the operators cannot fit into it or around it, if they cannot pull the levers or push the buttons with sufficient force for a sufficient length of time, or if they cannot reach the controls in the first place, then the mechanical system is at least effectively useless, and at worst potentially dangerous.

CHAPTER 4

Man–Man Communication: Words and Symbols

In many respects information is the keystone of contemporary civilization. Without it governments would be unable to govern, generals would be unable to command their troops or put them in a position to win battles, and industrial organizations would be unable to function. However much we might complain about the bad news which is heard and seen continually in the media, as individuals we would be lost without constant, accurate, updated information about the state of our world.

The transmission of information between individuals, therefore, plays an extremely important, indeed a fundamental, role in everyday living and working. However, before the ergonomics of man–man communication can be discussed in more detail the concept of information transmission needs to be considered in its widest context, and not simply in terms of the written or spoken word. At work, for example, an operator is likely to need information about how a machine works, how to operate it safely, what to do when it breaks down, and so on. Much of this he will obviously receive from his colleagues, from his supervisor or from a manual. However, information is also passed to him from the machine via its displays telling him, for example, how hot it is, how fast it is moving, or what parts of a process have been reached. In addition the operator imparts information to the machine via his controls, 'telling' it what he wishes it to do—to go up, to go down, to turn right, or to increase speed.

Since information is so important in the working environment, it should be apparent that the way in which the information is transmitted is as crucial as the message being conveyed. This suggests, therefore, that the efficient communication of information, in addition to simply the information *per se*, is also fundamental to the working situation.

Put in its simplest terms, any communication system may be conceived of as being composed of a transmitter who (or which) communicates information through some medium to a receiver. But communication can only be said to have occurred if the receiver interprets the information in the way in which the transmitter intended. The point may perhaps be illustrated by an example in which two people who speak different languages are trying to converse. The difficulties which are experienced arise largely from the receiver being unable to translate correctly the ideas conveyed by the transmitter. In such cases 'sign language' possibly helps the communication process by placing the message in a different medium—but one which can be understood by both. The problems could be increased, however, if either one can speak a smattering of the other's language but interprets the message incorrectly. In this case the communication system is likely to break down because misunderstandings occur.

Because effective communication is so important for efficient working, and because inefficient communication can lead to a breakdown within the man–machine system, one of the ergonomist's major functions is to design systems which will enhance the chances of the correct message being received and understood correctly. For this reason the present and following two chapters will consider different aspects of this communication process.

Since either a man or a machine can act as the transmitter or the receiver, four possible combinations of transmitter and receiver can be envisaged: a man communicating with another man; a machine communicating with a man; a man communicating with a machine; and a machine communicating with another machine. Since the first three of these options involve the operator interacting with his environment, they are within the realms of ergonomics and will be discussed in this and in the next two chapters. The fourth option (machine–machine communication) lies more in the domain of the cyberneticist or the engineer and will not be discussed in this book.

The present chapter, therefore, considers the problems presented by man-to-man communication—primarily those involved in the written communication of information and instructions. In Chapter 5, the question of communication between the machine and the operator is considered, while the channels for reply (from man to machine) will be discussed in Chapter 6.

MEANING, COMMUNICATION AND THE MESSAGE

The process underlying communication can perhaps best be understood when one realizes that the linguistic origin of the word is the Latin *communis*, meaning 'common'. One clear characteristic of meaning in human communication, therefore, is its 'commonness'—an essential prerequisite being a common understanding by those involved in the communication process. Fisher (1978), however, points out that the quality of commonness does not necessarily mean that all participants in the communicative process have to have identical understanding of the symbols or thoughts to be transmitted, just so long as some understanding is common to them all. This implies, therefore, that communication can take place even if both the transmitter and receiver do not 'speak the same language'—as long as the basic ideas are received accurately.

The role of ergonomics, then, is to arrange a situation which enhances the chances of the transmitter and the receiver having a maximum common understanding. However, before considering the ways in which this state of affairs may come about it is interesting to consider briefly some of the reasons why the two may not share a common channel for communication—why meanings and messages may be misinterpreted.

First, personalities might be involved. If the receiver dislikes or mistrusts the transmitter, he could be unreceptive to what he has to say. Indeed his understanding of the message could well be distorted by his opinion of the transmitter or by what he thinks are his motives. Second, even if the receiver is receptive to the message it may become distorted by his own personal preconceptions (his set). As Bartlett (1950) has suggested, one of the chief functions of the active mind is to 'fill up gaps', that is to try constantly to link new material with older material to make it more meaningful. So a message might be distorted by the receiver placing it in a context in which it is not meant to be placed, simply because of his 'set' to the message. Third, there is the problem of interest and attention. If the receiver is not attending to the message, perhaps because of boredom, part of it may easily be missed or misinterpreted. Finally, messages can become distorted because of our limited capacity to process the information which is being transmitted. This concept has already been discussed briefly when dealing with our ability to store different stimuli, but it also applies to the storage of complex, meaningful material such as ideas and concepts. If the message is too detailed or too long, it may overload our memory system and become distorted, lost or simply unable to be retrieved. In summary, therefore, the common channel for communication may be distorted by social influences, personal attitudes and expectations, boredom and a lack of interest, or information overload.

Messages which have to be transmitted from man to man may take many forms, and Miller (1972) divides them into three groups according to the physical appearance of the stimulus. These he terms verbal (including written words and signs), vocal (including spoken words and voice variations), and physical (including gestures and movements) stimuli. The remainder of this chapter considers, primarily, the influence which ergonomics can play in facilitating the transmission of 'verbal' stimuli (including written communication). As the reader will become aware, many of the principles involved with verbal stimuli may also be applied to vocal stimuli but, because other aspects of the environment are likely to be important (for example environmental noise, or social and organizational factors), the specific problems of speech communication are dealt with in Chapters 5 and 10. The whole area of the role of non-verbal stimuli in the communication process (body signs, gestures, etc.) is somewhat, too wide for the scope of this book, but texts such as those by Argyle and Cook (1976) or Morris (1977) may prove interesting.

TYPOGRAPHICAL ASPECTS IN WRITTEN COMMUNICATION

For written information to be communicated efficiently, the message needs not only to be read (and interpreted) correctly but absorbed in the shortest time possible. Fast reading is important from the point of view of economy of time and, perhaps more importantly, it ensures that our long-term storage memory capacities are not overloaded. For example, the longer that a reader needs to decipher a word or a symbol on a page (perhaps because of bad handwriting or because he has the fifth carbon copy or a faint photocopy), the slower will be his rate of comprehension. Furthermore, it seems likely that such factors play quite a large part in determining whether the material will be read at all (McLaughlin, 1966). It is only comparatively recently, however, that those involved in presenting written information have considered, in any systematic way, the question of typographical factors such as the type and quality of the print and the page layout, despite the fact that it is these aspects which convey the message.

One industry which appears to have a fairly good record on this is the newspaper industry, probably because they have to capture their reader's attention very quickly. Most people who take a daily newspaper do not read it systematically from front to back, like a book, but probably only glance at the headlines and read only the paragraphs of news below which looks interesting. Even the paragraphs of news may not be read systematically if the reader is in a hurry or is only mildly interested—he may simply run his eyes down a paragraph looking for news of a particular sort.

One of the questions which needs to be answered by typographical designers, therefore, is how to capture the reader's eye. In most written communications this is done by headlines and by subheadings.

In 1946, Paterson and Tinker investigated the relative values of upper and lower-case letters in newspaper headlines. They presented their subjects with one headline at a time for a limited period and measured the number of words read in the time available. At normal reading distance, their results indicated that on average more words were read when using lower-case letters than when using headlines printed all in capitals of the same type face and point size (in this case 24-point Cheltenham Extra condensed). These results were replicated by Poulton (1967) who showed a 9 per cent advantage of lower-case over upper-case headlines in reading ability. A similar advantage was obtained by mixing upper and lower-case letters, although these were not read any more reliably than headlines printed entirely in lower case.

Poulton (1969a) suggests that the reason for the superiority of mixed over upper-case letters probably lies in the shape of the envelope surrounding the whole word as presented (rather than the individual letters). Words presented in capitals do not have any distinctive shape since all the letters are the same height. The shapes of various lower-case words, however, are more likely to be different because of the ascenders and descenders on different letters. For example the shape of the envelope enclosing the word *dog* is different from that enclosing the word *cat* due to the extensions of the *d* and *g* in *dog* and the *t* in *cat*. The envelopes around *DOG* and *CAT*, however, are not widely different, and this lack of distinctive shapes between different words in capitals means that the reader has to examine some of the intermediate letters to identify the words, thus increasing reading time.

In addition to the form of the letters which make up the printed word, their size in an important consideration. Type size is conventionally specified in terms of the height of a line of print, the unit being the 'point' which is 0.0138″ (about $\frac{1}{3}$mm) high.

Using two measures of readability, the amount read and the degree of comprehension, Burt (1959) compared the readability of passages composed of different sizes of Times Roman print (designed in 1932) ranging from 8 to 14 point. Unfortunately he did not provide any data to support his claim, but was able to state that the 10 point prose was the most legible. Most students found the 9 point print equally as legible as the 10 point, while older people often did best with 11 or even 12 point. These results were replicated by Poulton (1969b) when he asked housewives to search for particular words in lists of ingredients printed in 10, 7.5, 6 and 4 point lower-case type.

Deciding on the appropriate type size, however, is only part of the battle of designing a printed page for easy reading and comprehension. The layout of the page, the use of paragraph indentations, the number of columns, etc., all play an important part in producing an aesthetic as well as a readable product. Poulton, Warren, and Bond (1970) provide an interesting summary of the appropriate points, some of which are expanded below.

Headings

Headings serve a number of different purposes. Most obviously they assist readers who are looking for particular sections, and their inclusion in a table of contents may make the search task still easier. Perhaps less obviously, headings are also extremely valuable to those who are reading the entire report since they give the reader a structure which assists him in integrating the information as he reads. Each of these may help to increase the reader's comprehension and his memory of what he has read, as Dooling and Lachman (1971) have demonstrated. Unfortunately, however, no specific research evidence is available to suggest either the type of text which would benefit most from headings or the form that headings should take. As Wright (1978) has pointed out we cannot yet answer questions such as 'What kinds of reading purpose are most dependent upon headings', 'Which texts need headings', or 'What are the characteristics of a good heading'?

Numbering Headings

Numbering can serve two purposes. It may help to make clear to the reader the way in which the sections are nested together (thus replacing, perhaps, the need for subheadings), and it also enables the reader to refer in the index to specific sections which are smaller than a page. In this respect Perry (1952) has suggested that Arabic numbers are preferable to Roman numbers, although this is possibly so only as the numbers become larger and less frequently used (compare, for example, the ease of interpreting the sets of symbols iv and 4, and xxviii and 28). In addition Wright (1977a) suggests that the use of numbers for indexing is probably more useful than using letters of the alphabet, and the larger the sequence the greater is the advantage for numbers. For example people, are more certain that 8 precedes 10 than H precedes J.

Sectioning Prose

The use of sections has two advantages for the reader. First, it provides structure to the prose by informing him where one set of ideas end and a new set begins, and secondly it gives him a chance to 'collect his thoughts'. Regarding the design aspects, Hartley and Burnhill (1976) have suggested that leaving an empty line between paragraphs is a more effective cue for the reader than just indenting the first line of the new paragraph. A further possibility suggested by Wright (1977a) is to colour either the print or the background. She points out that a coloured background can sometimes be a useful way of providing supplementary indexing information, for example, as with an appendix. It enables the reader to turn quickly from the text to the beginning of the appendix without having to look for page numbers.

Cueing

In many cases significant words, phrases or passages in a text may be highlighted—perhaps by underlining, by italicizing or by the use of asterisks in the margin. This is known as typographic cueing and it helps the reader to pick out the salient points of the prose. If a cueing service is not provided by the author it may be added by the reader. In a sample of used textbooks, for example, Fowler and Barker (1974) found that the readers had inserted some form of cueing such as underlining or asterisks in almost 92 per cent of the books.

Various methods of cueing may be available to the printer, although italics are the most often used. However, the little evidence which is available suggests that the occasional use of italics for emphasizing significant points may be no better than plain text for ease of comprehension, and a complex cueing system may actually impair study.

The research on underlining as a means of cueing is not entirely conclusive. For example Cashen and Leicht (1970) showed that students who had received offprints with relevant passages underlined in red were better at answering questions both on the underlined statements and on the adjacent, uncued passages. It would appear, however, that the advantage of underlining is only apparent if the reader is not under a time stress to complete the passage. For example, in a study in which all subjects were allowed the same time to read the material, Crouse and Idstein (1972) found that cueing led to higher scores when readers were given 25 minutes to study a 600-word text, but not when they were given a much shorter time (either 2.5 or 5 minutes) to study a shorter (200-word) passage. Furthermore, the use of underlining as a cueing technique is likely to make the underlined parts of the text harder, rather than easier, to read because it causes the spaces between the lines of the text to be filled in.

Another cueing aid available is to highlight parts of the text by colouring the background of the particular line or part of a line. Using this technique Fowler and Barker (1974) found that subjects who were questioned a week after having read material were more likely to answer correctly the questions which related to the highlighted parts of the text than the non-highlighted material.

Finally, Foster and Coles (1977) considered the relative merits of using capital letters and bold type as cues. Their results indicated that the bold type was a better all-round technique of cueing. Although the capital letters led to higher scores on the cued material, when the subjects were tested on the uncued material they performed worse than those who had had no cueing at all. The authors suggest that when readers used the capital letter cueing technique, it hindered their reading performance, perhaps by affecting their reading behaviour, so that they had less time available for taking in the uncued portions of the text.

In summary, therefore, the results of the studies which have considered the value of cueing are inconclusive. In some cases cueing can increase reading performance, probably because the salient points are emphasized so reducing the time needed to search the material and the amount of information which has to be stored in memory. In other cases, however, cueing appear to have resulted in no benefit to the reader and even, in some cases, to have caused reduced overall performance because of its interference with normal reading patterns. The resolution of this conflict, therefore, might need to wait until research has been carried out in the light of the subject's own preferences for cueing.

ALTERNATIVE WAYS OF PRESENTING INFORMATION AND INSTRUCTION

The large number of books, journals, magazines, manuals and newspapers which are available to us every day would seem to point to the fact that we crave for the printed word. Whenever a new product is marketed it needs to be advertised, with words and phrases being used to present it in the best light. It will also normally carry an instruction manual to tell the operator how it works and how to work it, and a maintenance manual giving details of what to do if it breaks down. In addition to each of these, the machine itself is likely to carry small pieces of symbolic information to help the operator to work it. So following the production of a new lathe, for example, a glossy leaflet to interest the buyer may also be produced; the operator is likely to receive a comprehensive instruction manual; the machine will possibly have small information labels dotted over it saying 'danger', 'on/off' or 'position control'; and the maintenance man will be given a maintenance manual. Each of these communication media will attempt to reach a different audience, so will provide different levels and types of information. Each, therefore, has its own problems of information presentation, but some general points do apply and these are discussed below. However, as with most design recommendations, it should be remembered that the most appropriate means of communication for one situation may not be so for another.

The first question to be answered is how should the information be presented? For example as Wright (1977a) points out, people who keep information and refer to it when necessary (for example when using reference manuals) may have different requirements from those who, having studied the information, will be retrieving it from memory. Similarly, those who read a report from beginning to end will use the material in a very different way from those readers who only wish to understand specific information, or answer specific queries, and do not intend reading the entire document. Even among those who do read all of the information, there may be some who read it in order to take a decision but, once the decision is made, they no longer have any interest in the technical details.

Studies of alternative ways of presenting complex subject-matter have shown that some non-prose formats can increase the reader's ability to use the information. As an example, Dwyer (1967) provided subjects with information on the structure of the heart either orally or by simple line drawings. He showed that the line drawing was more effective when the subjects were subsequently tested on their memory of the heart's structure. However for tests which examined the subjects' 'understanding of the hearts' function, the oral presentation was most effective. Dwyer's results appear to suggest, therefore, that simple information is better presented visually, but when any action is to be taken, perhaps to integrate the information, it may be better to present it orally, with all the different emphases that the voice can attain.

In addition to the use to which it is to be put, the type of information to be communicated is also important when deciding on the appropriate communication mode. Booher (1975), for example, showed that if the information is 'static', that is merely informing the operator about, say, the location of controls on a machine, then it is better to present it as a series of pictures. 'Dynamic' information, on the other hand, for example informing the operator how to switch the machine on, was better presented as pictures and words.

The problems facing the ergonomist (or indeed the designer of any technical information), therefore, is to decide on the most appropriate way of presenting his information (or instructions) to his predicted audience. In doing so his first decision is to choose through which of the two senses, visual or auditory, his information is to be presented. Even this choice, however, can be limited, for example by environmental constraints. Thus, spoken instructions may be more costly to produce and are certainly less effective in, say, a noisy environment. Since more is said in the next chapter about using the auditory channel for information presentation, the remainder of this chapter deals with the visual presentation of information and instructions. In this case the designer is likely to have to choose between prose or some pictorial presentation such as graphs, barcharts, pictures, flowcharts and tables as his information channel. These are further discussed below.

Graphs and Barcharts

Graphs are used in a number of situations to present numerical information. They have the distinct advantage that they give the reader both the numerical data itself and indicate any apparent trends in the data. Furthermore, it two or more functions are plotted on the same graph, the differences and similarities in their relative trends can be easily seen.

Some data are available to help the designer to draw his graphs to their best advantage. For example Milroy and Poulton (1978) have shown that if more than one curve is drawn on the same graph, and each have to be labelled, then labelling each curve directly (rather than putting labels elsewhere on the graph or even under the legend) is likely to result in the quickest readings without any loss of accuracy. This is probably because direct labelling requires the reader to make fewer visual scans of the graph and so puts less strain on his short-term memory.

The number of curves that can be included on the same graph without any loss of intelligibility will clearly depend on the degree to which the curves overlap each other. As a simple rule-of-thumb Wright (1977a) suggests that three curves may well be the maximum for clarity if too much overlap is apparent. Indeed, it may also be advantageous to break graphs down into smaller units rather than to have too much information presented on a single graph.

Finally, Wright reports other work which has indicated that presenting the axes of a graph as a square (that is, extending the x and y axes on the top and right hands respectively) increases the ease with which the values of extreme points on the graph may be read. This is possibly because the squared axes

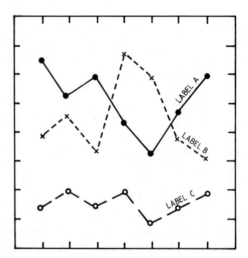

Figure 4.1   A hypothetical graph illustrating the principles of direct labelling and extending the axes

help to produce the perspective of a 'whole' figure, along the principles which are discussed below. Some of these suggestions are illustrated in Figure 4.1.

Despite the fact that they are used extensively, particularly in the advertizing industry, little comparable work has been carried out to investigate similar aspects of barcharts. Wright (1977a), however, has summarized their advantages and disadvantages in the following way:

One of the a priori advantages of bar charts would seem to be the ease to which they can be used to present many differently related variables. Some alternative ways of doing this are shown in Figure 11 [4.2]. Experimental comparisons among these different formats are hard to find but some of the formats will be more suitable for a specific purpose than others. For example, Figure 11a [4.2a] conveys the most information in the least space and very clearly shows the growth in total expenditure over time; but comparisons may be error prone either between items in the same year (e.g. in 1951 more was spent on defence than on education, but this is not obvious in Figure 11a [4.2a] or between different years for the same item (e.g. defence in 1968 and 1973). Indeed there may be two different ways of displaying the information to facilitate these two different comparisons. For example Figure 11c [4.2c] enables a comparison of items within the same year but comparisons between years are still very difficult; on the other hand, Figure 11b [4.2b] facilitates comparisons between the years but the distance between different items of expenditure may cause errors for these comparisons (e.g. education versus housing plus roads in 1968). Figure 11b [4.2b] makes it easier to compare the change in expenditure on particular items over time, whereas the lack of alignment between the different items makes this difficult in Figure 11a [4.2a].
(Reprinted by permission of Elsevier Scientific Publishing Co.)

Flowcharts

The flowchart (sometimes called an 'algorithm' or 'logical tree') presents the information to the reader as a set of choices and pathways and looks very similar to the computer programmer's flowcharts so much in evidence these days. On the surface at least, it appears to offer the reader information in an easily 'digestible' form, illustrating the various relationships (pathways) between different aspects in the information.

Rather than aiding the reader, however, if otherwise easily comprehensible information is to be transmitted, the research available has indicated that, under some circumstances, flowcharts may actually create problems. For example, although Wright and Reid (1973) have demonstrated that presenting difficult problems as flowcharts produced fewer subsequent errors than when the material was presented as prose, short sentences or tables, if easy problems were given flowchart presentation produced no advantage. Furthermore, if the material had to be remembered, the performance of those subjects who had been presented with flowcharts deteriorated faster. It would appear, therefore, that flowcharts are better if complicated information needs to be acted on immediately, but have little value if the information is otherwise easy to understand.

As with graphs, a few research findings are available to help in the design of a flowchart. For example Kamman (1975) compared the efficiency of two types of flowchart design of part of a telephone directory. The first was the

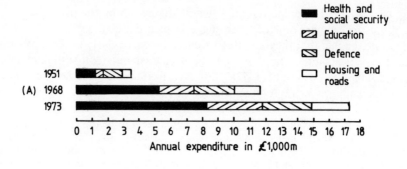

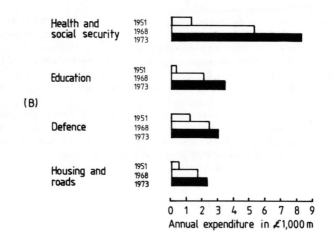

Figure 4.2 Alternative ways of displaying information using bar charts (from Wright, 1977a, reproduced with permission from *Instructional Science*)

normal vertical, binary, branching structure consisting of a series of question boxes leading, via arrows, to further question boxes or finally to 'action' boxes. The second consisted of a streamlined flow of short captions arranged horizontally. Each caption was embodied in white 'rivers' on a grey background, and each decision point could branch into three or even four pathways (see

Figure 4.3). The differences between these two charts lay in the maximum number of choices required to reach an action box (seven in the first chart and three in the second), and the number of words on the chart (chart 2 required about half of the number of words used in chart 1). Comparing these two forms and the normal telephone directory format, Kamman showed that the second (streamlined flow) chart produced significantly fewer errors than the first, and when subjects used either form they were less error prone and performed the task quicker than when they used the normal telephone directory.

Kamman's study illustrated one way of reducing the number of choice points. However, in some cases it may not be either possible or practical to do so, and in such cases Gane, Horabin, and Lewis (1966) have suggested forming small subcharts each of which deals with separate sequences in the operation. A more radical solution has been advanced by Jones (1968) who suggests that the material can be rewritten in a 'branching list' structure. In this case the information is listed vertically and the reader is required to jump to other questions. For example question 1 might read 'Is the motor turning? If YES continue with question 2. If No go to question 15'. Wright (1977b) argues that such a format may be useful in providing specific answers to specific questions, but it loses some of the other advantages of flowcharts—particularly the visual appearance of the decision making-structure.

In addition to reducing the number of choice-points and words, Wright suggests other ways of increasing flowchart efficiency. First, it is often the case that the various outcomes are not equally likely. The designer may therefore decide to design a flowchart so that the most frequently used decision path is the shortest possible—rather than reducing the overall average path length. Secondly there may be occasions when it makes better sense to emphasize particular aspects of the choice. For example, the fact that some of the outcomes may be more important than others might be an argument for putting the important decisions at the top of the tree; or different choices may have different levels of uncertainty attached to them, so making it reasonable for the decisions to be made in the order from the most to the least certainty.

Pictures

Illustrations are widely used in modern communication systems, particularly when technical terms are being referred to, and pictorial information is increasingly being used to communicate instructions for using equipment. This is particularly so when the equipment is to be used by people of different nationalities or with low levels of literacy. In addition, as Szlichcinski (1979a) observes, pictures can also be used during problem solving to answer specific questions which arise and they can influence the type of problem-solving strategy adopted. This, in turn, will determine what information is sought.

It must be emphasized, however, that pictures do have their limitations. For example Vernon (1953), reviewed the results of wartime research on the visual presentation of statistics and concluded that people needed special training to understand diagrams and make proper use of them. She also argued that

68

Design 1

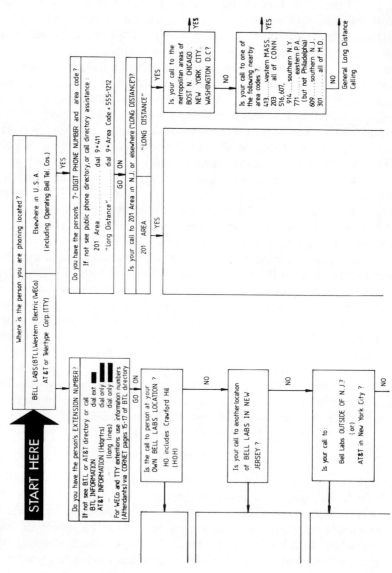

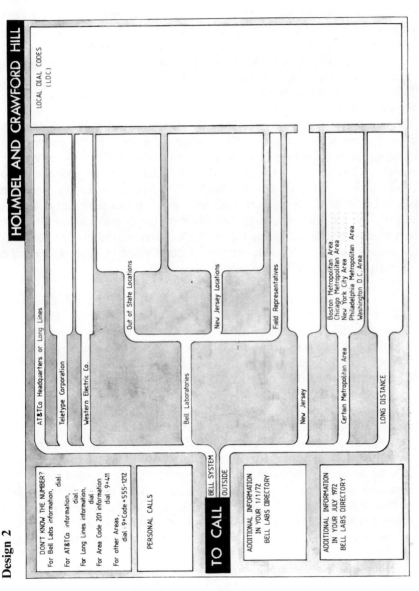

Figure 4.3 Two flowchart designs compared by Kamman (1975). Problems set using design 2 were read quicker and with fewer errors than with design 1; reproduced by permission of the Human Factors Society Inc.

diagrams and pictorial charts are likely to be less effective than prose if the concepts being presented are too complex for the readers to understand. As an example of the drawbacks of pictures, Szlichcinski (1979b) asked subjects to perform an operation involving turning a control, pressing a lever and throwing a switch, and were given only pictures as instructions. His subjects experienced great difficulty in performing this task correctly.

Despite these problems, however, pictures will continue to be used for communicating information and instruction and it is up to ergonomists to try to produce guidelines for their efficient production.

In essence two types of pictures may be used: discrete symbols or more complex illustrations.

The first of these, discrete pictographic symbols, are already in widespread use on equipment, for example vehicle controls, and for public information, for example the small signs indicating mens' and womens' toilets. Unfortunately, however, little research has been carried out to consider the appropriate design of such symbols. Nevertheless, Easterby (1970) has suggested that the three separate but related issues which are most important in their design are the perceptibility, the discriminability and the meaning of the pictures. Thus the observer must be able to see and distinguish different symbols and realize to what the symbols refer. To enhance these factors Easterby suggests a number of design principles; the more important of which are:

1. A solid boundary to the figure: a figure which does not have such a boundary lacks contrast and form and does not attract attention well; Figure 4.4a illustrates this point.
2. Simplicity: a simple shape is perceived more readily than one which contains too much detail (see, for example, Figure 4.4b).
3. Closure of the figure: in many respects our perceptual system tends to integrate information derived from the senses to produce a 'whole' figure. For example, looking across the room the image falling on the retina might be composed merely of three or four horizontal 'planks' holding a few vertical, rectangular shapes. If asked to draw the image, however, one might produce a recognizable set of bookshelves. This ability of our visual system to produce a 'whole' figure or *gestalt* can be incorporated in the symbol design if a recognizable boundary is produced for the figure. Thus the first of the symbols in Figure 4.4c is perceived more as a 'whole' than the second. (It may be remembered that Wright advocated this procedure for the production of graphs, by suggesting that the axes should be extended to form a square, p. 64.)
4. Stability: because our perceptual system attempts to impose form on the incoming sensory information, some types of figures are inherently 'unstable'. For example, most people will know of the picture showing two faces in silhouettes (Rubin's vase-profiles pattern). Depending on the 'set' of the observer he will see either two faces or a Grecian vase and, once the possibilities are seen, they tend to keep changing from one to the other. The

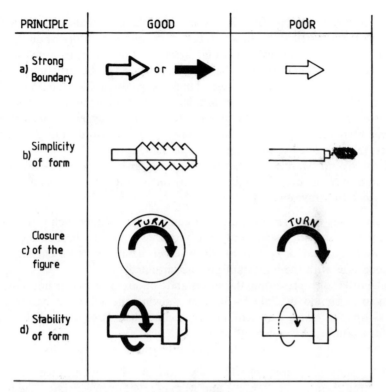

PRINCIPLE	GOOD	POOR
a) Strong Boundary		
b) Simplicity of form		
c) Closure of the figure		
d) Stability of form		

Figure 4.4   Some of the principles important in symbol design; reproduced from Easterby (1970), *Ergonomics*, **13**, 149–158, by permission of Taylor and Francis Ltd·

same effect may be seen in Figure 4.4d in which the arrow appears to be indicating alternately a clockwise and an anticlockwise motion. Because of this instability the meaning of the picture becomes ambiguous, whereas removing part of the arrow reduces this ambiguity.

5. Symmetry: to make recognition easier, the various elements in the picture should be as symmetrical as possible, providing the production of a symmetrical figure does not mislead the observer. Szlichcinski (1979a), however, suggests that this principle is controversial. Thus he points out that when children are asked to choose which of several picture orientations are most like the original object, a three-quarter view is more often chosen than a symmetrical view showing a single side. Similarly, subjects located controls on a panel and operated them more rapidly from slides showing the more symmetrical front or side views. Szlichcinski suggests, therefore, that an asymmetrical figure may often provide more information than one which is symmetrical.

Although discrete symbols can be used to represent a great deal of information, their usefulness is mainly restricted to denoting various features of the environment. However it is sometimes the case that illustrations have to be used for instructions, in which case more complex information has to be given than can be conveyed in single symbols. For this reason pictures that describe actions which the user has to perform need to be structured in some way to form a sequential message. Thus the pictographic instructions must fulfil many of the functions of natural language, so that the ways in which the pictures are arranged and the shapes from which they are formed need to convey their meaning in a similar fashion to the way words and grammatical structure determine the meaning of simple sentences.

Barnard and Marcel (1977) carried out a fairly comprehensive study to determine how pictures should be designed and arranged to convey information. In essence they investigated:

1. Part versus whole representation of equipment: whether each picture in the sequence should display only the important part to be considered (for example a dial or a switch), or whether it should represent the whole apparatus with the important part accentuated.
2. The order of representing the status and actions: whether or not individual pictures should be linked—the first showing the action to be taken (for example, throwing a switch) and the second the consequence of that action (for example, showing an illuminated light).
3. The use of insets to illustrate important aspects of the actions.

Using subjects' reports of what they thought different sets of pictures were instructing them to do, Barnard and Marcel (1977) demonstrated the need to include details of the relevant parts of the machine in each picture (in other words, the 'whole' representation). Surprisingly, however, the use of insets resulted in poorer understanding than did comparable instructions without insets. Overall, the action-state representation yielded the best results.

Tables

Tables probably represent the most usual means of presenting information. However, perhaps because they are in such common usage, it is often the case that little thought is given to their design.

A table is little more than a systematic arrangement of different items of information, which can be numerical, as in timetables, or non-numerical, such as the table of chapters in the front of this book. The designer has a variety of ways available to arrange the information. For example the items could be listed alphabetically or in numerical order, or they could be grouped with reference to a common feature. Furthermore, as Wright and Fox (1970) point out, it is always possible to provide aids such as spatial cues by aligning items vertically or in rows, or to introduce typographic aids such as variations in the size, style or colour of the print.

The many forms of tables which can be seen in present day usage differ, primarily, in the amount and kind of work which they require the user to perform. In this respect Wright and Fox (1972) refer to explicit and implicit tables. Explicit tables are those which give the reader all the information he wants, for example a bus timetable which lists each time the bus will arrive. An implicit table, on the other hand, gives some information only and leaves the reader to infer any additional information that he may require. The bus timetable which tells the reader that the bus will arrive 'every 20 minutes' after 11.00 am is an example of such a table.

For the designer, the main advantage of the implicit over the explicit tables lies in the space which is saved. However, they are disadvantageous as far as the reader is concerned since he needs to carry out some further calculations, to which he might possibly obtain the wrong answer.

With regard to the explicit types of table, Wright and Fox (1970) suggest that there are two main variants: linear and two- (or more) dimensional (the latter are sometimes called matrices). In the linear version the authors argue that the reader searches through what is effectively one long list to find the information that he requires—as in, for example, a tax code or a telephone directory. With the matrix version on the other hand, the reader has to make two or more searches for different parts of the information required, and then has to coordinate the results of this search in a particular way. For example, when reading a timetable to determine the time at which his train will arrive at a particular station, a passenger will first scan the rows to find his train, then scan the column to find his station. The time of arrival is given at the intersection of the row and column.

In many respects, of course, the linear type of explicit table may be viewed simply as an implicit version of the matrix table. Take the telephone directory as an example: If it were to be used as a linear table to find the telephone number of, say, a Mr H. K. Williams, one would start at the beginning of the directory and work through until the required name is read. This is clearly not the way in which a directory is used—in fact a number of dimensions are (implicitly) imposed. Thus one first looks for the initial of the surname (dimension 1). The second dimension would be the second letter (or even the second couple of letters), etc. In this way the search time is considerably reduced.

From the research which has been carried out, it is difficult to draw general guidelines as to which type of table should be chosen, since users and the type of information to be conveyed differ markedly from situation to situation. However, from their studies Wright and Fox (1970) suggest that some people were completely unable to use matrices, and their difficulty persisted even when the experimenter worked through examples with them. In addition conversions were made slightly more rapidly, but much more accurately, with the explicit than with the implicit table. Thus they argue that linear, explicit tables are generally better than either matrix or implicit tables, particularly if they are to be used by the general public.

Nevertheless there are many situations when a multidimensional table is more desirable, perhaps because of space limitations or because the user population is known to be more capable. In such cases a few design guidelines are available,

A. Two decisions on two dimensions

	E1	E2	F1	F2
G1	–	–	–	–
G2	–	–	–	–
H1	–	–	–	–
H2	–	–	–	–

B. Four decisions on two dimensions

		1		2	
		E	F	E	F
1	G	–	–	–	–
	H	–	–	–	–
2	G	–	–	–	–
	H	–	–	–	–

Figure 4.5 Two ways of arranging four alternatives along two dimensions used by Wright (1977b), *Frgonomics*, **20**, 91–96, and reproduced by permission of Taylor and Francis Ltd.

while some of the typographical aspects (such as print styles and typeface size) were discussed above.

The first general rule suggests that the number of decisions to be made should be reduced to a minimum. Wright (1977b), for example, showed that subjects made far fewer errors and performed quicker when only two dimensions were given than when the table required four decisions. Thus the arrangement of Figure 4.5a was preferred to that of Figure 4.5b. This is probably because, whereas the reader merely has to scan in one direction for each dimension in Table (a), he needs to scan in two directions for Table (b). To find alternative F2 in Table 4.5b, for example, the reader first needs a horizontal scan to 2, then a vertical scan to F, so potentially increasing the time taken and the errors made. Once again, therefore, Wright has demonstrated that using only one dimension was preferable to using two.

Second, it has been found beneficial to group items together rather than to have all items in the table equally spaced. Tinker (1960), for example, has shown that grouping items into blocks of five was more helpful to the reader than grouping by tens, but the latter grouping was better than no grouping at all.

Finally, ruled lines appear to be as effective as spaces for marking distinctions between columns. In this respect Wright and Fox (1970) make the following comment regarding the size of the gap between the grouped items:

Although there seems to be no direct experimental evidence on the point, it is a common observation that when there is a large gap between the item that is looked up in the table, and the information which is as a consequence read off from the table, it is easy for the eye to misalign the two columns and so err by reading off information either directly below or above that required. Indeed it was to the lessening of such difficulties that Tinker attributed the beneficial effect of grouping items within columns. Therefore it would seem advantageous to have the space between corresponding items within a table as small as possible. . . even if lots of space within a table makes it easier on the eye, too much space in some places may make it harder on the user.
(This material appeared in Vol. 1, pp. 234–242 of *Applied Ergonomics*, Published by IPC Science and Technology Press Ltd., Guildford, Surrey, UK)

Prose

The use of prose for instruction and information has purposefully been left until last to give the reader a chance to consider the value of alternatives such as tables or flowcharts. However, it is the case that most information which is transmitted from man to man is carried out by combining words (either verbal or visual) into (sometimes) meaningful sentences.

In many cases misinterpretations of the words used can lead to communication breakdown. This may be either total or partial but, in either case, is undesirable. If there is a total breakdown then the necessary work will not be done, but if it is partial then the work might be done incorrectly. Misinterpretation and misunderstanding, therefore, may be costly and even dangerous. For example, Conrad (1962) reports the case of a boy who died needlessly because his mother could not understand the instructions to operate a telephone. Furthermore Wright and Barnard (1975a) have reported that in at least one London borough several local authority staff are employed full-time to assist people in completing their forms for benefits. As they point out, if to such staff costs are added the administration time and the postage that normally accompanies returning incompleted or wrongly completed forms, some idea can be obtained of the total cost and wasted time which can accompany poor communication.

Chapanis (1965b) has called attention to the problem of words and their meaning in language, and has suggested four ways in which the intended meaning of the transmitter may not be adequately received: ambiguity, incomplete information, misleading information, and verbose information. These are not, of course, mutually exclusive. For example, information may be ambiguous because it is incomplete; or it may be misleading because it is too verbose.

Ambiguity

The meaning of the message may be open to more than one (sensible) interpretation. This problem often arises because of a need to conserve space by reducing apparently redundant words. For example, the author has seen people apparently hesitating near a lift bearing the sign 'In case of fire do not use lift'. Most people may well surmise that the instruction suggests that the lift should not be used *when* a fire breaks out, but it is not difficult to imagine the meaning as 'a fire will break out if the lift is used'. By adding the word 'the', to make the instruction read 'In *the* case of fire do not use lift', this ambiguity is removed.

Ambiguity arises not only because apparently redundant words are removed, but it can also occur when prose written in one language is translated into another. Slight changes in the arrangement of translated words can radically alter the meaning or the comprehensibility of the message. As was pointed out in Chapter 3, understanding the characteristics of other cultures is becoming more important as the amount of trade carried out between different countries increases. This also means being able to communicate adequately with people from foreign lands.

Although adequate translations will often be beneficial to the communication process, in some cases poor translations may actually produce poorer results than no translation at all. Subtle differences in the meaning of words of different cultures may lead to misunderstandings. To an English car mechanic, for example, 'gas' implies exhaust whereas his American counterpart would think of petrol; 'boot' in the United Kingdom is the same as 'trunk' in the United States, and so on.

As an example of this problem, Sinaiko and Brislin (1973) asked Vietnamese airforce mechanics to perform various maintenance tasks using maintenance manuals which were written in either English (American); or in high quality, lesser quality, or poor quality translations of the original. Although their Vietnamese subjects made many errors when using the original American material (61 per cent), their errors rate increased to 89 per cent when they were given the poor Vietnamese translations from which to work.

Incomplete information

Quite clearly, a communication system is in great danger of breaking down if some important aspects of the information are withheld from the receiver. In many cases this lack of information is withheld not by intent but by accident, and this can happen in such media as instruction manuals when the author may leave out what to him seem to be irrelevant or unnecessary pieces of information. To the reader, however, who may not be conversant with the machine which he has to operate, some of the omitted information could be vital. Omitting to tell the operator that the machine first needs to be switched on, and where he can find the switch, is a common example of this type of accidental incomplete information. Another example is given by Chapanis (1965b) who describes the instructions to operate a mechanical timer so that different combinations of input and output relationships could be attained. Unfortunately, however, important supplementary information was omitted, without which the timer would not operate!

The transmitted information can also be incomplete if unfamiliar terms are used but not explained, in which case although the information is available it cannot be used. As any novice who has attempted to understand computer manuals will appreciate, for example, instructions containing unfamiliar terms are a source of annoyance and poor performance. The problem can often be solved, however, with the liberal use of examples.

Misleading information

Information which is deliberately incorrect or misleading will clearly lead to communication difficulties—if the operator is told to do the wrong thing he is unlikely to carry out his task efficiently. However, it should also be realized that misleading information sometimes may not be given deliberately, but could occur if the information is ambiguous, incomplete or arises from rumours. Even the simplest message can become distorted due to slight additions or alterations being made as it is passed on from person to person.

Too verbose information

This is perhaps the opposite problem of the ambiguity which arises from incomplete or abbreviated information. In this case the meaning of the message is not altogether apparent at the first (or second) reading (or listening) because either too much, too wordy or too jargonized information is provided. The problem appears to be particularly prevalent in government, legal and even military circles as the following example given by Chapanis (1965b) illustrates: 'This radio uses a long life pilot lamp that may stay on for a short time if radio is turned off before radio warms up and starts to play'. As Chapanis suggests, this 29-word statement could have been more explicitly put as 'the pilot lamp stays on for a while after you turn the radio off' (16 words).

Some Communication Rules

Wright and Barnard (1975a) have produced a useful guide for the production of efficient instructions and information. Since they dealt mainly with the design of forms, many of the points which they raise have been dealt with earlier (for example, typographical aspects) but some are important in the present context.

Use short sentences

The purpose of a sentence is to allow the reader to take a short pause before taking in further information. Since we have a limited capacity to process information, however, if the sentence is unduly long the reader will either being to forget some of the earlier information in the sentence or will impose his own structure. In this case he will break up the long sentence himself, perhaps in the wrong places, and obtain a slightly different meaning to the one intended.

Wright and Barnard suggest that to be readily understood each sentence should contain only one clause. In order to implement this they suggest that writers should scrutinize sentences with more than one verb, looking for alternative ways of expressing the information.

Use active sentences

It is often possible to express a sentence, particularly an instruction, in either the active or the passive sence. For example the command 'Bolt down the hatch' implies a more active state than the passive instruction 'The hatch is to be bolted down'. Reviewing this field Greene (1970) concluded that active sentences are more easily understood and remembered than are the equivalent passive forms.

Use affirmative sentences

Just as a sentence may be written in the active or passive form, it may also be written in the negative or in the affirmative. In her review of sentence construction, Greene (1970) also concluded that sentences with negative elements are more difficult to understand than affirmative sentences. Words such as *not, except, unless* each have negative elements, so that an instruction 'Voters may not vote for more than one candidate' may be better understood in its active, affirmative form of 'Vote for one candidate only' (Wright, 1975).

Wright and Barnard (1975a) also point out that words such as *reduce* have negative connotations too, although they lack the specific negative prefixes such as *un, dis-, in,* etc. Thus 'If the dial falls below 15, reduced output will occur' could be rewritten as 'Output will increase if the dial rises above 15'. (It should be pointed out, however, that this approach is only valid if the two forms of the sentence have the same meaning. For example, because 'output falls below a dial reading of 15', it may not be the case that 'output rises above a dial reading of 15'.) In a similar way to the word *reduce*, the terms *more* and *better* are more easily understood than the terms *less* and *worse*.

Use familiar words

This point has been made earlier. If the reader is unfamiliar with the jargon being used he may not understand all the information.

Organize temporal sequences

If a series of actions is either to be described or performed, it is better if they are written or spoken in the order in which they are to be carried out. Broadbent (1977) illustrates this point with the following example: if an action involved pushing in the carburettor control and then starting the engine, it is better to write the instruction 'depress the carburettor plunger before starting the engine' than 'before starting the engine depress the carburettor plunger'. In the first case the words are matched to the actions, in much the same way as, in ergonomics philosophy, the environment is matched to the man.

READABILITY

Having finally decided on the appropriate wording of instructions and information, a designer may wish to check that his prose is the most readable he is able to devise. In many respects, readability is largely subjective, with individual styles of writing being able to be discerned in different pieces of work. As Klare (1963) points out, however, although there are rules for efficient writing, careful adherence to them does not guarantee good writing. Writing is an art not a science. However, he points to three general principles which the writer must bear in mind. First, to know something of the type of reader for whom he is writing—his educational level, motivation and reader experience. Second, he should consider his own purpose for writing—what he wants to do: he may want

Table 4.1   The range of Flesch 'reading ease' scores (from Flesch, 1948)

Score	Description of style	Typical magazine	Syllables per 100 words	Average sentence length in words
0 to 30	Very difficult	Scientific	192 or more	29 or more
30 to 50	Difficult	Academic	167	25
50 to 60	Fairly diffcult	Quality	155	21
60 to 70	Standard	Digests	147	17
70 to 80	Fairly easy	Slick-fiction	139	14
80 to 90	Easy	Pulp-fiction	131	11
90 to 100	Very easy	Comics	123 or less	8 or less

to help the reader read more efficiently; to judge whether the material is acceptable; to read for comprehension, learning and retention; to understand orally presented material; or to accomplish some combination of these and other purposes. Third, as emphasized above, he should select his words carefully. It is the words used and their construction into sentences which make the prose readable.

To produce some standard in the readability of material, various readability formulas have been proposed, which operate by analysing the words used in selected passages in the prose. Klare (1963) describes many such formulas, but perhaps the most popular was produced by Flesch in 1948.

To compute Flesch's 'reading ease' score, a 100-word sample passage of the prose to be analysed is first chosen. Then the number of syllables (wl) in those 100 words are counted in the way in which they are read aloud. For, example 1980 will have four syllables (*nine—teen—eigh—ty*). Second, the formula requries the average number of words per sentence (sl) to be computed. With this data reading ease can be calculated as:

$$206.835 - 0.84\,wl - 1.015\,sl$$

This score puts the piece of writing on a scale between 0 (practically unreadable) and 100 (easy for any literate person). As a more detailed guideline, however, Flesch produces a table of typical scores for different types of prose as shown in Table 4.1 above.

SUMMARY

This chapter considered communication in its most basic—the transmission of information and meaning from one man to another. After considering the meaning of meaning, and the ways in which the communication system may be corrupted, the remainder of the chapter discussed the various processes involved in transmitting information using the printed page: typographical aspects and the use of various communication media such as tables, graphs and pictures. Finally, some of the problems involved in the efficient communication of prose were considered.

CHAPTER 5

Machine–Man Communication: Displays

The concept of communication was introduced in the last chapter, when aspects of the first link in the working chain—communicating information and instructions from one man to another—were discussed. Once the initial instructions have been given to him, however, most of the information which the operator subsequently receives will come not from another human being, who can perhaps be questioned further in the case of misunderstandings, nor from written instructions which can be read a few times to ensure that they are understood, but from single or composite instruments which display information about the state of the system. Since the operator is only able to make an appropriate response on the basis of the information which he receives, it is clear that this aspect of the system needs to receive careful consideration, and that the display should be designed with both the worker and the work in mind.

Although many types of display are commercially available, the choice will often be narrowed to only a few for any one particular task or situation. Thus it falls on the ergonomist to choose the most appropriate display by considering the requirements of the situation and the various uses to which the information produced by the display will be put.

In practice the 'best' display is normally chosen by trading off the criteria of speed, accuracy and sensitivity in communicating the important information. Since the communication act requires the receiver to interpret correctly the message which originates from the transmitter, however, these criteria refer just as much to the performance of the operator as they do to that of the machine. It is for this reason that the needs of the task and the man have to be made explicit. In some cases it may be that the speed with which the information can be absorbed is more important than the other criteria, for example when a pilot reads the rapidly changing height information from his altimeter on takeoff or landing. In other cases accuracy may be as important—for the avoidance of errors or ambiguity. It is of little value, for example, for the pilot to be able to read his altimeter quickly if either he reads it inaccurately or it is inaccurate. In other situations the display may need to be very sensitive (for instance, to detect the slightest change in the variable which is being measured). For example, a heart-rate monitor in an intensive care unit should respond immediately any change occurs in the patient's condition. It would be useless as a machine if the nurse was able to read (quickly) the fact that the patient's heart had stopped beating (accurate) 3 minutes ago (insensitive).

Speed, accuracy and sensitivity, then, are the primary criteria on which a display's value is judged, but they are not independent criteria. A display which can be read quickly is valueless if its readings (either the readings which it gives or those which are recorded) are inaccurate; a display which communicates slight changes in machine state is worthless if it takes a long time to read, etc. For this reason it must again be emphasized that both the worker's and the system's requirements need to be considered before the appropriate display is chosen.

The display, therefore, represents the sole means by which a machine is able to communicate information about its internal state to the operator. As Rolfe and Allnutt (1967) put it, 'The display translates what is at first imperceptible to us into perceptible terms'. As such it can only function in one of the man's five sensory modes: vision, hearing, touch, taste or smell, although under normal circumstances only the first two (or sometimes three) of these senses are used. To understand fully the efficient matching of the display's requirements to the man's capabilities, therefore, much of the information regarding the human sensory system which was discussed above in Chapter 2 needs to be considered.

Before discussing visual and auditory displays in more detail, it is useful to compare the two types and to consider briefly the situations in which they are more usefully employed.

In general, *visual displays* are more appropriate when:

1. The information is presented in a noisy environment. Under such conditions the auditory displays may not be perceived.
2. The message is long and complex. Compare, for example, a written sentence from a teleprinter (visual) and the same information presented from a taperecorder (auditory). Because the eyes are able to scan the visual material more than once, the short-term memory capacities are not overloaded. Unless the taped message is translated into written material, the words which are decoded somehow have to be stored in memory while other words in the message are being decoded.
3. The message needs to be referred to later. Visual information may produce a permanent record—unless auditory recording equipment is used, acoustic information is stored only in memory.
4. The auditory system is overburdened—perhaps because of too many auditory displays or (as in (1) above) in a noisy environment.
5. The message does not require an immediate response.

Visual displays are more appropriate, therefore, for providing continuous information to the operator. On the other hand, *auditory displays* are more appropriate when:

1. The message requires an immediate response. It is for this reason that warning

messages are normally presented in the form of a klaxon or a bell—they are more attention-getting.

2. The visual system is overburdened—perhaps because of too many visual displays or in conditions with too high a level of ambient lighting.

3. The information needs to be presented irrespective of the position of the operator's head. The drawback of visual displays lies in the fact that the operator needs to be looking at them before they are able to communicate the information. Auditory displays, however, do not have such restrictions. It is for this reason, too, that they make such acceptable warning indicators.

4. Vision is limited—for example in darkness, at night, or when it is likely that the operator will not have had time to adapt to the light or dark conditions.

VISUAL DISPLAYS

Types of Display: Digital and Analogue

Visual displays are perhaps the most commonly used instruments for communicating information from machine to man. They are also, however, most commonly badly designed—with sometimes disastrous results. For example, Fitts and Jones (1974a) analysed 270 errors made by pilots in reading and interpreting their instruments, each of which could have led to an accident. From their subsequent analysis, they felt that the various types of errors committed could be classified into nine major categories. These include 'simple' misinterpretations of the instruments (for example, misreading the altimeter by 1,000 feet (304) m); reversal errors (for example, interpreting the bank indicator as showing a left bank when, in fact, the aircraft was banking to the right); misinterpreting visual warning signals; substitution errors (confusing one instrument with another) and illusory errors (for example, misreading the altimeter because of an illusory conflict between body sensations and the instrument indication). It is clear, therefore, that there are many ways by which the information from the machine, although accurate, can be misinterpreted, which makes it effectively worse than useless (particularly if the misinterpreted information leads to a totally incorrect control action and hence a serious mistake).

In essence, visual displays take two forms: digital and analogue. The digital display, which has become more apparent in recent years with the advent of pocket calculators and digital watches, presents the information directly as numbers. With the analogue display on the other hand, the operator has to interpret the information from the position of a pointer on a scale, from the shape, position or inclination of a picture on the screen, or from some other indication which is analogous to the real state of the machine. For example, a clockface is an analogue scale, since if the minute hand points to '9' it has travelled three-quarters of the way around the face and indicates a position which is analogous to three-quarters of the way through the hour; the pilot's pictorial display of the artificial horizon (which indicates the attitude of the aircraft relative to the horizon) is an analogue display since the positions of the respective 'aircraft' and 'horizon' indicators in the display are analogous to the respective positions of the aircraft and ground in the real world. Finally, a warning light is an analogue display since the state of the light (on or off) is analogous to the state of the machine to the world.

The Use of Displays

In many respects the use to which a display is to be put determines the type which is chosen although, of course, exceptions to this rule exist as will be discussed below.

Displays may be used:

1. To make quantitative readings—that is, to read the state of the machine in numerical terms: the temperature in degrees centigrade, the height in metres, the speed in kilometres per hour, etc.
2. To make qualitative readings—that is, to infer the 'quality' of the machine state, for example whether the machine is 'cold', 'warm' or 'hot', rather than its precise temperature; whether the aircraft is banking 'shallowly' or 'steeply' to the right or to the left, rather than its precise angle, etc. Also subsumed under this heading is the use of displays to make check readings, in other words, to compare the state indicated by one display with that shown by another.
3. In combination with controls to set the machine or to track (maintain) a steady machine state.
4. To warn the operator of danger or that a specified machine state has been reached.

Murrell (1971) points out that when deciding on the type of display to be used and what its characteristics should be, it is important to consider what information the operator needs to do his job effectively, and then to question how this can be given quickly and unambiguously. For example, the same display can be used in some cases to indicate whether the machine is on or off; at other times a direct numerical (for example, voltage) reading may be required, while in other cases the operator may wish to compare the information with that obtained from other displays. Only when the ergonomist has determined the proportion of time for which the display is used in each case, and has 'weighted' these figures by a factor of the importance of each operation in the particular situation, will he be in a position to be able to decide on the appropriate type of display.

Displays for Quantitative Readings

Both digital and analogue displays may be used for making quantitative readings, although it is only in recent years with the increase in electronic sophistication that digital displays have been a viable alternative to analogue displays. A mechanical type of digital display which is operated by a system of reels (as may be seen, for example, in the milometers of most types of car) has two major drawbacks: First, a number may not always be fully visible as it moves around on a continuously revolving wheel. Instances can occur when all that may be seen is the bottom half of one number and the top half of the next. Second, because the number is moving, the image will not be in the same place on the retina. This is not so with electronic numeral displays, in which the number is altered by illuminating different segments or dots.

Given the fact that electronic displays are now available, therefore, the question arises under what circumstances should the analogue or digital display be used? In this respect, the majority of work which has been carried out has concerned the design of a suitable aircraft altimeter, but the principles which have evolved may be extrapolated to any situation in which a man needs to interact with his machine.

The early experiments which were carried out to resolve the question tended to point to the digital display as producing less reading errors and faster reading times than their analogue counterparts. For example Murrell and Kingston (1966) describe results from an experiment carried out on skilled journeymen using a graduated and digital micrometer. Whereas about 3 per cent of the readings were in error using the graduated micrometer, the same men made only 0.05 per cent errors when they used a digital micrometer. Similarly Zeff (1965) compared the conventional and digital clockface for the speed of reading

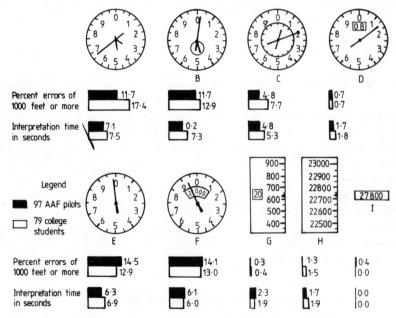

Figure 5.1 Speed and accuracy in reading altitude from different types of instruments (from Grether, 1949)

and errors; he showed the digital clockface could be read faster than the conventional clock, with one-tenth of the errors.

One of the earliest investigators to consider the value of dials and counters was Grether (1949), who used photographs of altimeters bound into booklets to compare the reading errors produced by trained pilots and college students. Each photograph showed a different height on one of eight dials or a counter (Figure 5.1). His results indicated that the standard three-pointer altimeter used extensively in aircraft (one pointer indicates 10 000s of feet, another 1000s of feet, and the third 100s of feet) had the highest error rate and longest reading times. The digital display, however, had the lowest error rate and reading times.

It is, perhaps, useful at this point to realize the implications of making reading errors of this nature. For example, the skilled journeymen in Murrell and Kingston's experiment misread their graduated micrometers by up to 100 000ths of an inch, a not insignificant amount in precision engineering terms. Perhaps a more sobering example is provided by Rolfe (1969a) who quotes the reports of investigations into two airline crashes, of which the following is an example:

'On the evening of 28th April 1958, a B.E.A. Viscount airliner fitted with modified three-point altimeters was making a final approach to land at Prestwick Airport at the end of a flight from London. The captain reported to the ground that he was 14 500 feet and descending. As he descended he reported he had passed through 12 500 feet. Soon after the last report the aircraft struck the ground. The subsequent investigation... showed that the captain had misread his altitude by 10 000 feet and had perpetuated his misreading error until the aircraft struck the ground and crashed.
(This material appeared in Vol. 1, pp. 16–24, of Applied Ergonomics, Published by IPC Science and Technology Press Ltd., Guildford, Survey, UK)

From this example, therefore, it is clear that a misreading error from a badly designed display may not necessarily simply be a 'one-off' event. Often a 'set' is introduced to perpetuate the error on subsequent readings—not only did the captain misread his altimeter once, he *continued* to misread it by the same amount until he hit the ground.

The impression may have been gained, so far, that numerical information should always be presented digitally. Indeed some authors tend to argue strongly for this. For example Nason and Bennett (1973), after showing that counters were superior to dials when precision reading was required, concluded that 'Counters are superior to dials for all quantitative reading tasks.... There may be few, if any, situations, from the standpoint of performance, where dials should be employed.' Digital displays, however, do have their drawbacks, particularly when the task requires some degree of check reading, when the display is used as a predictor in addition to providing numerical information, when the rate of change of information is particularly fast or, as discussed earlier, when mechanical digital displays are used.

Check Reading

This will be discussed in more detail in Chapter 7. Many functions of a display involve providing not only precise, quantitative information, but allowing the operator to compare display readings or to check that a certain value is or is not being indicated. For example Murrell (1971) discusses data from a survey of HM ships in 1952 which indicates that, on electrical and steam equipment, only 18 per cent of the dials were used for quantitative readings whereas over 75 per cent were used for some form of checking. Similar conclusions were reached by White, Warrick and Grether (1953) who report that the average duration of a pilot's visual fixations when flying on instruments is approximately 0.5sec, suggesting that aircraft instruments are typically read in a check reading manner.

Because they lack the additional perceptual information that pointers provide (for example, the angle between the pointer and 'north', or the difference in angle between two pointers) the only way that check reading can be carried out using digital displays is for the operator to compare numbers. For this reason digital displays may increase both the time needed to read the instruments and the number of errors made or this type of task. For example, Connell (1948) mounted four displays side by side on a panel. She found that the average time taken to determine whether or not the same reading appeared on each of the four digital displays was $\frac{1}{2}$ sec longer than when the subjects were presented with four, circular scale and pointer instruments. The number of errors also increased using digital displays.

The use of spatial cues for prediction

It is very rare for an operator to simply read the values indicated on a display. In the majority of cases, although possibly he may not realize it, he uses the information which he receives to make predictions, perhaps about the state of the machine or about the machine's future behaviour. This point has been emphasized by Graham (1954) who, considering the use of displays for quantitative readings, has argued that 'the criteria of ... speed and accuracy ... is only half of the problem. Very often the operator has to translate what he sees into an appropriate action.'

This unconscious prediction can be illustrated by the following example: An operator knows that the machine is likely to explode when the display indicates 10 'units'. He is asked simply to read the display and it reads 9.5 units. At this point his understandable reactions (possibly of panic) will be due to his prediction of the machine's future state!

A second example may be given by considering the conventional (analogue) watchface: The large (minute) hand may point at 7 and the small (hour) hand somewhere between 11 and 12, indicating that the time is 11.35. In the same glance, however, the position of the minute hand provides spatial cues to enable the reader to predict, quickly, that 25 minutes are left of the hour. Digital displays do not give this information so readily—the operator needs to carry out fast numerical computations (with all their attendant problems) to achieve the same result.

The increased efficiency of analogue displays in providing such spatial cues has been demonstrated by Simon and Roscoe (1956). They presented subjects with display information intended to represent an aircraft's present altitude, the predicted altitude after 1 minute, and the final altitude to be reached. One of four types of display were presented: (a) a vertical (strip) display incorporating three pointers; (b) a circular (dial) display with three pointers (similar to the three-point altimeter); (c) three separate five-digit counters; and (d) three separate circular, single-pointer displays.

From this information subjects had to decide:

1. whether they were diving, climbing or flying level,
2. in order to reach and/or maintain their final altitude, whether they should climb, dive or continue flying level,
3. if they should climb or dive, whether they should increase, decrease or maintain their present rate, and
4. whether they should eject!

In terms of the time taken to complete the task, the results showed the vertical displays to have the lowest average time in a series of ten problems (56.3 sec), the combined circular display coming second (64.2 sec), the digital displays third (74.6 sec), and the separated circular displays fourth (79.7 sec). In terms of errors, the digital displays had the highest (approximately 7 per cent), and the vertical strip displays the lowest proportion (approximately 3 per cent).

When discussing their results, therefore, Simon and Roscoe argue that the poor showing of the counter display was related to its failure to provide any *direct* spatial cues.

Blurring

Many of the earlier experiments carried out to compare the efficiency of digital and analogue scales for quantitative readings presented the stimuli statically. Thus each separate reading was shown to the subject either for a specific short period or under conditions in which he was timed. As already discussed, under these circumstances the digital display produces fewer reading errors in faster times than does the analogue display.

In the dynamic (real-life) situation, however, the values indicated by the display will normally be changing constantly, sometimes very quickly. Under such circumstances the image may become blurred, and this can have serious consequences if the numbers have to be read with some precision. As anyone who has tried to read the fast changing times provided by a digital clock at a sports meeting will know, the blurred image makes reading the information practically impossible. In such cases the spatial cues which are provided by the pointer in an analogue display may help the operator in making his reading.

In summary, therefore, it would appear that for recording quantitative information, digital displays make the task easier for the operator (faster reading time and less errors) than do analogue displays. The superiority of the digital displays is reduced, however, in situations where the values change quickly, where some degree of check reading is required, and the operator might need to predict the future 'conditions' of the machine.

Displays for Qualitative Readings

In some situations the operator may use his display not to record precise readings but to indicate the qualitative state of his machine. For example, rather than needing to know the machine temperature in degrees centigrade, he may simply need to know whether it is 'hot', 'neutral' or 'cold', or whether it is 'safe', 'dangerous' or 'critical'. This task, therefore, can be conceived of as a form of check reading. For this reason and because no numerical value needs to be recorded, in these circumstances the analogue display may be more effective than a digital display.

Visual coding

As McCormick (1976) points out, the optimum designs of displays for qualitative readings depend on what needs to be read. If the whole range of machine states can be divided into a limited number of 'levels', then the best way of representing these levels is by separately coding them on the dial face. This is normally carried out by coding the different areas in some way. A number of visual coding methods are available to the designer to enable him to distinguish between different sections on the dial face, and some of the coding methods which Morgan et al. (1963) list include the use of colours, shapes (both numbers, letters and geometric shapes), and different brightnesses.

In an attempt to compare the efficiency of various coding methods, Hitt (1961) presented subjects with information coded according to five types of code: numbers, letters, geometric shapes, colour, and position. With this information the subjects were asked to make various inferences: to identify whether a particular object was present, to locate it on a grid, to count the number of different objects indicated, to verify the function of the objects indicated by each code, or to compare the type of objects in one display with another. These types of tasks, of course, are carried out normally in most working situations. For example an operator may need to identify that the machine has reached a 'ready' state, to locate a particular fault, to verify that the machine is safe, to count the number of machines in operation or to compare the state of machine 1 with that of machine 2.

For each of the different operations, Hitt demonstrated that numerical and colour coding were equally the most effective. In all cases coding according to position was the least effective method. Similar results were obtained by Smith and Thomas (1964) who demonstrated that a colour code consisting of five different colours was more effective than any geometric shape code that they used. In addition, Christ (1975) has reviewed a number of similarly conducted experiments and has shown that colour was superior to size, brightness and shape coding for accuracy of identification. However, in some instances the use of letters and digits were shown to be more efficient, but this is not altogether surprising since the tasks were in the nature of identifying objects. When tasks which involved searching were reviewed, colour was superior in all cases.

Colour coding, therefore, would appear to be the most useful 'all-round' coding method for qualitative displays. As a coding system, however, colour does have its limitations, primarily when coloured ambient illumination is used or if the operator is colour-blind.

Since a colour is perceived as a result of a particular wavelength of the light being reflected from a surface, it is clear that if coloured light is used to illuminate a coloured patch, the resultant colour perceived will not be what would be expected under white ambient illumination. This may cause the colour coding system either to be lost entirely or to be changed from what was learned previously.

A further limitation in the use of a colour coding system is if colour-blind operators are likely to be used. As was discussed in Chapter 2, approximately 6 per cent of healthy, adult males have marked reduced sensitivity to colours, most having difficulty in distinguishing between red, green or blue colours.

Pictorial displays

The pictorial display is a particular type of qualitative display which, as its name suggests, presents the operator with a pictorial representation of the machine state. Examples of this type of display may be found in many large process industries when the operator needs to know which stage of the process has been reached. Large panels may often be seen which give the operator a 'picture' (usually a representational picture in terms of interconnecting lines) of the system. Parts of the display may be colour-coded and/or illuminated to indicate where the process has reached, or any other information.

Another example of a pictorial display is the artificial horizon seen on most pilot's control panels, which indicates the position of the aircraft with respect to the horizon. The display consists of a horizontal bar and an outline symbol of an aircraft, and the relationship between the aircraft and the horizon is indicated by the relative positions of these two aspects of the display.

The overriding principle in the design of pictorial displays is to ensure that the picture which they give is as realistic to the real-life situation as possible. For example, if a section of railway track is represented pictorially by a series of interconnecting lines on a signalman's panel, the length of the lines should, as near as possible, be in proportion to the relative lengths of the sections of the track, the points should be in their relative correct positions, and the speed of the 'train' (perhaps indicated by a moving light) should be proportional to the real train speed. If principles such as these are adhered to, a pictorial display will provide useful information to the operator.

In some cases, however, it may be difficult to decide on the form which a 'real-life' representation should take. For example, with the artificial horizon in the pilot's cockpit the question has arisen whether it is more true to life for the aircraft symbol to move and the horizon bar to remain static, or vice versa. From the point of view of a person standing on the ground watching an aircraft bank, there is no doubt that the plane moves with respect to the horizon. From the pilot's position looking out of the aircraft, however, the horizon appears to move while his world (the aircraft) remains 'stationary'.

The many studies which have been carried out to investigate this specific type of pictorial display have concluded that the moving aircraft type of pictorial display is interpreted quicker and with less errors than the moving-horizon display (Johnson and Roscoe, 1972). The reason for this may be understood when one considers that the pictorial display represents a figure–ground configuration, with the figure being whatever is the moving part of the display and the ground the stationary part. When a pilot is looking out of the cockpit window the horizon, being a normally accepted, stable frame of reference, is the ground against which his and other moving aircraft are figures. However, when he moves his eyes to accommodate on the panel containing the artificial-horizon display, the panel itself becomes the ground with the moving-horizon bar becoming the figure. In this way it is quite possible for him to misinterpret the elements of his display and make a control movement that aggravates the condition. As with the symbols discussed in Chapter 4, therefore, it is important to determine the figure–ground relationships of the elements in the display before an adequate pictorial display is designed.

Displays for Check Reading

This use for displays has already been discussed, when it was shown that an analogue display helps the operator more in his task than does a digital display. As was discussed, this is probably because the check reading judgements are made as much on the basis of the position of the pointer relative to the whole dial as to the actual reading.

Displays for Setting and Tracking

Little work has been carried out to find the optimum type of display to use when settings have to be made. It may be suggested, however, that as long as the display and control movements are compatible (see Chapter 7), the precise design of a display for this function is not too critical.

Comparing the merits of analogue and digital displays for this type of task, it would appear that analogue displays are easier to use. For example, Benson, Huddleston, and Rolfe (1965) asked subjects to maintain a constantly changing display reading at a steady position (in other words, to carry out a tracking task). At the same time they were required to react to a light which was illuminated at irregular intervals. (This use of a secondary task ensured that the subject's total attention was employed.) Two types of display were used:

Table 5.1 The respective merits of analogue and digital displays

Function	Analogue display	Digital display
Quantitative reading	Best if a precise reading is not required, or if the task contains predictive or checking components	Best for accurate reading of slow changing values; poor if the task includes any predictive or checking component
Qualitative readings	Best for warnings, checking and prediction; useful to have visually coded areas.	Poor
Setting and tracking	Best	Poor

a combined analogue and digital display, or a single digital display. Their results indicated that whereas the subjects' performance on the two displays was not significantly different, their performance on the secondary task was worse when the digital display alone was presented than when they were presented with the combined analogue and digital display. This, coupled with the fact that the subjects' heart rate, sweat rate, muscle activity and respiration rate all increased when the digital display alone was used, suggests that the digital display made the subjects work harder to maintain their performance.

The possibility that analogue and digital displays may differentially affect the quality (rather than accuracy) of an operator's performance on a setting task was also considered by Rolfe (1969b). His subjects were asked to alter their 'altitude' using either a digital or a combined analogue/digital (counter/pointer) display. Again, no significant differences in either speed or accuracy were reported between the two types of display when the experiment was carried out in the presence of a secondary task. Differences in the *type* of error, however, were apparent. On the combined display, 62 per cent of the 21 errors made were caused by the operator 'levelling out' too early (for example, levelling at 25 000 feet (7600 m) when being asked to ascend to 26 000 feet (7900 m). Using the numerical display, however, the majority of the 26 errors (69 per cent) were caused by the subject 'overshooting' the required setting (in other words, exceeding it by more than 150 feet (45 m) and then returning to the correct setting).

Again therefore, it would appear that adding the analogue component to the display alters the operator's behaviour, to make him more 'conservative' and, perhaps, cautious in his setting.

Table 5.1 summarizes the relative advantages and disadvantages of analogue and digital displays as discussed above.

Visual Display Design

From the foregoing discussion it is clear that if the efficiency of the system is to be maximized, it is extremely important to choose the correct type of display.

To do this one needs a full analysis of the task that the operator has to undertake, which can then be matched against the capabilities of the display to carry out these requirements.

Unfortunately, however, knowing which type of display is useful in which circumstances is only half of a solution to the problem. Just as the previous chapter illustrated how man–man communication can go hopelessly wrong if, for example, the words are not put together in a suitable fashion, so too can the right type of display give a wrong (or misleading) communication if it poorly designed.

The purpose of this section, therefore, is to consider the second half of the problem—the design of visual displays. Naturally very few people are in a position to design suitable displays, but many are in a position to purchase them. It is the intention of this section, therefore, to provide guidelines for making a realistic choice.

Display design for quantitative readings

Analogue displays (dials)

The design aspects of a dial which an operator would use simply to take readings of the machine state have been investigated from most possible aspects. These studies were carried out in the early years of ergonomics, with some of the recommendations being embodied in British Standards (for example, BS 3693 parts 1 and 2, 1964; 1969), and they have been discussed at length by Murrell (1971).

Possibly the two most important criteria to consider when choosing a dial are the ease and the accuracy of making a reading. Translated into design features this implies 'simple, uncluttered, bold design' (McCormick, 1976).

The first requirement to consider is that the dial should be large enough to be read comfortably by an observer positioned at some distance (the reading distance) away. This distance, of course, may change with the position of the dial on the console. For example, a dial which is placed above the operator (not a recommended position as will be discussed in Chapter 7) will probably be further from his eyes than one directly in front of him. In addition, since the information which the operator needs will be obtained from the scale itself, the size and length of the scale will play a large part in determining the size of the dial. For all practical purposes, the British Standards Institute (1964, 1969) has suggested that the length of the scale (L) is related to the reading distance (D) by the formula $D = 14.4L$. For example if the dial is to be read at a distance of 90 cm from the operator, the scale base length (the distance around the scale between 'minimum' and 'maximum') should be at least 6 cm).

Having chosen a dial of the appropriate size, the second question which a potential purchaser should ask concerns the number and sizes of the scale divisions and markers.

The distance between the scale markers represents possibly one of the most fundamental factors which determines the readability of the scale. This, for any particular scale base, will be a function of the number of markers used. Thus as the space between two successive markers is reduced, while the distance between the dial and observer is maintained, the visual angle which the space subtends at the eye will also decrease in proportion. If this visual angle falls below a critical value, then the observer begins to make reading errors, and these increase in frequency linearly with the logarithm of the decreasing visual angle. Murrell (1958) has demonstrated that, for most reading applications, a visual angle of 2 min of arc is the critical point. This would imply, therefore, that for a reading distance of D units a gap of approximately $0.0006\,D$ units between the markers is required—in other words, a minimum scale gap of less than 1 mm for a reading distance of 1 metre. It must be remembered, however, that the studies reported by Murrell were carried out under optimum conditions. Under poor illumination, for example, the critical visual angle would be increased and require a larger scale gap. Practice and poor vision also play an important role in acuity tasks of this nature. For these and other reasons, McCormick (1976) recommends minimum scale gaps of between 1.25 and 1.75 mm.

The desirable length and thickness of scale markers have been the subject of a number of investigations which have indicated that within limits some variations can be accepted without influencing reading ability to any great extent. However, it is obvious that the major scale markers (those associated with particular values) should be emphasized, and the British Standards Institute (1964) suggest that each major marker should be twice the length of the minor markers. Regarding the thickness of the markers, Morgan et al. (1963) suggest that the major markers should be between 0.125 and 0.875 mm for good to poor illumination conditions, and McCormick (1976) appears to agree with these dimensions.

Regarding the arrangement of scale markers, the relative merits of different numbering systems have been investigated by several research workers. The body of opinion which has evolved appears to suggest that a system which progresses in 1s or 10s (1...2...3, or 10...20...30, etc.) is the easiest to use. A system increasing in 2s would appear to be slightly less effective, whereas progressions of 4s, 8s, 25s or decimals is not recommended. (It may, of course, be the case that these results simply reflect our practice in counting in 10s rather than in 4s). The minor markers, should help the observer's accuracy in interpolating between major markers, and so for this reason the divisions need to be sensible. Since people can be fairly accurate in interpolating into 5ths or even 10ths of a scale division (Cohen and Follert, 1970), however, the number of minor markers may not be too critical.

No readings could be made without the use of a pointer, and some details are available to give guidance about appropriate pointer design (Spencer, 1963). In summary, the pointer should be pointed rather than blunt, and the tip should meet, but not overlap, the base of the scale markers.

If the pointer is placed away from the surface of the dial, problems of parallax may occur. These can perhaps be demonstrated by envisaging reading a dial first from the left, then in front and then to the right of the dial. In each case a slightly different reading will be made because of the different lines of sight between the dial number, the pointer and the eye. To ensure that these parallax errors do not occur, some dials have a small, thin mirror positioned behind the pointer. Only when the observer cannot see the reflection of the pointer in the mirror will he be positioned in front of the dial and a true reading made.

In summary, therefore, a potential purchaser should pose the following questions:

1. Is the scale base length large enough to be read at a 'normal' reading distance?
2. Is the gap between scale markers at least a millimetre?
3. Are the major markers able to be differentiated easily from the minor markers?
4. Is the numbering system in 1s or 10s?
5. Is the pointer designed adequately?
6. Does the dial, as a whole, look uncluttered and easy to use?

Digital displays (counters)

Digital displays may take any one of three forms: mechanical, electronic and computer-generated using a cathode ray tube (CRT). In the mechanical type of display, the numbers are printed on either a revolving drum or on pieces of metal which 'flip' over. However, with the discovery of crystals and diodes which become illuminated when a current is passed through them, the mass production of cheap electronic digital displays has become available. Computer-generated displays are, of course, a form of electronic displays but their advantage lies in the easy generation of letters as well as numbers, so the ergonomist must also consider the production of alphanumeric displays.

Mechanical counters Two aspects of the counter are important: First the way in which the numbers change over and secondly the design of the numbers themselves.

If the numbers do not change in a 'snap' action but revolve around a drum, an undesirable situation will sometimes arise in which only parts of the number are shown in the display window, for example the bottom half of the number 8 and the top half of 9. Naturally this will increase the time needed to read the number and may possibly lead to errors.

With regards the design of the numbers, the primary problem is to ensure that no two or more numbers can be easily confused. For example, the numbers 0, 3, 6, 8 and 9 are often confused with each other since they usually have curved or rounded outlines. This dilemma can often be increased when letters are introduced—for example B, 8 and 3 are commonly confused. This has led to a number of investigations to design an appropriate set of alphanumerics, perhaps the earliest of which was carried out by Berger in 1944. He was primarily interested in the design of car numberplates and varied such aspects as the angle with which the vertical of a 7 met the top bar, and the diagonals of the 8 intercept; where the central intersection of the 3 should be placed; the length of the top bar of the 5; and whether the 4 should be open or closed.

Berger also compared the legibility of white numbers on a black background and black numbers on a white background. Interestingly, he found a difference between the two forms when he considered the optimum stroke-width (quite literally the thickness of the stroke of the number) to height ratio. For white numbers on a black background the optimum ratio was approximately 1:13, whereas if the numbers were black on a white background a ratio of about 1:8 was preferable. These relationships are shown in Figure 5.2, from which it is clear that the white characters on a black background can be seen from a greater distance than when black characters were used. The difference in optimum stroke-width of white and black characters has been confirmed by other investigators, and is likely to be caused by an irradiating effect of the white background 'spreading out' (that is, tending to blur) into the black areas, but not the converse.

The effect of different number founts (shapes) on legibility (as opposed to confusibility) does not appear to be too great. Most conventional alphanumeric founts can be read with reasonable accuracy under normal conditions where

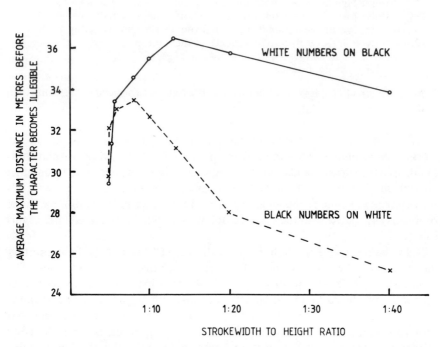

Figure 5.2 Average distances in metres at which numbers having different stroke–width: height ratios can be read (adapted from Berger, 1944). Character size 42 × 80 mm

size, contrast, illumination and time permit. For this reason the only recommendation which Buckler (1977) makes is that fount styles are kept clear and simple.

As McCormick (1976) points out, however, adverse viewing conditions may create significant differences in the legibility of different styles. For this reason he recommends using letters produced either by the (US) Navy Aeromedical Equipment Laboratory (NAMEL) or according to the (US) Airforce–Navy Drawing 10400 (AND). The shape of the British Standards Institute recommendation (1964) for numerals is similar to the AND fount.

Although the character's shape may not be too important to legibility, its size certainly does matter. This depends on such factors as the reading distance, contrast, illumination and the time allowed for viewing. Peters and Adams (1959) have proposed a simple formula for determining the optimum character size in terms of the viewing distance, the importance of the number and the level of ambient illumination:

thus
$$H = 0.0022D + K1 + K2$$
where
H = height of the letter in inches
D = viewing distance in inches
$K1$ = a correction factor for viewing conditions:
    0.06 for high ambient illumination; favourable reading conditions,
    0.16 for high ambient illumination; unfavourable reading conditions,
    0.16 for low ambient illumination; favourable reading conditions,
    0.26 for low ambient illumination; unfavourable reading conditions,
$K2$ = a correction factor for the importance of the number:
    0.075 if the number is very important
    0.0 for other importance

McCormick (1976) has pointed out that the figures produced correspond well with other recommendations.

Electronic displays Electronic displays now play an important part in the working environment. They range from the computer linked visual display units (VDUs) to the small crystals which can be seen in digital watches and calculators and which illuminate when a current is passed through them. Because VDUs have some ergonomic problems associated with them which are additional to the problems associated with the crystal displays, they will be discussed later in this chapter.

With the advent of the liquid crystal display (LCD) and the light-emitting diode (LED), there is less need for mechanical display presentations. The advantage of the electronic display are threefold: First, because any particular number is produced by the illumination of different 'segments' of the display, the character is always in the same part of the observer's retinal field. No problems with numbers 'flicking' or rolling over will occur. Second, the LED display generates its own illumination, enabling it to be read in the dark without additional light sources. However, this is not the case with the LCD display— since the segments appear black, background illumination is needed for adequate contrast. Third both the LED and the LCD displays can be driven by a computer or other electronic device, and this provides the designer with greater flexibility in terms of both the design and the content of the displayed information. Finally, with the advances in electronics, multicolour LCDs have been produced (Matsumoto, 1977) which will allow the designer to incorporate colour coding into his products.

Many of the design considerations which were discussed above apply also to electronic displays. However, the segmented characters produced by LEDs or LCDs do have further problems, particularly due to the fact that the appearance of the character, because it is composed of a number of segments (usually seven), differs from that of a drawn number. As Figure 5.3 illustrates, the main difference lies in the fact that the normal curves are lost. A secondary problem concerns the spacing between different numbers. Depending on the number which is to be read, because the character is formed by illuminating different segments, the spacing between numbers will vary, as is shown Figure 5.3. For these reasons, Plath (1970) has cautioned against the overuse of segmented numbers.

Van Nes and Bouma (1979, 1980) have considered the design of segmented numbers in more detail and have suggested that:

1. The smaller the number of segments from which a digit is built up, the better it is recognized. For example the number 1 (which is composed of two segments) is recognized more quickly than 9 (which is composed of 6 segments).
2. The probability of confusing two different numbers decreases as the difference between the number of segments which make up the two numbers is increased; and
3. Not all segments appear to be equally important for perception.

On the basis of these arguments, mainly the third, they suggested a new set of

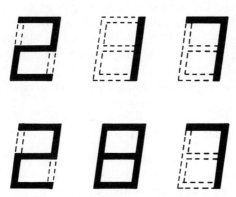

Figure 5.3 The different spaces produced by different combinations of electronic, seven-segment numbers

segmented numbers which have variable segment thicknesses. The top and the two right hand segments of a seven-segment number are about two-thirds the thickness of the other four segments. The authors do emphasize, however, that these new designs are still at the experimental stage.

A combined analogue and digital display for quantitative readings

When discussing earlier the relative merits of analogue and digital displays for making quantitative readings, it was shown that each had their advantages and disadvantages in different situations. For example an analogue display proved to be more valuable when it was recording fast changes in the machine state, whereas the digital display was more appropriate when making static readings; the digital display, on the other hand, was more useful when having to read precise values, etc. A sensible compromise, therefore, would seem to be to produce a display which has both analogue and digital components.

Only one detailed study reported in the literature has considered this compromise. After producing such a display, Rolfe (1969a) carried out both a pilot opinion survey and a laboratory experiment to investigate whether the digital display should be separate from (that is, above) or an integral part of (that is, inside) an analogue altimeter display. Using static trials (subjects were presented with photographs of each altimeter showing different readings), no difference was obtained between either of these two combined forms, although a significant difference was obtained between these two displays and an analogue-only display. Thus the average reading time for both counter–pointer displays was consistently about one-third that of the analogue display. No errors were made on either of the counter–pointer displays while 20 per cent of the readings made on the analogue display were in error by 100 'feet' (30 'metres') or more.

Dynamic trails, however, pointed to the 'counter-inside' display being more valuable than the 'counter-outside' display. This was also reflected in the pilots' opinion survey which suggested that the counter-outside display tended to appear as two independent displays demanding an alternation of attention. Remembering Easterby's suggestion that pictographic symbols should have a recognizable boundary, that the display should appear as a 'whole', it is not surprising that the counter-inside display was read more efficiently.

Display design for qualitative readings

All of the available evidence indicates that an analogue (dial) display represents the most efficient means of presenting qualitative information to the operator. The pointer position, and its speed and direction of movement, provide the operator with relatively low-level information of the machine state which, coupled with coded areas on the dial, will enable fast qualitative judgements to be made.

Colour coding

When discussing above the various types of coding available to the designer (p. 88), it was apparent that colour coding was the most efficient of the coding systems, and this is possibly reflected in the fact that most indicators which are used for quantitative readings are colour coded. Two important questions need to be answered, however, before an appropriately colour-coded display can be produced. These are: what colours should represent which type of machine state, and how many colours should be used overall?

The maximum number of colours which ought to be used in the coding system would appear to be around ten. For example Jones (1962) suggests that the normal observer can identify about nine or ten surface colours, primarily varying in hue. On the other hand Morgan *et al*. (1963) suggest a maximum of 11 different colours. However, it is possible to increase this number quite significantly if different dimensions are combined, for example size, luminance levels, hue and colour purity (Jones, 1962)

Having decided on the number of colours to be used, the next question relates to *which* colours should be used. In this respect it was pointed out earlier that many colours are already associated with different moods or conditions, so it is important that the colours chosen should be compatible with the operator's stereotyped ideas of what they represent. Unfortunately little research has been carried out to answer this problem, so most of the colours used in modern displays are chosen on the basis of 'commonsense'. In this respect Morgan *et al*. (1963) have suggested the meanings and uses of a number of different colours (Table 5.2).

Finally, Poulton (1975) has pointed out that colour coding can often be used to indicate sizes, for example some shops code cloth sizes according to colour; and the value of a resistor has been colour coded for some years (black $= 0, \ldots$ white $= 9$). On the basis of a small survey in which he asked people to say what colour they thought represented the largest size, etc., Poulton was able to

Table 5.2 The suggested meanings of different colours (adapted from Morgan *et al*., 1963, and reproduced by permission of McGraw Hill)

Colour	Meaning	Use
Red	Hazard	Fire: Alarms, extinguishers, hose Danger: Symbols Stop: Signs on machinery, road signs Emergency
	Hot	
Orange	Possible danger (but not immediate hazard)	Dangerous parts of machines and guards
Yellow	Caution	
Green	Safety	First-aid equipment
Blue	Caution	
	Cold	
Purple	Radiation Hazards	

recommend an 'ergonomic colour code' for sizes in which red represents the largest size and white the smallest. Intermediate sizes should be represented in colours arranged in rainbow order.

Pictorial qualitative displays

The use of pictorial displays which illustrate the state of the machine was discussed above (p. 89). In this respect it was shown that the overriding design principle is that the display should be as 'true to life' as possible.

Display design for check reading

Although comparing two readings (comparison), or ensuring that a value has or has not been reached (checking), is a task which frequently has to be carried out, very little research appears to have been conducted to help design an appropriate check reading display. This is possibly because for check reading extensive use has been made of colour coding to differentiate the important areas of the dial, although this may not be the most appropriate technique available. For comparison readings the most useful technique is to arrange the

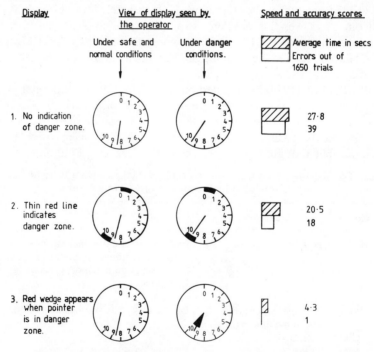

Figure 5.4   The value of using a warning flag appearing on a display to indicate 'danger' (from Kurke, 1956. Copyright 1956 by the American Psychological Association. Reprinted with permission)

various displays in a particular way, as will be discussed later in Chapter 7.

An interesting design for a check reading dial face has been proposed by Kurke (1956), who suggested a dial face which was so arranged that when the indicator is pointing to, say, a 'danger' reading, there appears on the dial face a high-contrast wedge-shaped 'flag' which is not present when the machine is operating in 'safe and normal' conditions. Comparing this type of dial with one which had a colour-coded (red) area permanently on the dial face and one which had no obvious aids, Kurke demonstrated significantly reduced errors and operation times when using the dial which had the warning flag (Figure 5.4). Using this display, therefore, the operator would only have to perceive the presence or absence of the flag. Using a display with colour-coded portions he needs to seek out the position of the pointer and then to decide whether or not it is in the red area. Finally, if no aids are available he needs to determine the pointer position, make a reading, and then to decide on the meaning of the reading. Clearly having to decide only whether a flag is present or not is likely to be a considerably easier task.

Cathode Ray Tube and Visual Display Units

The cathode ray tube (CRT) and its more computer-jargonized relative, the visual display unit (VDU), offers the designer a more flexible man–machine communication system than either the analogue or the digital display alone. Being computer linked, in many respects it represents an amalgam of all types of display, allowing alphabetic characters, numbers and even diagrams to be displayed. Indeed Chorley (1973) suggests four main advantages of these displays: First, they are sometimes the only suitable means of presenting certain types of information (for example radar or television). Second, with limited panel spaces and with the increase in the amount of information to be shown, some type of time-sharing display becomes possible. Third, using appropriate software the versatility of the electronic display allows flexibility of formats, in the order in which information is presented (for example, emergency signals can be programmed to override normal communications) and in the way in which different information is emphasized. Fourth, the VDU generates its own illumination and can therefore be used in the dark.

Despite the advantages of VDUs, however, a number of factors need to be considered from an ergonomics standpoint before the display is able to be used effectively by the operator. These factors may be subsumed under two headings: those important in the production of the display itself, and those important in producing the displayed information.

Display design

To those who need to compare different types of VDUs, four factors are important when considering the 'engineering' of the display. These are the display luminance, the regeneration rate, the resolution, and the colour.

Luminance

Gould (1968) points out that the amount of light reflected from the usual paper pages which people read these days under ambient house and office illumination is around 50 mL (the definitions and units of illumination are discussed in more detail in Chapter 11). He therefore suggests that this level may be used to estimate the recommended luminance for symbols on CRTs or VDUs. As is discussed later in Chapter 11, if the luminance departs significantly from the optimum level, visual fatigue may occur. Comparing the specifications of different VDUs of the time, Gould considered that most fell within this loose specification.

Coupled with overall luminance is the luminance contrast between the characters on the display and the background. Thus the brightness of the display symbols must be significantly higher than that of the background to ensure that the symbols are readily and accurately identified. Unfortunately, however, the problem is not as simple to solve as it would appear. Whereas, for example, the illumination of any one of the characters on this page is constant, a symbol (or light spot) produced on a VDU is composed of a spread of illumination levels with the brightest part in the centre and gradually becoming dimmer towards the edges. This means, therefore, that the contrast between the edge of the symbol and its background is not as high as between its centre and the background, resulting in a slightly 'hazy' symbol. To overcome this slight 'blurring' effect and to increase the contrast difference, Gould (1968) suggests both reducing the level of ambient illumination (and thus the amount of illumination reflected from the background) and adding a darkened filter to the screen.

The question whether light symbols should be viewed against a dark background (positive contrast), or vice versa (negative contrast), has been asked recently by a number of investigators. The evidence which has emerged appears to demonstrate, fairly conclusively, that performance is increased using the *negative* contrast (dark symbols against a light background) technique (Bauer and Cavonius, 1980; Radl, 1980). These findings, of course, are in contradiction to those of Berger (1944) who investigated the contrast direction of car numberplates (see p. 95). The reason for these different conclusions, however, may lie in the particular task normally confronting the VDU operator—namely the transcription of information from the printed page to the screen or vice versa. Thus, as Radl (1980) points out, using a negative contrast image the adaptation conditions for the eye are better for this type of task which require eye movements between the VDU screen and the paper sheet. Furthermore, having a light background reduces the possibility of glaring reflections occurring on the screen which tend to mask the displayed symbols (see p. 240). This only emphasizes, once again, the need for investigations to be carried out within the system under consideration.

Regeneration rate

The regeneration rate refers to the speed in which symbols are produced on the screen which, if it is not fast enough, causes them to appear to flicker. Clearly, therefore, to reduce flicker (and thus visual fatigue) the regeneration rate must be equal to or greater than critical fusion frequency. As was discussed above in Chapter 2, many variables influence critical fusion frequency including the luminance level and the position of the screen relative to the observer. Gould (1968) recommends a regeneration rate of faster than 60 c.p.s., which is the rate of television receivers. On this basis over 60 per cent of the VDUs he surveyed had regeneration rates which were lower than his recommendation.

Resolution

Two basic methods are available to produce symbols on a VDU—either cursive (line writing) or raster. Using cursive regeneration the spot is deflected in such a way that each symbol of the display is traced out in turn, with the spot being blanked out between symbols. This method enables conventionally shaped letters and numbers to be employed, and it makes use of the maximum possible resolution of which the display unit is capable.

Using a raster method the electron beam scans the whole area of the screen in a regular pattern, but only 'turns on' when it is appropriate to form a character. (This is how television screens operate.) This method gives slightly less well-defined characters since the resolution is lower because of the finite spacing between the raster lines. However, provided that a large enough number of lines is used (the domestic television system uses 625) the line structure will not be visible.

Colour

With advancing technology, the use of coloured symbols on VDU screens is now a viable proposition. For example Engel (1980) points out that, in addition to enlivening the image and improving the legibility of displays, colour can also be used to structure and emphasize parts of the text. However, he also cautions against the overuse of colour in visual display units. Words or groups of words which are composed of different colours may break up the natural grouping by emphasizing the 'wrong' aspects of the 'phrase', so interrupting the word's *gestalt*. The need for the ergonomist to decide which aspects of the text are important, and to consider some of the principles discussed in Chapter 4 is therefore clear.

Because the use of colour in VDU displays is a very recent advance, little published work is available to suggest which colours should be used. Using 30 subjects, however, Radl (1980) conducted an experiment to investigate operator performance using different coloured symbols. White and six other colour combinations were used and subjects were asked to transcribe the letters from the VDU screen to white paper. His results illustrate the performance advantage of yellow characters over other colours.

Designing the displayed information

Discussing the display design leads, naturally, to a consideration of the ways in which the display symbols are produced. If a raster generation method is used, for example, the display resolution will influence the size and typography of the displayed characters.

A number of authors have considered the problem of the number of raster lines for legibility, and all of the available evidence points to an optimum of ten lines per character. The legibility of a character, however, also depends on its height with respect to the observer (that is, the angle which the symbol subtends at the observer's eye), and Hemingway and Erickson (1969) have drawn together different studies to produce a family of curves relating lines per symbol and the angle subtended at the eye for 80, 90 and 95 per cent correct detection (Figure 5.5).

As the typographical design of letters was shown (Chapter 4) to affect their legibility, the same is true for the design of characters on a VDU. Modern displays use a number of different character founts, of which some have been shown to be less readable than others.

Vartabedian (1971) reported a study in which he compared the speed of subjects' recognition and the number of errors made for different CRT displays. In essence two forms of display were used—either a dot matrix or stroke display (these are produced by either a raster or cursive operation method). The dot matrix appears as its name implies: each character is formed by a series of dots

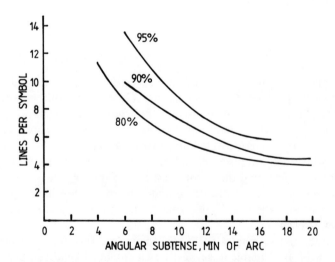

Figure 5.5 The relationship between the number of VDU raster lines and the angle between the lines subtended at the eye for 80, 90, and 95 per cent correct detection (from Hemingway and Erickson, 1969, and reproduced by permission of the Human Factors Society Inc.)

in a matrix, with the various dots being either turned on or not turned on to form the symbol. The stroke symbol was formed from the normal cursive method.

Six display were compared: a 5 × 7 dot matrix, using either circular or elongated 'dots'; a 7 × 9 dot matrix, using either circular or slanting 'dots'; an upright stroke; and a slanted stroke. The results indicated, fairly conclusively, that the 7 × 9 circular dot matrix produced less errors and faster reaction times than did the other forms. Whereas the 5 × 9 circular dots produced roughly similar numbers of errors as the 7 × 9 matrix, the characters took rather longer to recognize. The two slanting displays (7 × 9 elongated, and stroke) produced a higher proportion of errors and a slower reaction time than their upright counterpart. Finally, Vartabedian also demonstrated that elongating the 'dot' also adversely affected legibility.

On the basis of this work Vartabedian (1973) produced a set of alphanumeric characters for a 7 × 9 circular dot display. A similar matrix was used by Huddleston (1974), although his display was composed of square rather than circular dots. In addition Maddox, Burnette, and Guttmann (1977) have produced recommendations for a 5 × 7 dot matrix display. Each of these displays are shown in Figure 5.6.

Finally, it should be remembered that the ergonomic factors important in a computer generated visual display unit do not end with the characters generated on the screen. For example, the design of the software which controls the display format also needs to be considered carefully. At this point we re-enter the realm of man–man communication, and all the principles discussed in the

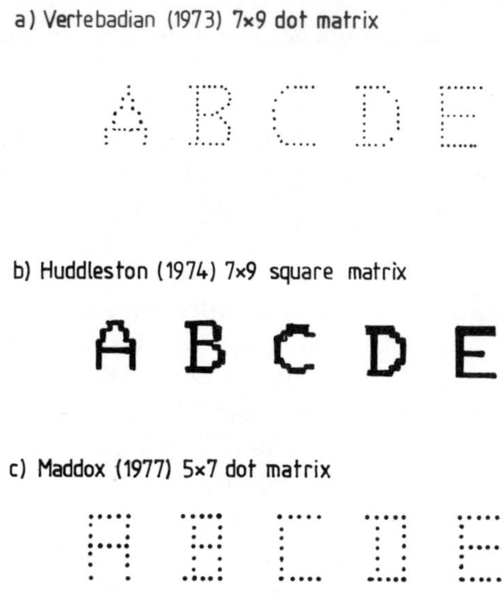

a) Vertebadian (1973) 7×9 dot matrix

b) Huddleston (1974) 7×9 square matrix

c) Maddox (1977) 5×7 dot matrix

Figure 5.6   Three-character founts produced for visual display units

previous chapter (layout, spacing, highlighting, headings, etc.) need consideration.

AUDITORY DISPLAYS

Although the visual modality is the most extensively used medium for presenting information to the operator, auditory displays do have their value—particularly if the visual system is already overloaded, or if the operator needs to have the information irrespective of where he is looking at the time. In addition, as Colquhoun (1975a) has pointed out, if combined with visual displays, auditory displays can often make monitoring performance superior to the use of visual displays alone. Auditory displays, therefore, are suitable primarily as warning devices although they are used in some circumstances to display information about the machine state. These aspects are discussed below, but it should also be remembered that in some cases quantitative information can be dispalyed in the auditory mode. The hourly chimes of a clock or the dots and dashes of morse code are examples of such uses.

Warning Displays

A warning sound is probably the simplest type of auditory display since it conveys information of an 'on-off' nature. Either the system is 'safe' or 'unsafe' and its state is indicated by the absence or presence of the sound. To be effective, therefore, a warning sound should be both perceptible and attention getting.

Unfortunately, little scientific data exists to guide the designer in his choice of suitable warning displays. Murrell (1971), for example, suggests that to be effective the sound intensity should be at least 10 db higher than the background noise, although he offers no supporting evidence for this assertion. As will become clearer in Chapter 10, because of masking effects the warning sound should be louder than the background noise, but by how much is not clear since the level of masking depends on such factors as the frequency and duration of the signal. Unfortunately no authors other than Murrell suggest a minimum level.

As long as it can be heard, a criterion which is possibly more important is the need for the warning display to be attention-getting. Again little design data is available, although it would appear sensible to use sound to which the ear is maximally sensitive—that is, in the range 500 to 3000 Hz. McCormick (1976) has drawn together most of the available evidence regarding the qualities of the sound which are suitable for auditory warning signals. In addition to using signals in the range 500 to 3000 Hz, he suggests that if the sound has to travel far, then frequencies below 1000 Hz should be used, and if they have to 'bend' around major obstacles or pass through partitions, then the frequency should be reduced to below 500 Hz. He also suggests that a modulated signal (1–8 beats per second or warbling sounds varying from 1–3 times per second) are different enough from normal sounds to demand attention. Finally, McCormick

Table 5.3 Types of auditory alarm— their characteristics and special features (from Morgan *et al.*, 1963,) and reproduced by permission of *McGraw Hill*

Alarm	Intensity	Frequency	Attention-getting ability	Noise-penetration ability	Special features
Diaphone	Very high	Very low	Good	Poor in low-frequency noise Good in high-frequency noise	
Horn	High	Low to high	Good	Good	Can be designed to beam sound directionally Can be rotated to get wide coverage
Whistle	High	Low to high	Good if intermittent	Good if frequency is properly chosen	Can be made directional by reflectors
Siren	High	Low to high	Very good if pitch rises and falls	Very good with rising and falling frequency	Can be coupled to horn to directional transmission
Bell	Medium	Medium to high	Good	Good in low-frequency noise	Can be provided with manual shutoff to ensure alarm until action is taken
Buzzer	Low to medium	Low to medium	Good	Fair if spectrum is suited to background noise	Can be provided with manual shutoff to ensure alarm until action is taken
Chimes and gong	Low to medium	Low to medium	Fair	Fair if spectrum it suited to background noise	
Oscillator	Low to high	Medium to high	Good if intermittent	Good if frequency is properly chosen	Can be presented over intercom system

argues that 'high-intensity' sudden onset signals are often desirable for alerting an operator, and these could be presented dichotically (alternating the signal from one ear to the other) if earphones are used. Overriding all of these considerations, however, is the requirement that the quality of the warning sound is different from any other sound which the operator is likely to experience in his workplace.

Table 5.3 describes the relative advantages and disadvantages of different types of auditory warning displays.

Other Qualitative Auditory Displays

Just as a visual display can present the observer with relatively low-level qualitative information—for example, whether the machine is 'hot', 'cool' or 'cold'—so too can auditory displays. In this case the coding is usually made in terms either of the pitch of the tone or some other quality.

The present-day telephone provides an ideal example of this use. As soon as the receiver is raised a tone is heard which indicates that the system is working and is ready to be used. On dialling or keying a number a different tone will be heard depending on the new state of the system: number ringing, engaged or unobtainable. Other examples of the use of auditory information in this way can be conceived but, in all cases, it is essential that the tones indicating the different machine states are easily distinguishable.

Tracking Displays

Auditory tracking aids have been used for some time to help pilots maintain a steady course. Perhaps the simplest of these was the A/N signal system, which consisted of a continuous 1020 Hz tone being heard when the pilot was flying on course. If he deviated to the left, the tone becomes more of a morse A (dot–dash) signal, whereas an N (dash–dot) signal (which interleaves with the A signal to form the continuous 'on-course' tone) would become apparent with a deviation to the right. Although some success was claimed with this system, it does depend, again, on the distinguishability of the two (A and N) codes. As McCormick (1976) points out, however, under adverse noise conditions the difference between the two signals may not be properly identified, and the pilot may think that he is to the right of the beam when actually he is to the left, and vice versa.

Hofmann and Heimstra (1972) report an auditory tracking system which proved to be superior to a visual display. In this case the subject was required to maintain a particular random course using a hand wheel, in which he heard no noise if he was on target. If he deviated by more than 5 per cent however, the subject received a noise signal to either the right or the left ear, depending on from which side of the track he was deviating. Using this system, more time was spent on target using the auditory than the visual display.

From these two examples, therefore, it is clear that acoustic displays can be used to provide simple, one-dimensional tracking information, although great care needs to be taken to ensure that the various displayed states are easily distinguished.

SUMMARY

This chapter discussed the various ways by which information is presented to the operator by the machine—through its displays. Although many types of display are available, it soon became apparent that the most appropriate display is highly dependent on the nature of the task for which the information is to be used. In the visual modality the choice is essentially between an analogue and a digital display, although a number of different ways of presenting these two basic types were discussed. In the auditory modality, the designer's choice is more restricted.

The following chapter considers the third main sensory system which is used at work—the sense of touch. Although the main theme of the chapter is to discuss ways of enhancing the information flow in the other direction—from the operator to his machine via his controls—it should not be forgotten that each time the operator operates a control, for example using a push-button, the control shape and dimensions are also passing information from the machine to the man.

CHAPTER 6

Man–Machine Communication: Controls

Controls represent the final link in the man–machine closed-loop system and are very much the complement of displays. Indeed the value of a well-designed display, perhaps in terms of a reduction in the number of reading errors or in a faster reading time, could be seriously reduced if the many features important in the design of the control with which it is associated are not considered. Display design was discussed in Chapter 5, so the present chapter considers the factors which are important from the point of view simply of designing the operator's controls and tools.

A survey carried out by Fitts and Jones (1947b) well illustrates the fact that poorly designed controls alone may lead to inefficiency and breakdown in the man–machine system. In a complementary study to their analysis of aircraft display reading errors, they analysed 460 'pilot error' experiences in operating aircraft controls; 68 per cent of the errors could be ascribed to poor control design. The remainder were due either to mistakes occurring because of a lack of compatibility between the display and the control (6 per cent) or to poor placement of the control on the cockpit panel (26 per cent). (These considerations are discussed in Chapter 7.)

Controls, therefore, are important components in the system. However, many factors need to be considered before an effective control system can be designed that will match the operator's abilities and behaviour. The operator's task needs to be analysed to determine the degree of accuracy, force, precision and manipulation, etc. which it requires; this then has to be compared with the operator's abilities to carry out such tasks. If his abilities do not much the requirements, then changes in the mechanical part of the system ought to be considered—perhaps involving different types or designs of controls and control systems.

This chapter, therefore, first considers the types of controls which are available for carrying out different types of operations. Second, some of the factors which are important in the design of controls and aspects of the system which could affect control effectiveness are examined. Finally, the discussion considers the extent to which such factors have been incorporated in the production of specific types of control.

TYPES OF CONTROL

Controls are commonly classified into two groups according to their function. The first type includes those which are used to make discrete alterations in the machine state, for example switching it 'on' or 'off' or switching to different levels of machine activity. The second type includes those controls used for making continuous settings—for example, a radio volume control allows the user to increase the volume gradually, and to stop at any of an infinite number of intensities within its operating range. McCormick (1976) further subdivides these two functions into:

Table 6.1 Types of controls and their functions

		Discrete		Continuous	
Control type	Activation	Discrete setting	Data entry	Quantitative setting	Continuous control
Hand push-button	Excellent	Can be used will need as many buttons as settings—not recommended	Good	Not applicable	Not applicable
Foot push-button	Good—	Not recommended	N/A	N/A	N/A
Toggle switch	Good—but prone to accidental activation	Fair—but poor if more than three possible settings to be made	N/A	N/A	N/A
Rotary selector switch	Can be used but on/off position may be confused with other positions	Excellent—provided settings are well marked	N/A	N/A	N/A
Knob	N/A	Poor	N/A	Good	Fair
Crank	Only applicable if large forces are needed to activate—e.g. open/close hatch	N/A	N/A	fair	Good
Handwheel	N/A	N/A	N/A	Good	Excellent
Lever	Good	Good—providing the are not too many settings	N/A	Good	Good
Pedal	Fair	N/A	N/A	Good	Fair

Discrete

1. activation—for example, turning a machine on or off,
2. data entry—as on a keyboard to enter either a letter or a number,
3. setting—switching to a specific machine state.

Continuous

1. quantitative setting—setting the machine to a particular value along a continuum, for example, turning a radio frequency control to receive a specific radio station,
2. continuous control—continuously altering the machine state, for example to maintain a particular level of activity (commonly known as tracking).

Because of these varying activities, different controls will be more appropriate for some purposes than for others. The respective advantages of various controls for different activities are shown in Table 6.1. Although it might be suggested that some general design of control could be produced for most activities, the evidence suggests that it is more appropriate to choose a control which has specific advantages for specific situations. This was illustrated by Chambers and Stockbridge (1970) who, after reviewing the literature concerning different types of push-button design recommendations, suggested that in terms of operating speeds the most efficient activation control was the push-button followed by the rocker switch, the slide switch and the toggle switch. However, because the push-button is usually associated with a 'ballistic-like movement which resulted in increased speed but reduced accuracy', the order for accuracy was the reverse that for speed.

Finally, as McCormick cautions, the usefulness of any control can be limited by such features (if relevant) as the ease to which it can be identified, its location and size, its relationship to the appropriate display, and the type of feedback which it gives to the operator. The next section considers some of these important factors.

FACTORS IMPORTANT IN CONTROL DESIGN

Feedback

The concept of feedback is one which has been discussed already in previous chapters. It refers to the information which the operator receives from both his environment and from his own body, which helps him to assess the spatial position of both himself and his body parts. In relation to hand controls, for example, the feedback which he obtains from his eyes, shoulders, arms, wrists and fingers tells the operator by how much a control has been moved and its final position. In addition, feedback from the more sensitive pressure receptors in the skin provides the operator with information concerning the nature of the control which he is operating, for example its size, texture and any tactile coding characteristics. Feedback, therefore, relates to the 'feel' of the control.

Burrows (1965) has pointed out that control 'feel' arises from two separate sources. Firstly it occurs as kinaesthetic feedback from the muscles—for example, telling the operator where his arm is at the time and the speed with which it is moving through space. As was discussed in Chapter 2, this is a very efficient form of feedback, particularly in learning different skills. Secondly, 'feel' is determined by the control itself in terms of the amount of resistance to movement which is built into it, its looseness, etc. To these sources should, of course, be added any tactile, visual or auditory feedback loops which may help the operator (for example, 'clicks' or marks on the control surface).

Possibly the main feedback cue arises from the control's resistance to movement. In the majority of cases, particularly when continuous settings are to be made, some inbuilt resistance is desirable since it allows the operator to make his settings with a certain degree of precision. In addition, resistance will often help to guard against the accidental activation of the control. If too much resistance is incorporated into the control, however, or resistance of the wrong type, performance may be reduced and the operator could experience fatigue. Understanding the nature of different types of resistance, therefore, should make it possible to choose those controls which have resistance characteristics which will minimize any possible negative effects, while at the same time maximizing performance.

Control resistance takes four main forms, and their advantages and disadvantages are shown in Table 6.2. From this table, it would appear that static friction is most appropriate for discrete setting controls since it reduces the possibility of accidental operation. For continuous setting controls, however, elastic or viscous resistance will allow greater precision due to the nature of the kinaesthetic feedback which it provides.

Regarding the level of resistance to be introduced, it is difficult to set any maximum figure since it will be related to the type of operator, the location of the controls, and the frequency, duration, direction and amount of control movement required. Clearly, however, the maximum level set should lie within the range of abilities of the operating population (see Damon, Stoudt, and McFarland (1971); and Roebuck, Kroemer, and Thomson, 1975). For minimum levels Morgan *et al.* (1963) suggest that, for all hand controls except push-buttons, resistance should not be less than 2 to 5 lb (9 to 22 N), since below this level the pressure sensitivity of the hand is very poor. If the full weight of the arm and hand rest on the control, the minimum resistance should be 10–12 lb (44 to 53 N). Table 6.3 summarizes some of the data available for maximum resistances for different types of control.

Size

The size and dimensions of the control clearly need to be related to the anthropometric dimensions of the limbs to be used. Thus the diameter of a

Table 6.2 The characteristics of static and coulomb, elastic, viscous and inertial control resistances

Type of resistance	Example of incidence	Characteristics	Advantages	Disadvantages
Static and coulomb	1. On/off switch 2. A 'stuck' control	The resistance is maximal at the start of the movement, but falls considerably with further force, i.e. the control slips	Reduced chance of accidental activation	Little precision control once the control has begun to move
Elastic	Spring-loaded control	Resistance is proportional to control displacement	1. Kinaesthetic cues may be maximally effective 2. Control returns to null position	Because control returns to neutral, operator's limb needs to be constantly active
Viscous	Plunger	Resistance is proportional to the velocity of the control movement	1. Good control precision—particularly rate of movement 2. Reduced chance of accidental operation 3. Operator can remove his limb and control remains in position	
Inertia	Large crank	Resistance is caused by the mass of the control	1. Allows smooth movement 2. Reduced chance of accidental operation due to high force required	1. May cause operator fatigue 2. Does not allow precise movement because of danger of overshooting

Table 6.3 Minimum resistances required for different controls (adapted from Morgan *et al.*, 1963, and reproduced by permission of McGraw Hill

Control	Minimum resistance
Hand push-button	10 oz. (2.8 N)
Foot push-button	4 lb (17.8 N) if foot does not rest on control; 1.25 lb (5.6 N) if foot rests on control
Toggle switch	10 oz. (2.8 N)
Rotary selector switch	12 oz. (3.3 N)
Knob	0–6 oz. (0–1.7 N) depending on function
Crank	2–5 lb (9–22 N) depending on size
Handwheel	5 lb (22 N)
Lever	2 lb (9 N)
Pedal	4 lb (17.8 N) if foot does not rest on control; 10 lb (44.5 N) if foot rests on control

push-button should be at least that of the fingertip (approximately 16 mm); the size of a knob on a lever equal to the breadth of grip (49 mm), etc. Garrett (1971) provides a set of these various dimensions for the human hand but it should be remembered that these dimensions will be altered, sometimes considerably, if the operator is wearing gloves.

Not only is it important to relate the size of the control to the dimensions of the limb which is used to operate it, it is also important to consider the type of action which is required of the operator, since all controls require some degree of manipulation. For hand controls, therefore, different types of manipulative tasks will require different control dimensions, depending on the part of the hand used to operate the control.

Most manipulative tasks can be placed along a continuum of 'gripping' to 'non-gripping' activities. In gripping activities the fingers and parts of the palm form a closed chain and act in opposition to each other to exert compressive forces on the object to be gripped. In non-gripping actions the forces are exerted either through the whole hand or through the fingertips in an open chain. In addition to the amount to which the fingers are closed, a second manipulative dimension can be related to the degree of hand/object contact. From such a two-dimensional classification it is possible to determine which anthropometric dimensions are required for any particular task, and also the forces and torques needed. Some of these dimensions, with some examples, are illustrated in Figure 6.1.

Weight

The weight of many controls becomes important only when the inertia is enough to cause undue resistance, for example as with a crank handle, otherwise the weight will be supported by the machine itself. However, some controls are often used away from a machine (particularly as hand tools) and in these cases the weight of a tool may obviously play an important part in its effectiveness.

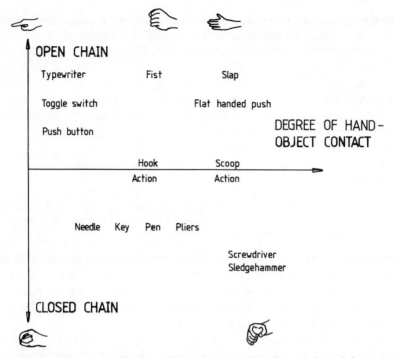

Figure 6.1 A classification of hand control functions (adapted from Grieve and Pheasant, 1981, reproduced by permission of Cambridge University Press)

In addition to the overall weight of a tool, its weight distribution is also an important consideration. When the hand is held in a relaxed, neutral position, a rod held in the hand makes an obtuse angle of approximately 102 degrees to the forearm (Tichauer, 1975). Major deviations from this will cause static load to be placed on the wrist muscles. If most of the weight is distributed towards the front or the back of the tool, therefore, so that the wrist needs to work to maintain its natural posture, the static load will soon cause fatigue. The ideal weight distribution will be one which places the maximum weight over the place where the tool is held, and maintains the 102-degree angle.

Control Texture

Since the control acts as the interface for information flowing between the operator and the machine, it might be thought obvious to point out that the quality of the control action will depend largely on the extent to which the operator's limb is able to remain in contact with it. As with many aspects of the working situation, however, ensuring that this occurs is not so simple. For example, it is obvious that the surfaces of hand-held controls should not be so smooth as to make it difficult to grip firmly. This is particularly important if the hands are likely to be moist from sweat. In addition, a highly polished surface may cause glare, perhaps adversely affecting the operator's performance on a visual task. On the other hand, however, surfaces which are to be grasped or which are likely to be rubbed against should be free of any abrasive properties (rough sufaces are often contaminated with sand or dirt, etc, so it is likely that an abraded wound would eventually become infected). A balance must be struck, therefore, between the two extremes, so the question becomes to what extent the control should be textured. Many of these problems are solved by using a non-reflective, rippled coating, but the ripples should not be raised to the degree that they cause painful pressure spots.

When the operator applies a force to a hand control, the direction of the force may act either transversely across the palm as in the case of a steering wheel, or longitudinally as with a lever. (A foot, too, will exert longitudinal forces along a pedal.) In both cases the rippled texture described above may help the designer in his task of minimizing the possibility of the hand slipping. However, the two types of force suggest that the directions of the ripples need to be considered in so far as they should be so arranged as to be at right angles to the likely direction of force.

Control Coding

The value of highlighting different aspects or areas of the machine to ease identification was discussed in the last chapter. For qualitative displays colour coding different areas was shown to produce increases in performance and a reduction in errors.

It is, of course, possible also to colour code controls. However, because they are usually operated by a limb it is probably more appropriate to code them along some tactile dimension, thus allowing the eyes to be released to accept other, visual, incoming information. Unfortunately, however, touch is a less accurate sensory mode than is vision for perceiving differences, which may lead to uncertainties in the operator's actions and probably slows down discrimination time. For this reason Moore (1976) recommends that tactile identification of controls should only be used as a final check on the identity of the control, rather than being the primary coding method. Adequate labelling, he suggests, is probably as efficient. Like colour coding, however, labelling also needs to be seen to be effective and, furthermore, requires careful design—as was discussed in Chapter 4.

In many cases controls may have to be positioned in places where labels or colours are not easily seen, and in such cases shape, texture, size and location coding, or indeed any combination of these, may be employed. For each of these techniques, the aim is to produce groups of controls such that the controls within each group are rarely confused with each other.

Coding by shape

An operator is able to distinguish different shapes primarily by the pressure differences, caused by various protrusions in the shape, which are produced over the hand tissue. For this reason shape coding is normally useful only to ungloved operators.

A number of sets of unconfusable shapes have been produced by different investigators over the years (for example, Hunt, 1953; Jenkins, 1947; Moore, 1974). In general the work has demonstrated that simple forms are easier to discriminate than are more complicated ones. Furthermore, the learning of their use can be simplified and made more efficient if, in addition to being individually discriminable by touch, the controls have shapes which, either by design or by convention, look similar to the part of the machine which they operate.

Coding by texture

In addition to the control shape, its texture may also be coded; for example the edge of a control may be rippled or knurled. As Bradley (1967) demonstrated, as long as the textures are distinct enough, confusions will not occur. He produced smooth, fluted and knurled textured control wheel edges which would not be confused. Furthermore, appropriate training and practice may significantly improve an operator's ability to distinguish between textures (Eckstrand and Morgan, 1956).

Like shape coding, but perhaps to a lesser extent, the stimulus perceived by the operator when using textured controls is related to the degree of deformation produced over his fingertips by the different textures. It is this aspect of the control which the operator learns to recognize. Any impediment to the perfect transmission of this information from the control to the operator, therefore, could result in errors being made. Gloves provide a good example of this problem. Even if the gloves are thin enough to allow the operator to manipulate his controls under normal circumstances, they may nevertheless impair (or perhaps more importantly alter) his perception of the surface texture. Dirt and grime collecting in the textured surface may also cause similar problems.

Since the important stimulus is the degree and pattern of skin and tissue deformation, texture coding need not be confined solely to 'smoothness', 'rippledness', or 'knurledness'. Moore (1974), for example, describes an experiment designed to investigate discriminability between different raised surface shapes on push-buttons. By asking subjects to feel (blindly) the surface of different push-buttons and to compare them with pictures, he was able to produce a set of six discriminable surface textures.

As far as tactile shape coding is concerned, Moore recommends the following five principles:

1. Shapes to distinguished by touch should have as gross a shape as possible, covering an area which can be touched by one finger.
2. Geometric shapes, numbers or letters should be formed from outlines rather than from solids.
3. Shapes should be made to vary along as many tactile dimensions as possible.

4. If at all possible, the buttons should be designed or chosen to ensure that the shapes remain in the same orientation at all times and do not revolve.
5. The shapes should not be uncomfortable or difficult to use.

Coding by size

The size of the control itself may provide useful visual or tactile cues, but size alone generally is not as useful for coding as is shape or texture. Again the different sizes used should be such that they are discriminable from one another. In this respect, Moore (1976) suggests that the sizes used should follow a logarithmic progression with at least a 20 per cent difference between each size.

Coding by colour

Colour coding has previously been discussed in relation to visual displays, and similar principles exist for control colour coding. Unlike visual displays, however, controls are not normally used to present visual information, so they may not be in a position which allows the operator unrestricted lines of sight. Indeed, they are often sited in positions which enable them to be operated while the operator is looking elsewhere, perhaps at the display. In these circumstances, colour coding—an essentially *visual* aid—is often of little value.

FACTORS WHICH AFFECT CONTROL EFFECTIVENESS

In many respects the majority of the topics discussed in this book will affect control effectiveness: the quality of the information reaching the operator; his position with respect to the control; his environment (both social and physical); fatigue; stress, etc. However, this section considers a few aspects of the man—machine interaction which are not discussed elsewhere and which are specific to control design—namely operator handedness, the wearing of clothing (in particular gloves and shoes), and the positional relationships between the control and the operator's limb.

Handedness

For everyday usage, an individual's handedness (or, more accurately, hand preference) may be classified as being 'left' or 'right' simply on the basis of the writing hand. However, when many different actions are taken into account this simple dichotomy clearly becomes insufficient. For different types of actions a single individual may have different hand preferences and this could cause problems from the point of view of determining which hand should operate which control. Taking the action of tightening a screw as a very simple example of the problem, a right-handed person would need to rotate his wrist with a movement of his palm upwards, in other words, twisting his wrist away from his body (supination). The operator with a left-hand preference, however, will need to protinate his wrist (palm moved downwards—wrist twisted towards his body). Unfortunately for the left-handed person, however, supination allows a greater torque and range of movement than does protination (Damon, Stoudt and McFarland, 1971).

The problem of left-handed operators, however, does not simply rest with considering the strengths and types of movement of the left and right hand. Controls, and particularly tool handles, are often designed for use by the right-handed operator, and when used by a left-handed operator they will be found to be difficult or uncomfortable to use. Such difficulties may well lead to fatigue and possibly to accidents.

The solution to the problem of left-handedness, however, is not simple. Various estimates of the incidence of left-handedness have been made and vary from 2 per cent to 29 per cent (Hardyck and Petrinovich, 1977), although a general rule of thumb is that less than 10 per cent of any large national population is left-handed (or at least has a left-hand preference for most activities) (Kimura and Vanderwolf, 1970). Since ambidexterity, in the sense of equal preference, is extremely rare, the question must be posed whether it is possible to accommodate fully the requirements of the left-handed operator. Naturally, sometimes right and left-handed tools could be made available, in which case the different handle shapes and directions of movement will need to be considered. In other cases, however, for example the question of where controls should be placed with respect to the operator's own position, it may not be possible (or economical) to produce both right and left-handed versions. In such cases, therefore, it is even more crucial to assess the requirements of the job and to match them to the capabilities of the operator.

The Presence of Clothing and Protective Clothing

The types of clothing which are most likely to interfere with efficient control action are gloves covering the hand and shoes on the feet. Both may increase the necessary dimensions of the control and both are likely to affect the operator's ability to use the control adequately.

Although gloves are designed to protect the operator's hands they may have a number of undesirable consequences, particularly in relation to the operator's ability to manipulate and to obtain feedback from his control. As an example, the normal sensation of 'grip' probably results from the pressure perceived when the flexed fingers around the gripped object press against each other. If the working glove happens to be too thick in these regions, high pressures can be generated between the fingers before the hand is firmly closed around the tool handle or equipment control, which may result in an insecure grasp. Furthermore a thick glove can also obstruct the fingers from wrapping around the handle sufficiently for a firm grip. On the other hand, if the operator is aware of these problems he may grip the control unnecessarily tightly and firmly so increasing fatigue in his finger and other muscles.

After carrying out a series of experiments to determine the degree to which gloves interfere with control manipulation and operation speed, Bradley (1969a) concluded that the efficiency with which instrument controls may be operated by a gloved hand depends on the glove characteristics, the physical character-istics of the control, and the type of control operation. Specifically, snugness of fit and resistance to slipping were shown to be the two most important glove parameters, and under some circumstances a snug glove which did not slip over the controls actually improved performance. In many circumstances, however, gloves are worn for protection against injury to the flesh and then snugness and even resistance to slipping may be absent. In such cases, therefore, the size of the controls should be increased to allow adequate manipulation and, as discussed earlier, the control ought to be textured to reduce the possibility of slipping.

In addition to interfering with grip, gloves can often impede the perception of any coded texture differences on various controls. As was discussed earlier, such texture differences cause different pressure patterns on the observer's skin and it is these which could be occluded by the gloves.

In a similar fashion to gloves interfering with the operation of hand held controls, shoes sometimes affect the efficiency of foot-pedal operation. Heavy, protective footwear, for example, may not allow the feet to be moved with the required precision since the necessary feedback may be either missing or of poor quality.

A second impeding feature of shoes which is often forgotten is the height of the heels. As will be discussed later, most foot-pedals are designed for the forces and angles produced by a foot having 'average' dimensions, and the presence of high heels, for example, can alter these parameters. As an example of these effects, Warner and Mace (1974) demonstrated that the average brake response time of 17 female drivers was increased by 0.1 sec when wearing platform (having a heel higher than 2 in. (5 cm) and a sole $\frac{1}{2}$ in. (1.5 cm) thick) as opposed to 'normal' shoes—despite the fact that the subjects were used to wearing platform shoes. (This is simply a further example of training being unable to overcome the effects of poor equipment design.) Translated into vehicle-stopping distances, this represents an increase of 1.42 feet (0.5 m) for each 10 mph speed increase, which could mean the difference between safe and unsafe stopping.

Control Shapes

Control shape can have an important influence on the way in which an operator uses the control, which may in turn affect the posture. Awkward postures can then put undue stress on the musculoskeletal system, causing fatigue over long periods of use. This applies particularly to the use of hand controls.

Many tools are used which, under normal operational conditions, require the wrist to be bent either downwards or upwards. The effect of this action is to cause the tendons which connect the finger muscles to the forearm bones in the elbow region to bend and to become subject to mechanical stress. Generally, no ill-effects are experienced when such a tool is used infrequently but, under continuous operation, the effect will be to cause muscular fatigue and thus loss of efficiency. As Tichauer (1975) points out, it is much safer to bend the tool than to bend the wrist. Perhaps a general recommendation would be to ensure that hand tools are designed to allow the device to be operated with the hand and forearm longitudinal axes as nearly aligned as possible.

Shape is also important when considering the cross-sectional configuration of a hand tool. If high grip forces are required, the handle should distribute the forces to as large a pressure-bearing area on the fingers and palm as possible, while still being small enough to allow the fingers to wrap round the handle. To this end, Pheasant and O'Neill (1975) have demonstrated that muscular strength is maximal when using handles of around 5 cm diameter. Below 3 to 4 cm and above 5 to 6 cm both the amount of contact which the hand has with the handle, and the amount of torque able to be exerted, falls dramatically.

THE DESIGN OF SPECIFIC TOOLS AND CONTROLS

When discussing the development of modern ergonomics, many sceptics describe the era of the 1950s and early 1960s as being the period of 'knobs and dials' investigations. By this they imply that ergonomists were interested mainly in the design aspects of specific types of displays or controls—rather than considering the functioning of these pieces of equipment as parts of the total work system. Although a cursory examination of the published literature of this period tends to support this observation, it could well be argued that basic research of this nature needed to be carried out before the advantages and disadvantages of different types of display or control in the system could be considered.

Because it is important to be able to recognize good 'ergonomic' control designs, this section provides a very basic description of the design recommendations for different types of control as elucidated by the research of the period. However it should be realized that some of the data are based on experiments using comparatively small numbers of subjects. Given the wide variability of performance abilities shown by individuals, therefore, it is not possible to state figures for diameters, loads and so on with absolute certainty. As Murrell (1971) points out, however, 'At the best (the figures) give a reasonable estimate of the design parameters concerned, at the worst they are better than guesswork on the part of the designer who, if he follows them, may rest assured that the equipment which he is producing is being designed in accordance with the best knowledge available at present'.

Hand Controls

Naturally the basic considerations of all hand controls concern the anthropometric and biomechanical capabilities of the operator's fingers hands and wrists.

Knobs

The knob is a cylindrically shaped control which is operated by grippi
thumb and forefinger around the circumference and moving them in oppos.
They may be used for making fine, continuous adjustments or as rotary selector
switches.

It is important that the diameter of this type of control is not too small to
prevent it from being gripped and turned easily. On the other hand, panel space
should not be wasted by using controls which are larger than those required
for efficient operation. For this reason Bradley (1969b) asked 48 subjects to
make various clockwise and anticlockwise 'standard' operations using knobs
with diameters ranging from 0.5 to 3.25 in. (1.2 to 8 cm) in 0.25 in. (0.6 cm)
increments. Two levels of shaft friction (torque) were used: 77–85 inch-grams
(0.5 to 0.6 Nm) (normal) and 171–181 inch-grams (1.2 to 1.3 Nm) (heavy). Using
turning times as his main measure, Bradley demonstrated that a knob diameter
of approximately 2 in. (5 cm) was optimum for both normal and heavy friction.
As the shaft friction increased the turning time was magnified significantly as
the knob diameter deviated from the optimum.

As control panels increase in complexity, more and more instruments may
have to be crowded into a limited panel space. One possible means by which
this may be accomplished is by grouping controls together. In the case of
concentric controls, this would mean ganging several control knobs perpendicu-
lar to the panel by mounting them on concentric shafts. Unfortunately, however,
ganging control knobs probably increases the chance of inadvertently operating
adjacent knobs. While turning one of the knobs the operator's fingertips or
knuckles might scrape against the face of the knob immediately behind it, or
his fingers or palm may rub against one of the knobs in front of it, so altering
the setting of the knob which was inadvertently operated.

Bradley (1969c) investigated the optimum dimensions for such controls. In
a very similar experiment to his determination of optimum knob diameters he
demonstrated that, for three ganged controls, if the diameter of the centre knob
is 5 cm (optimum knob diameter) the diameter of the front knob should be
less than 2.5 cm and that of the rear knob about 8 cm. His results also indicated that the front and middle knobs should each be 8 cm thick, whereas
the back knob could be as thin as 0.6 cm. Bradley further argues that
concentrically ganged knobs should only be used if enough shaft friction is
present to prevent accidental operation.

Push-buttons

Push-buttons are small, single-action controls which operate in one direction
only. They are usually activated by the fingers but are sometimes foot-operated.
They range in size from comparatively large 'on' or 'off' buttons for heavy
machine tools to small, individual finger-controlled buttons such as seen on a
modern electric typewriter keyboard. Three types of push-button are commonly
available: 'latching' (push-on, lock-on), 'momentary' (push-on, release-off) and

'alternate action' (push-on, push-off). All three modes are found in hand-operated buttons, but the latching mode is seldom found in foot-operated buttons.

The important physical parameters of push-buttons are their size, separation, shape, operating force, the provision of feedback, and the separation between the buttons. Of these factors the shape of the button has already been discussed, and the separation between buttons is discussed in the following chapter.

Regarding size, the limiting factor must be the dimensions of the fingers which are to operate the controls. In this respect, a minimum diameter of 1.5 cm is often suggested for finger operated controls and 2 cm for those operated by the base of the thumb (for example an emergency-stop button). Unfortunately little data is presently available to enable a comparison to be made between these dimensions and the size of the average fingertip. However Chambers and Stockbridge (1970) quote data for mean 'finger widths' of four male subjects which suggest that the above 'recommended' dimensions are at least 3 mm too small if the whole of the fingertip is to operate the button.

The resistance required for these controls depends largely on the type of operation to be carried out. Moore (1975), for example, suggests a range of resistances from 283 to 1133 g if one finger is used, 140 to 560 g if different fingers are used, and 283 to 2272 g for large, thumb-operated controls. Chambers and Stockbridge (1970) emphasize, however, that caution must be given against reducing the resistance by too much since it becomes easier to operate the button inadvertently. A hair trigger on a gun, for example, requires a safety catch to prevent accidents.

As has already been discussed, effective feedback is essential if the control is to be operated efficiently; the operator needs to know that the control has functioned correctly. Although this feedback may be visual, in the case of push-buttons it is most usefully given by feel or by an audible click. The eyes can then be released for other work. However the resistance needed to produce such a click should not exceed the maximum resistance suggested above or fatigue may occur after continued use (users of some modern calculator keyboards will have experienced this fatigue). Finally, modern control panels often use indicator lights as feedback devices but, as Moore (1975) rightly points out, they are only useful if they can be seen by the operator. He also suggests that the value of indicator lights usually bears an inverse relationship to the number alight at any one time; the more that are alight, the less each new one will mean to the operator.

Switches

Switches will normally be one of two shapes; either rotary selector switches, which have the appearance of knobs but which are used to make discrete settings, or a toggle type of switch which has the appearance of a miniature lever. Toggle switches will generally have two positions—'on' or 'off'; speed and ease of operation may be sacrificed with more positions.

Very little data is available on which to base the design of toggle switches, although Morgan *et al.* (1963) suggest that the maximum length should be 2.5 cm with a minimum of 3 mm. Murrell (1971), however, would allow up to 5 cm in length but agrees with the minimum dimensions. Both authorities agree that the switch should 'snap' into position and should not allow the possibility of intermediate positions.

The advantage of a rotary selector switch lies in the increased number of positions which may be used (Chapanis, 1951, suggests between 3 and 24). Most of the dimensions discussed for rotary control knobs are relevant to selector switches, apart from the resistance. As with the toggle switch, the resistance should be applied to enable the switch to 'snap' into position—in other words, it should be reduced at each position.

Perhaps the most important design consideration for rotary selector switches is the way in which the settings are indicated. This is usually effected by moulding either the whole switch or the top part of it into a pointed shape, or by making a mark on the switch surface. Whatever method is employed it is essential that there is no ambiguity regarding the position selected (see Figure 6.2).

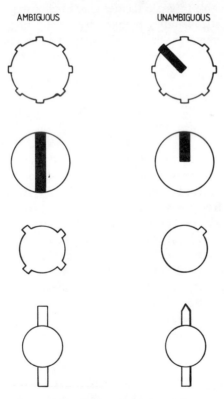

Figure 6.2 Some examples of ambiguous and unambiguous switch shapes and markings

Levers and joysticks

The difference between a lever and a joystick is simply that joysticks operate in two dimensions whereas levers only operate in one. For this reason, joysticks are used more often for complex tracking or positioning tasks (for example, vehicle guidance), whereas levers are used in situations in which only one dimension is to be altered (for example, vehicle speed or direction).

Because joysticks are used in situations in which precision adjustments are made, it is desirable that only the hand and fingers are used since these muscles are more densely supplied with nerves than are, for example, those in the arm. For this reason, joysticks will generally be smaller than levers. To aid precision they should have resistance in all directions with, perhaps, a return to centre position if the hand is removed. Morgan *et al.* (1963) further suggest that the joystick should be designed to enable the operator to rest his wrist while making the movements, and that the pivot point should be positioned under the point at which the wrist is rested. As is discussed later in Chapter 10, however, the value of resting the wrist while making tracking movements may be severely reduced if the joystick is to be used under vibrating conditions.

Little data is available to guide the designer in choosing the optimum length of either lever or joystick handles, perhaps because the important consideration relates more to the extent to which the display alters in relation to the control movement. This factor, the control–display ratio, is discussed more fully in the next chapter.

Foot Controls

Pedals

Pedals are frequently used when large forces need to be applied with relative speed, but they are rarely used for the primary control process—this function is usually reserved for hand controls. Historically this has arisen because it has been felt that the feet are slower and have less precision than the hands. However, as Kroemer (1971) points out, such assertions are based upon very little data. Indeed his own experiments suggest that an operator can be trained to use his feet with almost as much effectiveness as his hands (consider, for example, the precision with which an experienced motorist can control the speed of a car accelerator pedal). Since the hands always appear to be the overloaded control channel (like the visual system and displays), efficient use of the feet may reduce operator overload.

Apart from the pedal size, which is clearly related to the amount of space available, the important parameters of foot pedals are the position and angle of the fulcrum (if the pedal is hinged) and the maximum force required to operate the pedal. These factors are clearly interrelated as was demonstrated by Hertzberg and Burke (1971). They measured the force able to be exerted by 100 subjects using pedals set at different angles to the floor. As may be seen from Figure 6.3, an optimum angle of between 25 degrees and 35 degrees

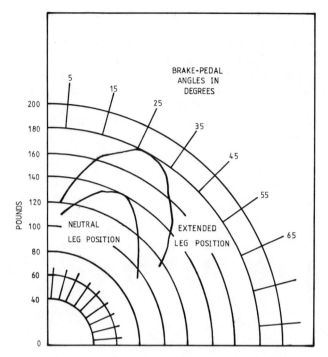

Figure 6.3 Maximum pedal forces able to be exerted in various leg and foot positions (from Hertzberg and Burke, 1971, reproduced by permission of The Human Factors Society, Inc.)

produced the highest forces. This was verified by asking the subjects to rate their ankle comfort at each angle; 80 per cent preferred angles between 25 and 35 degrees.

The angles which were suggested by Hertzberg and Burke are the initial pedal/floor angles—not the operating angle of the pedal. This should be related to the possible range of movement at the ankle, otherwise fatigue is likely to occur very quickly. In this respect, Damon, Stoudt, and McFarland (1971) suggest that 95 per cent of the population can attain an ankle angle of 20 degrees, so it would appear sensible not to exceed this angle by any large amount.

In their experiment Hertzberg and Burke used fit, male aircraft personnel as their subjects, who were asked to push their feet against the pedal as hard as possible. As can be seen from Figure 6.3, under these circumstances forces of approximately 140 lb (623 N) were obtained with a 'neutral' leg position. Mortimer (1974), however, has demonstrated that although 95 per cent of his male subjects were able to attain these forces, a similar proportion of his female subjects were able to press with a force of only 70 lb (311 N). Once again, therefore, this demonstrates the need for the design of the equipment to take account of the type of operator who is to use it.

SUMMARY

With the efficient transmission of information to the machine via controls, the communication link is completed. However, as this chapter has demonstrated, the choice and design of appropriate controls are not independent considerations, they are highly related to the type of task, the type of operator and of his clothing. Only when all aspects of the system have been considered can the information be transmitted from machine to man and from man to machine in an unimpeded fashion.

CHAPTER 7

Workspace Design

So far the discussion has centred around the principles which are important for designing specific aspects of the operator's immediate work environment, particularly various components of his machines. So he is given information (perhaps instruction) about his task or his machine by a colleague or by a manual (Chapter 4) or by the machine itself (Chapter 5). In Chapter 6 the ways in which the operator communicates with the machine via his controls were considered. Unfortunately, however, the important aspects in a worker's environment which need to be considered do not end here. Any advances gained by the ergonomic design of, for example, his displays and controls may be negated entirely if these individual components are inappropriately arranged 'in front of' the operator, in other words, in his immediate workspace. (The term *workspace* should not be confused with *workplace* which refers to the arrangement of different workspaces. This latter aspect is discussed in the next chapter.)

The aim of this chapter is to consider the principles governing the ways in which controls and displays ought to be arranged around the operator to guarantee their most efficient use. This means that at least four separate aspects could be considered:

1. The position of a control with respect to another control;
2. The position of a display with respect to another display;
3. The position of controls with respect to displays, and vice versa; and
4. The shape and dimensions of the operator's workspace.

Although much of the discussion will centre around the relatively concrete concept of a 'panel' with analogue or digital displays and particular controls, the extended concept of a workspace should always be borne in mind. For example, included in a car driver's immediate workspace will be his dashboard (which include displays and controls), and his steering wheel and foot pedals (controls). In addition to these displays and controls, however, he has others. For example, the windscreen can also be considered as part of the driver's display system since it displays to him a great deal of information regarding the operation of his machine (his car) *vis-à-vis* the road, and other displays will also be available outside of the vehicle, for example signals, road signs and even other vehicles. In the future discussion, therefore, notions such as 'display', 'control' or 'panel' should be taken in their widest sense.

Determining where the controls and displays should be placed within the operator's workspace is not a simple problem to solve. Not only do aesthetics and styling need to be considered, but also such factors as the operator's comfort and safety, the closeness of the controls for ease of use, the separation of controls for the avoidance of mistakes, the balance of work between limbs, the avoidance of overloading the operator, the need to satisfy a wide range of operator sizes, the layout of the components for ease of operator understanding, and many other factors which are possibly not quantifiable. Because the problem is so complex a few computer programs have been developed to help the designer to arrive at a solution. For example Bonney and Williams (1977) describe the CAPABLE (Controls And Panel Arrangement By Logical Evaluation) program which has been used in a few applied situations. Although they are still in their developmental stages, such computer applications may be useful in the future.

Even a computer program, however, has to be developed according to general principles or rules, and the purpose of this chapter is to discuss some of these principles.

GENERAL PRINCIPLES OF WORKSPACE DESIGN

When determining how controls and display should be arranged in front of the operator, the overriding consideration must be to ensure that they can be used quickly and accurately. For this reason an attempt is normally made to ensure that the arrangements of both sets of components are positioned to suggest to the operator the manner by which they should be used. This means, therefore, that they are arranged in terms of the sequence in which they would normally be used, their frequency of use and their importance. However, overriding these considerations is the basic requirement that the components should be accessible to the operator when he needs them. This implies the need to take account of the appropriate anthropometric data and of the position which the operator adopts when carrying out his tasks. Finally, any restrictions which are placed upon the operator's movements, by his clothing or by other equipment, will affect these considerations.

Sequence of Usage Principle

Time sequence

This principle suggests that if controls or displays are normally operated in some sequence, for example switching on a lathe, increasing rpm, moving spindles together, etc., then they should be arranged in that sequential order. All that the operator then needs to do is to alter his controls or read his displays one at a time in a particular sequence, rather than in an apparently random fashion.

A slight variation of arranging components according to a temporal sequence was described by Shackel (1962) when discussing how part of a computer

A. ORIGINAL W PATTERN

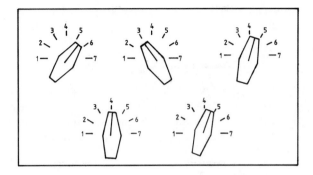

B. REDESIGNED SEQUENTIAL PATTERN

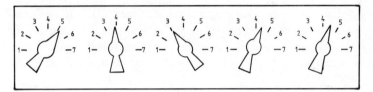

Figure 7.1 Rearranging five switches to be used in temporal sequence
(both settings read 54345)

console was redesigned using this principle. In the original prototype five switches, arranged on a console in front of the operator, were used to make a five digit setting (see Figure 7.1a). Unfortunately, however, the position of the switches did not suggest to the operator the way in which they should be used. The correct sequence was executed in the form of a capital letter W, with the top left switch setting tens of thousands, the bottom left setting thousands, the top middle hundreds, etc. Since a 'natural' setting sequence would be to set the first three digits using the top trio of controls and the last two the bottom pair, errors were clearly likely to occur. For his redesigned console, illustrated in Figure 7.1b, Shackel arranged the controls in a simple, linear, sequential order.

Functional sequence

Having arranged the panel components according to a temporal sequence, it is also possible to arrange them in terms of their function—either within the temporal sequence, or in terms of a temporal sequence of different functions. For example, the workspace around a pilot include components concerning his altitude, attitude, speed, radio contact, etc. It is common sense to suggest that all the controls and displays which are related to any one of these functions should be grouped together. However, it may also be the case that the different functions are used in a temporal sequence—for example the radio, then the speed, then altitude, and then the attitude, etc. In such a case it is sensible not only to group the components according to their function (perhaps in terms of a sequence of usage order within each group), but to arrange the groups in the order in which they are to be used.

Frequency of Usage Principle

As its name implies this principle suggests that controls and displays should be arranged in terms of how frequently they are used by the operator. Thus the more frequently used components should be placed within easy sight and reach of the operator. As will be discussed below, this usually means directly in front of him.

Importance

The frequency of usage principle is a useful guide to design. However, if it were taken to extremes, situations might arise in which a rarely used, but very important, component might be positioned well outside the operator's effective area. Obvious examples include emergency controls which, by their nature, are operated infrequently but which, when they *are* used, need to be operated quickly and accurately. The frequency of usage principle, therefore, must be tempered by importance.

THE POSITION OF CONTROLS WITH RESPECT
TO OTHER CONTROLS

Spacing

Having ensured that the controls have been placed in the optimum area in front of the operator (that is having regard to sequence and frequency of use, function and importance), the next problem facing the designer concerns the specific position of each control. In this respect, the amount of space allowed between each control is of paramount importance. Too much space is likely to make the operator move his limbs unnecessarily or, if a large number of controls is to be arranged, will result in a less than optimal spatial arragement. Too little space, on the other hand, may result in the wrong control being activated accidentally.

The minimum spacing required between each control is determined largely by the type of control to be operated (and thus the limb to be used), the way in which the control is to be operated (sequentially, simultaneously or randomly), and the presence or absence of protective clothing.

The importance of designing controls to fit the limb or part of the limb which is to be used to operate them should be clear. For example push-buttons, which are operated by the fingertips, will require less intercontrol space than pedals operated by the feet. Similar 'commonsense' can also be applied to the manner by which the control is to be operated. Using levers as an example: if two levers have to be operated simultaneously the space between them will need to be able to accommodate two hands (or at least the parts of two hands which overlap the lever handles). Sequential operation, however, is likely to be made using one hand only and this would require less intercontrol space. Again, if more than one lever is to be operated at once and by the same hand, the intercontrol spacing should be reduced sufficiently to allow the fingers to spread easily to each control.

Similar arguments may be advanced for other controls. For example, regarding concentric controls (knobs) Bradley (1969d) measured the time taken to reach and to turn four controls arranged centrally, and recorded the number of inadvertent touchings of adjacent controls made by 24 right-handed subjects. He varied the spacing between the knobs (0.5 to 1.75 in. [1 to 4 cm]), the knob diameter (0.5 to 1.5 in. [1 to 3.5 cm]) and the knob configuration.

Bradley's results indicated that performance increased rapidly as the distance between the knobs increased up to a distance of 1 in. (2.5 cm), after which it continued to increase but at a much slower rate. Interestingly, too, his results indicated that 'inadvertent touchings' of knobs depended upon their position. Figure 7.2 shows the frequency of touchings when one, two or four knobs were arranged around the knob to be operated. Whenever any knob was arranged horizontally to that which has to be operated, the right-hand knob was always inadvertently touched. In a vertically arrangement, however, the lower knob was most vulnerable. In many ways these results might have been predicted,

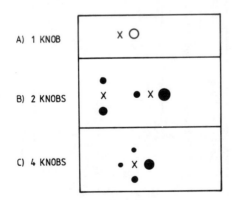

X OPERATED KNOB

● CROWDING AREA—REPRESENTS THE FREQUENCY
WITH WHICH THE KNOB
WAS TOUCHED

Figure 7.2 The effect of the number and orientation of adjacent knobs on the likelihood that they will be inadvertently touched (from Bradley, 1969d, and reproduced by permission of the Human Factors Society Inc.)

Table 7.1 Desirable separations between different types of controls (in cm) (adapted from Morgan *et al.*, 1963, and reproduced by permission of McGraw Hill)

Type of operation		Finger		Hand			Foot	
		Push-button	Toggle switch	Lever	Crank	Knob	Foot pedal between pedals	between centres
One finger/hand/foot	Simultaneously							
	Sequentially	2.5	2.5	12.5	12.5	12.5	10	20
	Randomly	5	5	10	10	5	15	25
Different fingers/hands/feet		1	1.5					

however, since right-handed subjects were used. Thus, using the right hand, the right control knob would normally have to be passed over to operate the central knob. Similarly with a vertical orientation an upward movement of the hand is likely to be used. The results are interesting, however, in so far as they illustrate the fact that consideration needs to be given not only to the positioning of one control *vis-à-vis* another, but with respect to the position and direction of movement of the operating limb. Table 7.1 indicates the desirable separation between different controls (from Morgan *et al.*, 1963).

Accidental Operation

This problem has already been introduced when discussing the spacing between controls. Thus the 'wrong' control may be activated if too little space is provided between adjacent controls, or if the controls are placed in less than optimum positions so that the operator's movements needed to operate one control requires him to pass over (with the danger of operating) another.

Even if the controls are placed in optimum positions, however, accidental activation still sometimes occurs. These mishaps need to be guarded against, and the designer has a number of techniques available:

1. recessing the control;
2. orienting the control so that the normal direction from which accidental activation may occur will not cause it to be operated. For example, if the operator needs to reach over a lever so that his arms move in a vertical direction, orienting the lever to operate horizontally may reduce the possibility of an accident;
3. covering the control with a hinged cover;
4. locking the control;
5. operationally sequencing a set of controls. If controls need to be operated in sequence, it is often possible to ensure that control 2 cannot be operated untill control 1 has been activated;
6. increasing control resistance.

The Position of Controls on the Console

Even if the controls have been arranged with ideal spacing between them, incorporating all the guards against accidental operation, one further consideration still remains—that of locating the controls on the console for the operator's optimum reach and performance.

Reach is clearly a problem which relates to the operator's anthropometric dimensions, and this is discussed further in the next chapter. Overall 'performance', however, depends on a number of factors including time and accuracy. One of these factors, performance time, was investigated by Sharp and Hornseth (1965). They placed three types of control (knob, toggle switch and push-button) in different positions on a console situated about 30 in. (76 cm) from the seated

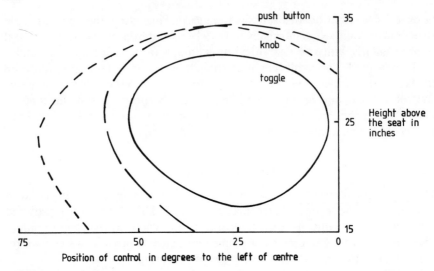

Figure 7.3 10 per cent reduced performance contours for the toggle, knob and push button controls placed in front and to the left of the operator (from Sharp and Hornseth, 1965)

operator. Their results, shown in Figure 7.3, demonstrate different performance 'maps' for the three different types of control. Thus the knob, for example, is able to be placed in a much wider range of positions than the toggle switch before the same decrease in performance is observed, implying that locating the toggle switch is more critical than the knob. Other, similar, types of conclusions may also be drawn from Figure 7.3.

Specialized Groups of Controls—The Keyboard

Although controls are often considered in terms of single, or sometimes pairs, of units used by the operator to alter a machine state (for example, a switch to turn the machine on or off or a knob to increase speed), these single units are sometimes used for extremely fast, sequential use in which information is presented to the machine is bursts. The typewriter, data entry and digital telephone keyboards are instances in which controls are used in this manner. Because the operator works with groups of either letters or digits, the arrangement of the keys on the keyboard needs to take account not only of the control spacing but of an optimum control grouping. If the arrangement is not ideal, since a full-time operator may press up to 75000 keys each day (Klemmer, 1971) fatigue may be induced in both the finger muscles and in the muscles maintaining posture in the back. (Ferguson and Duncan, 1974, for example, have demonstrated that the finger-loads using a standard keyboard layout are maldistributed, with the ring and little finger being overloaded).

The normal typewriter keyboard (named the QWERTY board because these letters occurs at the beginning of the top line) appears to have been accepted as a standard keyboard arrangement by most nations. This keyboards was originally designed by L. Sholes and his colleagues in 1874, to conform to the mechanical constraints of contemporary typewriters. The arrangement was such that the letter most likely to be typed next was obscured by the operator's hand, so reducing both the operator's typing speed and the risk of keys being jammed by pressing two at once (Martin, 1972). However, two alternative keyboard arrangements have been proposed in the past.

The first, developed by Dvorak in 1936, arranges the keys according to the frequency of letter-occurrence in common English usage. The vowels and most common consonants are placed on the middle row (that is AOEUIDHTNS), which means that nearly 70 per cent of common words are typed on this line alone.

The second layout, which has been suggested for 'non-typists', is to arrange the keys in three rows alphabetically, starting with A at the left of the top row and ending with Z at the right of the bottom. Klemmer (1971), however, suggests that the available research has demonstrated that, at least in America, there appears to be no group of potential users who perform better on this type of arrangement than on the QWERTY board.

Controversy presently exists as to the relative merits of the QWERTY and the Dvorak boards. For example a US government-sponsored study in 1956 demonstrated little difference between the advantages of each arrangement. Martin (1972), however, has suggested that (unreported) novice training experiments carried out in Great Britain have demonstrated a 10 per cent saving in training time using the Dvorak board. However, if for no other reason, the high cost of converting machines and retraining typists to use the Dvorak board will probably ensure the continuation of QWERTY board.

The story of the development of data entry keyboards is rather different, perhaps because the great advances in computer technology which have occurred arose at a time when ergonomics was able to be of some value.

With data entry keys, the problem deals simply with the ten keys 0 to 9 and concerns the most appropriate arrangement to suit the operator. In essence, there are three ways of arranging the numbers:

1. in one row across the top, as with the QWERTY typewriter board,
2. the ten numbers (0 to 9) arranged in a $3 \times 3 + 1$ matrix starting with 1,2, 3, on the top row and ending with 0 below the third row (as on a digital telephone keypad), or
3. the same $3 \times 3 + 1$ matrix but having the arrangement 7,8,9; 4,5,6; 1,2,3 and 0 (as on a calculator keypad).

Conrad and Hull (1968) compared keying efficiency using the last two arrangements. No significant differences were obtained in terms of the speed of data entry, but significantly fewer errors were made using the telephone pad arrangement (1,2,3; 4,5, etc.) than with the calculator pad arrangement (7,8,9; 4,5,6, etc.) (6.4 per cent versus 8.2 per cent).

THE POSITION OF VISUAL DISPLAYS WITH
RESPECT TO OTHER DISPLAYS

The main consideration when placing visual displays on the operator's console concerns the physical relationship between the display and its associated control(s). This will be discussed in more detail in the next section. However, two specific aspects of placement which are peculiar to displays need to be considered. These are the visibility of the display and the ways in which the pointers are aligned when groups of analogue displays are used for check reading purpose.

Visibility Requirements

Since the value of a visual display depends entirely on the operator's ability to perceive it, making sure that displays are within his line of sight is an extremely important aspect of the panel design. However, it is an aspect which is sometime overlooked, since the problem can take two forms.

Firstly visibility might be impaired because the display is obscured (either fully or partially) by another component on the console. This is not an unusual occurrence, particularly when controls and displays are placed close together.

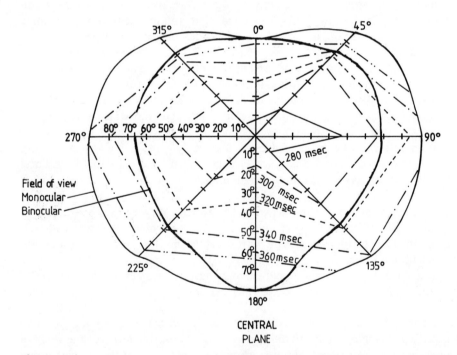

Figure 7.4 Equal response times for displays in the visual field (from Haines and Gilliland, 1973. Copyright 1973 by the American Psychological Association. Reprinted with permission).

For example, a display located towards the periphery of a console may appear to be partially obscured by part of a control which has been placed near to it, simply because of the operator's position with respect to the control and the display. This aspect is considered in more detail in the next chapter. In addition, reaching out to operate a control can easily cause a display to be obscured by the hand or arm.

The response to a display can also be impaired because, although the operator's visual field is quite wide, his speed and (under stress) accuracy of response depends largely on the position in the visual field at which the stimulus occurred. Haines and Gilliland (1973), for example, measured the speed of response of seven subjects to separate lights placed in different areas of their visual field. Figure 7.4 illustrates their results and shows the various boundaries of equal reaction time over the visual field. The fastest time, therefore, was obtained within an area bounded by about 8 degrees above and below the central sight line, up to 40 degrees to the right, but only 10 degrees to the left of the midpoint. This general 'oval but skewed to the right' trend can be seen in the other equi-response time boundaries—although the right side superiority reduces towards the edge of the visual field.

Haines and Gilliland suggest that Figure 7.4 can be used to aid the designer in the placement of important displays (particularly warning displays) in front of the operator. By superimposing the contours over a scaled-down model, the positions for the various displays may be pinpointed.

Grouping Display

For check reading

As was discussed in Chapter 5, displays are not only used for making quantitative reading but also for simply checking that the machine state is within certain (safe) boundaries. Since the operator make his readings largely in terms of the amount to which the pointer deviates from a particular position, rather than from the precise dial reading, his task would appear to contain a fairly large memory component. The use of colouring 'danger' areas on the dial helps to reduce the operator's load, as does the arrangement of the dial so that, when showing a normal state, the pointer is aligned to, say, 9 o'clock (White, Warrick, and Grether, 1953).

If a number of displays are used for check reading, however, they may be arranged so that each facilitates the reading of the other. Johnsgard (1953), for example, has demonstrated that pointer symmetry (that is, arranging the pointers to form a pattern) is possibly better than pointer alignment (arranging the pointers to point in one direction) for check reading. He compared four arrangements of a bank of 16 dials for check reading accuracy. In the first arrangement, the dials were each rotated so that, under 'normal' conditions, all pointers were aligned at the 9 o'clock position. For the other conditions, the dials were rotated so that the pointers produced different patterns under 'normal'

140

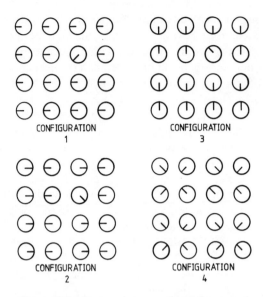

CONFIGURATION 1

CONFIGURATION 3

CONFIGURATION 2

CONFIGURATION 4

Figure 7.5 Four-pointer configurations used by Johnsgard, 1953 (one pointer, in each case, is indicating 'non-normal'). Copyright 1953 by the American Psychological Association. Reprinted by permission

conditions (Figure 7.5). Any deviation of any pointer would be highlighted by a break in the pattern.

Johnsgard's results clearly demonstrated that one configuration (4), when the pointers were arranged to point to the centres of groups of four dials, was less effective than any of the other three. However, arranging the dials so that the pointers were vertically symmetrical (configuration 3) produced significantly more correct responses than simply a horizontal alignment. Similar results were obtained by Ross, Katchmar, and Bell, (1955), although they were unable to support the contention that a symmetrical (patterned) alignment produces more correct responses than a simple, uniform alignment.

The use of patterns such as these was taken one step further by Dashevsky (1964) who extended the lines made by the pointers under 'normal' conditions on to the console (Figure 7.6). Comparing these extension lines, with the open (no-line) arrangement, he produced an 85 per cent reduction in error rate over a uniform alignment, and a 92 per cent error reduction when the dials were rotated as in Johnsgard's condition 4.

For flow diagrams

Different displays can often be grouped to provide the user with a 'working model', or flow diagram, of the process or machine, showing him in pictorial form the parts of the working system from which the information is arriving. For

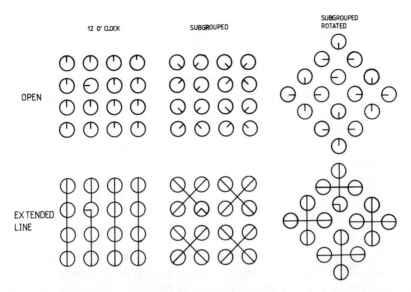

Figure 7.6 Groups of displays used by Dashevsky (1964) to investigate whether extensor lines between displays aid check reading. Copyright 1964 by the American Psychological Association. Reprinted by permission

example, large panels are commonly found in modern railway signal boxes which graphically depict the rail sector under the signalman's control, the position of the points, and the movement of the train through the sector. Such displays help the operator to interrelate the different aspects of the total system which are displayed, and allow him to locate faults or delays quickly. Although such composite displays are now more frequently used in large process industries, Morgan *et al.* (1963) caution that they are limited in terms of the additional space required for the connecting lines between displays. In addition much of the value of the display is lost if too much information is presented in too complex a form. Nevertheless, flow diagrams allow the operator to feel that he has some involvement with the system which he is asked to control.

THE POSITIONING OF DISPLAYS AND CONTROLS

The one theme which has been emphasized continuously throughout this book is that the information which flows from the machine to the man and back to the machine can be seriously disrupted if any part of the chain is deficient (for example the display, the operator's processing capacities, or the control). However, even if each separate component is operating at its peak efficiency, the process may still be interrupted if the information received from one component does not 'fit' the requirements of the next link in the chain. For example, information displayed at too fast a rate may overload the operator's perceptual and decision-making capabilities; a control with much inbuilt resistance may not be able to be operated due to a lack of muscle power in the operator's arm, etc. Taking this argument further it is apparent that, for information flowing in the display–man–control direction, the operator's workloads may be reduced if a display is compatible with its associated control(s). Then, just as a well-designed display or an appropriate control will help the operator in his task because it 'fits' his capacities, if the display and control are themselves compatible less work will be needed to complete the display–man–control link. In the same way it is important to ensure that the control–display relationships are compatible to allow an ideal flow of information the other way. These two aspects are considered below but, as was discussed earlier, the term 'display' must be taken in its widest sense throughout the discussion. So far the term has been used as having a specific meaning, that of a component on the panel, comprising either a set of numbers or a dial and pointer, which indicates the machine state. It should not be forgotten, however, that any aspect of the operator's environment may display information to him about the machine. For example, the way that the machine itself behaves (for example, the speed at which a drill moves into a piece of wood or metal) is information displayed, as is the behaviour of the machine with respect to the rest of the environment (for example, the way in which a car moves along a road is displayed to the driver through his windscreen).

Control-display Compatibility

In effect, a display and a control may be said to be compatible if one suggests the way in which the other should be operated. Compatibility effects can possibly best be exemplified by a simple experiment described by Welford (1976). Subjects were presented with up to eight stimulus lights, to any one of which they had to respond as quickly as possible by pressing the appropriate numbered key. Two conditions were given. In the first condition (highly compatible) each button was located below the light with which it was associated. In the second, the low compatible condition, the buttons were arranged in random order. The results demonstrated that a higher response time was produced when the lights (displays) and keys (controls) were incompatible than when they were compatible. As Welford explains it:

presumably in the first (condition), once the light was identified the key was also, whereas in the second arrangement the light had to be identified first and then its number used to locate the corresponding key. In other words, in the second condition the data had to be recoded from digital to spatial form before a response could be chosen, so that the translation mechanism had more 'work' to do than when each light was located above its corresponding key.

In the low compatible condition, in other words, the display did not 'suggest' the appropriate key response.

In addition to an increased speed of response which is likely to result from using compatible displays and controls, Murrell (1971) suggests two further reasons to aim for compatibility. First, learning time for the operation of equipment on which controls are compatible will be much shorter than if the controls are incompatible. In the experiment described above, for example, the subjects had to learn the new (random) arrangement of buttons before they could respond quickly. Second, when placed under stress, the operator's performance on equipment with incompatible display–control relationships will deteriorate as he reverts to the relationship which he expects should occur. The example has already been given of a press operated by a lever being lifted to make it descend. Whereas under 'normal' circumstances this incompatible operation was carried out efficiently, when a stressful occasion occurred and the operator needed to raise the press quickly, his 'natural' reaction was to lift the lever thus making it descend.

Fitts and Seeger (1953) suggest that this occurs because a response which a person makes can be considered to be a function of two sets of probabilities:

1. the probabilities (uncertainties) appropriate to the specific situation in which the operator finds himself—his training, instructions, past experience, successes, failures, etc; and
2. 'the more general and more stable experiences or habits based on the operator's experiences in many other situations'.

They suggest that training will nearly always lead to changes in the former but will have relatively little effect on the latter. Loveless (1962) continues this argument a little further by suggesting that the new (trained) behaviours do not replace the old behaviours which were learned as a result of past experiences and expectations, they merely overlay them. Certain situations may arise in which the old behaviours may come to the fore. 'During training the old response is weakened sufficiently to allow the new response to appear, and the latter is then strengthened by further practice; but the old habit has been suppressed rather than eliminated. The suppression is likely to be in part temporary, so the old response may reappear after a period away from the task'. He further suggests that 'it can also be predicted on theoretical grounds that habit regression will occur when the operator's motivation is decreased, when he is fatigued and when he is subjected to any novel change in the working situation'.

The suggestion is made, therefore, that compatibility arrangements are learned—that there are more examples of compatible than incompatible relationships in everyday living, and that the more one is exposed to these relationships the more likely will be the tendency for the operator to expect these relationships to occur. This begs the question whether there are developmental trends in the learning and durability of compatible relationships, but unfortunately little data is available to answer this question.

Two main ways of arranging compatibility between control and display exist. First *spatial compatibility*, in which the two components are compatible if the position of one on the console suggests that of the other. Crossman's lights

144

and buttons in the first condition had high spatial compatibility. Second *movement compatibility*, in which the movement of the control suggests the way in which its associated display is likely to move, and vice versa. For example, by turning a knob clockwise an operator would expect its associated display to indicate an increased reading. Relationships which are expected by the majority of the population are described as population stereotypes.

Spatial compatibility

As evidence that the need for spatial compatibility exists on all types of equipment, Chapanis and Lindenbaum (1959) investigated different control–display positions using four-burner gas and electric cookers, with the burners imagined as the displays. Four arrangements were investigated and the results indicated fewer errors as the position of the controls matched more that of the displays (Figure 7.7). Thus, in arrangement I, when the burners were staggered to be in the same sequential order as the controls, no operating errors were detected. Chapanis and Mankin (1967) followed this study with an investigation of ten control–display relationships. In this case, however, the controls were in

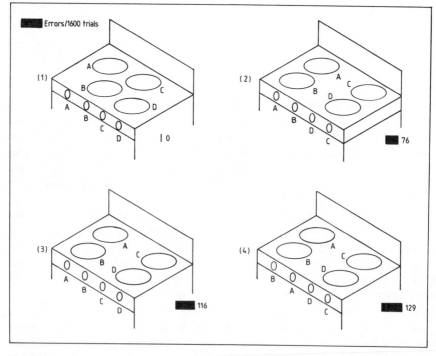

Figure 7.7 Four control–display arrangements used by Chapanis and Lindenbaum (1959); reproduced by permission of The Human Factors Society Inc.

the same plane as the burners (displays), an arrangement commonly used with split-level cookers.

Whereas population stereotypes exist for control–display movement relationships, as is discussed below, no such stereotypes seem to exist for spatial relationships. Thus Shinar and Acton (1978) asked over 200 subjects which sequence of controls they considered was most appropriate, using the arrangement of burners described by Chapanis and Lindenbaum (1959). Of the four most obvious arrangements (ABCD, ABDC, BADC, BACD), the first three were suggested by very similar numbers of respondents.

This being the case, it would appear that some further aid to spatial compatibility is required. Two suggestions have been made: First, to use linking 'sensor lines' between controls and displays (Chapanis and Lockhead, 1965) and, second, to colour code the display and its appropriate control (Pook, 1969). In both cases the experimental results were remarkably similar in so far as the aids produced no increase in performance when the display and control had maximum spatial compatibility. When the display and control were arranged incompatibly, however, large increases in accuracy and response time were obtained (up to 95 per cent reduction of errors on large panels using sensor lines, and approximately 40 per cent faster response using colour coding).

Although the discussion of spatial compatibility has so far centred around visual displays, displays utilizing other sensory modalities may also be arranged to be 'spatially' compatible. For example, Simon and Rudell (1967) demonstrated that information presented to the right ear was reacted to faster by the right hand, whereas a left-hand advantage was obtained for information presented to the left ear. Since we have fairly accurate abilities to localize sounds, particularly at the lower frequencies used for, for example, warning klaxons, these data suggest that the emergency controls associated with such an auditory display should be located in the same direction as the display.

Finally a study reported by Shackel (1959) illustrates the need to use the terms 'display' and 'control' very loosely. He describes a panel layout which consisted of an arrangement of 24 potentiometer switches and their associated jack plug sockets. In the case of an operator needing to put a plug into the appropriate socket for the potentiometer switch, in some respects the switch may be conceived of as being the display with the socket functioning as the control (or vice versa). Using a redesigned panel layout so that the position of the jack plug sockets were compatible with their associated switches, Shackel showed that the subject's response-times and error rates were reduced.

Movement compatibility

Controls and displays are composed of moving (or in the case of digital or auditory displays, apparently moving) parts. Since the two components are related both in the machine and in the operator's 'mind', it is important to ensure that the directions of movement of their moving parts are compatible with each other and conform to the operator's movement expectations. Again it must be stressed that the display represents the means from which the operator receives his information. A windscreen, for example, displays as much information as does a dial.

The commonly expected control–display relationships which might arise have been fairly extensively researched and documented, and can be divided into two subsets.

The first group include those relationships in which the control and display are in the same plane—perhaps on a console panel. In these cases the general movement rule, using either linear (for example, a lever) or rotary (for example, knob or switch) controls, is 'up or clockwise' for an increased or right-moving reading, and anticlockwise or down for a left or reduced reading. In addition, to make the operator's task easier, it has often been suggested that strip displays are associated with linear controls, and dial displays with rotary controls.

If the type of control and display are mixed, for example a rotary control and a strip display, curious effects which sometimes deviate from these generalizations may sometimes be noted. This was first documented by Warrick (1947) (and is discussed by Brebner and Sandow, 1976) who pointed out that the indicator is expected to move in the same direction as the point on the control closest to the display. Whereas the expected movements of vertical controls and displays may simply conform to Warrick's principle, if the display is vertical and the control rotary then Warrick's principle implies that the side of the display on which the control is situated is important. Thus a clockwise movement on the left of the display would be expected to produce a reduced reading since the part of the control next to the display is moving downwards. The same control on the right, however, would need an anticlockwise movement to produce an expected fall in the reading (Figure 7.8). Brebner and Sandow (1976)

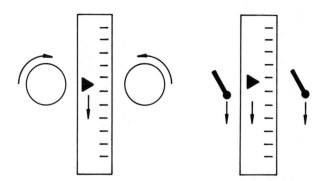

Figure 7.8 The different control actions required according to Warrick's principle, when the control is on the left and right sides of the display

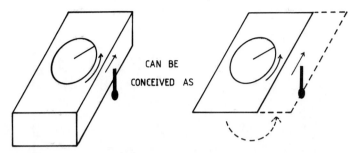

Figure 7.9 The application of Warrick's principle to cases in which the control and display are in different planes

note that this effect is enhanced if the scale values are placed on the side of the vertical display which is opposite that of the control.

The second control–display relationship occurs when the two components are situated in different planes, perhaps with the display positioned vertically in front of the operator with his control on a horizontal console. With this type of arrangement the appropriate movement relationship is that which would occur if the two planes were imagined as being one simple plane. Warrick's principle may then be invoked (Figure 7.9).

Control–display ratio

When discussing movement compatibility, an important aspect to consider is the degree to which the moving parts of the control and the display actually move. This has implications for the sensitivity to which the operator is able to operate and adjust his control. A small control movement which is associated with a large display 'deflection' means that the operator has to deal with a highly sensitive system; as with a 'hair trigger' on a gun, a small movement produces a large response. At the other end of the continuum, therefore, lies the low sensitive system in which a large control movement is required to produce a small movement on the corresponding display. (Like the terms 'control' and 'display', 'movement' must also be taken in its widest sense. This includes not only linear distance but, perhaps, degree of turn, number of revolutions, or even the intensity of a reaction.) The ratio between the control and display movements is, naturally, known as the control–display ratio and is an index of the system sensitivity. Thus a low control–display ratio (small control: high display movement) indicates a highly sensitive system, whereas a high control–display ratio suggests the reverse.

The optimum ratio for any one control–display system depends entirely on the system's requirements and properties, and so no general formulae exist to help the designer to decide the appropriate ratio to use. However the optimum ratio for a particular system can be determined experimentally, as was demonstrated by Jenkins and Connor (1949).

148

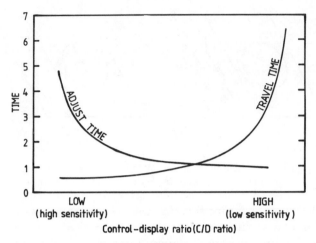

Figure 7.10 The relationship between adjustment and travel time for varying control–display ratios (from Jenkins and Connor, 1949)

Whenever a continuous control is used, the operator effectively performs two types of movement. First is a relatively crude, gross motor movement to position the control in the vicinity of the final setting, described by Jenkins and Connor as the 'travel phase'. Second, there appears a much finer movement, the 'adjustment' phase, in which the operator regulates his motor control to bring about the final control setting. The optimum control–display ratio, Jenkins and Connor argue, will be when both of these movements occur in the shortest time.

Figure 7.10 illustrates these relationships. Taking, for example, a system which has a low control–display ratio (small control: high display movement), the proportion of the control time used for adjustment will be higher than that for travel. However, as the amount of control movement needed to produce a corresponding display movement increases (in other words, as the control–display ratio increases or the system sensitivity decreases), the time spent travelling will also increase. At some point the two times will intersect, and Jenkins and Connor argue that this will be at the optimum control–display ratio *for that system.*

THE SHAPE OF THE WORKSPACE

By this point the reader should have a good idea of where the individual components of the operator's machine (principally its controls and displays) should be placed. The final consideration is how his machine, his panel, his console should be arranged around him.

Again little experimental data is available to guide the designer in his choice of shapes. This is probably because 'commonsense' and 'aesthetic' principles (which are by no means complementary) have arisen over the years.

The 'commonsense' approach to the shaping of the workspace argues that the appropriate shape is one which allows the operator to reach all of his controls easily and to see each of his displays. Thus full details are needed about the anthropometric dimensions of the user population: the arm reach, the hand sizes, the sitting height, the eye height, etc.

As Morgan *et al.* (1963) point out, the most appropriate shape to ensure that the operator can reach all of the controls easily is one with a continuously curved surface which effectively wraps around in front of the operator. However this is often difficult to produce, particularly if large, flat displays need to be mounted on the console face. To overcome this problem, therefore, a sectioned console may offer the next best alternative, with angled side panels which can, again, be located around the operator. For this type of design, Siegel and Brown (1958) demonstrated that side panels at 50 to 55 degree angles produced the lowest number of operator body movements.

SUMMARY

Having decided on the appropriate controls and displays, this chapter considered how they should be arranged around the operator—in his workspace. The overriding principles to consider are to arrange the components in terms of their sequence and frequency of use, and their importance. Having estimated these parameters, however, careful consideration needs to be given to where to place displays with respect to other displays, controls with respect to other controls and, finally, displays with respect to controls (and vice versa). Only when these aspects have been assessed will the link between man and his machine be able to work to its best advantage.

CHAPTER 8

Workplace Design

Having designed the operator's immediate workspace—his displays, controls and consoles—the final problem concerns the ways in which the various components in the total working environment (groups of both men and machines) should be arranged around the operator in his workplace. Just as the efficiency of a display or a control can be reduced by an inadequate console design, so can the usefulness of different machines if they are poorly sited for the operation. In this case, 'usefulness' probably reflects the ease of use rather than speed or accuracy. If speed is a factor, it is important more in terms of the operator's mobility around his machines, rather than the faster reaction times which have been considered so far.

When discussing where men and machines should be positioned in the workplace, the social relationships of the men themselves also need to be examined, an aspect which so far has not been considered. However, as was discussed in Chapter 4, the types of communication links which are set up between individuals, and between groups of individuals, may facilitate or hinder activity, and the existing social relationships can easily influence these communication links. The needs for privacy, peace and territory are various aspects of the social environment which are discussed later in this chapter.

Perhaps the first decision which needs to be made when designing the workplace, however, is whether or not the operator is to be seated. Specific aspects of seating design are discussed in the next chapter, but suffice it to say at this stage that although a seated operator will experience less fatigue, his mobility is likely to be reduced.

This chapter considers first the physical aspects of the environment, which are important when considering where men and machines should be placed, and then discusses some of the social factors.

PHYSICAL REQUIREMENTS IN THE WORKPLACE

Anthropometric Considerations

The importance of designing the environment to fit the anthropometric dimensions of the operator has been stressed throughout this book. They are, perhaps literally, the most basic considerations for ensuring that the operator is able to fit into his environment.

Anthropometric details are no less important when considering the ideal

150

arrangement of machines in the system. For example, if men need to walk down passageways between machines their distance apart will need to be at least equal to the operator's shoulder breadth so that they can accommodate both his arms. Such dimensions would naturally have to be increased in relation to the types and dimensions of clothing which the operator is likely to be wearing. For 99 per cent of the population, Damon, Stoudt, and McFarland (1971) suggest the following minimum dimensions for various kinds of passageways (including corridoors, aisles and tunnels):

Height, at least 195 cm or 160 cm if stooping is permitted (although this is undesirable),
Width, at least 63 cm.

If space is at a premium, the passageway can be trapezoidal in shape, 63 cm wide at shoulder level and 30 cm at the feet. For two or more people, Damon, Stoudt and McFarland suggest the same height but increasing the width by approximately 50 cm for each extra person likely to walk abreast.

Whereas these figures apply specifically to passageways which are bounded on each side by walls, it is likely that similar dimensions will also be required for gangways between two machines. Indeed a case might be made for increasing the width at least, since machines, unlike walls, may have protruding controls which could possibly be knocked by a passing operator, resulting in injury or accidental activation of the machine.

Anthropometric and biomechanical considerations are also important when the operator needs to move between different levels—using either stairs, ramps or ladders.

Corlett et al. (1972) compared the average energy required by eight male subjects to climb either stairs or ramps which had slopes ranging from 10 to 30 degrees. Using oxygen consumption, heart rate and maximum knee-joint angles as their comparison measures, the authors demonstrated that the physiological cost of stairs was always less than that of a ramp of equal slope. However they also point out that stairs impose a particular gait on the user which causes an increase in knee-bending. For old or lame people this makes ramps easier to use—regardless of their energy cost. In addition it should be remembered that ramps allow the operator to push or to pull large loads between levels (within the constraints discussed in Chapter 3). Stairs make this activity more difficult to carry out and, if the operator is carrying goods, ramps are likely to be safer than stairs.

Although for normal purposes stairs require less effort than ramps, they should still be designed to fit the user. Four aspects of the stair geometry appear to be important in determining their ease and safety of use: first, the height of the risers (the vertical distance between one step and another); second, the depth of tread (the distance between the front and back of the step); third, the steepness (slope) of the stairs; and finally the tread texture. The first three factors determine the amount of energy required, while the fourth affects the likelihood of slipping occurring.

Regarding the height of the riser and the depth of tread, many architectural codes of practice still embody a formula derived in about 1972 by Francois Blondel, who was director of the Royal Academy of Architecture in Paris. He concluded, from personal observation, that the normal pace in level walking (said to be about 60 cm or 24 in.) must be decreased by a regular and fixed amount to allow the foot to be raised in climbing stairs. His formula stated that the pace (60 cm; 24 in.) should be decreased by 5 cm (2 in.) for every inch or riser, in other words the depth of tread should be 60 cm (24 in.) minus twice the height of the riser.

Fitch, Templer and Corcoran (1974) argue that this formula has been passed down through generations of architects, despite the fact that in stairways which span either larger or smaller gaps than usual the rule produces either extremely narrow or extremely wide treads corresponding to the higher or smaller risers. In addition, as was discussed in Chapter 3, the average human body is bigger now than in the 17th century, and today's inch is shorter than that used by Blondel in his calculations.

As Corlett *et al.* (1972) and Fitch, Templer, and Corcoran (1974) demonstrate, all stairs impose an individual gait on the user due to the knee and pelvic angles required. From the proprioceptive feedback this soon becomes learned as he moves up (or down) the stairs, so that any irregularity in either riser height or tread depth will cause a misstep to occur. (This was well understood by castle builders in the Middle Ages who added an irregularly spaced 'trip step' near to the bottom of the stairs to catch the unwary intruder.) A worn step caused by continued use may be enough to make the user stumble.

Although the gait adopted when climbing or descending stairs can provide useful feedback cues to the user, if it is irregular or accentuated it can also cause discomfort and increase the energy expenditure. Corlett *et al.* (1972), for example, demonstrated that subjects used less oxygen when climbing a stairs with a 6 in. (15 cm) riser than with a 4 in. (10 cm) riser and, naturally, climbed the stairs faster. A larger riser height was also suggested by Fitch, Templer, and Corcoran (1974) whose data also suggested that increasing the height from 6 to nearly 9 in. (15 to 22 cm) reduced the number of missteps.

The ideal depth of tread is clearly related to the user's foot size and, as Ward and Beadling (1970) have noted, as the size of the tread decreases a point is reached where the shod foot can no longer fit on the step without being twisted sideways. This awkward movement induces a crab-like gait and a poorly balanced posture. If stairs are to be adequate for the largest people, therefore, Fitch Templer, and Corcoran (1974) suggest that the tread will need to be about 11.5 in. (27 cm) deep, allowing a nominal $\frac{1}{4}$ in. ($\frac{1}{2}$ cm) between the heel and the riser, and allowing the toe of the shoe to overhang the step by 1.75 in. (1.5 cm) (on descent). Finally all authors appear to support the contention that the steeper the stairs, the less energy is expended. Corlett *et al.*'s results, for example, showed that almost twice as much oxygen was used on a 30 degree slope than on a 10 degree slope.

Although the dimensions of a stairway may be designed eventually to fit the user, accidents can still occur if the tread texture is too smooth. This was illustrated in Chapter 3 when discussing how reduced friction between the walker and the floor can result in slipping. Kroemer's (1974) data, for example, illustrates how different combinations of shoe and floor material can aid or hinder slipping.

Apart from using lifts, which are space-consuming and extremely costly to install, the final practical means by which an operator can move from one floor to another is by using ladders. These are particularly useful when slopes greater than the 20 to 30 degrees discussed above are employed.

Ladders, however, are more dangerous to use than stairs. Data are not available which compare the probability of accidents occurring when ladders are used with those occurring on stairs, but the Royal Society for the Prevention of Accidents (1977) suggest that about 10 per cent of all falls from a height in industry involve ladders.

Dewar (1977) describes an analysis of one set of 248 ladder accident reports. 66 per cent of accidents occurred when the ladder slipped, either during climbing or when working on the ladder, while of the remaining 34 per cent a large proportion was attributed to the man misplacing his feet or otherwise stumbling when climbing. It would appear, therefore, that two aspects of the interaction between the user and the ladder are of major importance: first, the forces which restrain the ladder on the lower level, and second, the climbing action of the man. For this second factor Dewar suggests two further influencers: the angle of the ladder, which affects the awkwardness of the climbing action, and the stature of the man, which alters the knee and pelvic angles required to move up the ladder.

For an environment with fixed machinery it is likely that the ends of the ladder will also be fixed, thus ruling out any slipping of the ladder. For those situations without fixed or secured ladders, however, Hepburn (1958) suggests that the 'quarter length' rule should be used: that is, the horizontal distance from the foot of the ladder to a point vertically below its top should be one-quarter of the ladder length. This represents a gradient of 4 in 1 or approximately 75 degrees.

This suggestion, however, has been challenged by Dewar (1977). By recording the angles of the hip and knee-joints of both tall (average height, 181 cm) and short (average height, 167 cm) men climbing ladders set at 70.4 and 75.2 degrees (representing gradients of 3:1 and 4:1 respectively), he demonstrated that the steeper (75 degree) ladder caused the body's centre of gravity to be further back from the ladder than was the case for the 70 degree ladder, thus placing more reliance on the hands for support. Under such circumstances, therefore, he argues that if the hands slip there is less chance that the person would be able to regain his balance.

With regard to stature, Dewar's data indicated a slight, although non-significant, increase in the amount of leg movement associated with the shorter subjects. In other words, the taller subjects were ascending the ladder with a less 'natural', 'stiffer-legged' posture. This difference increased with the steeper ladder. Since a forced and unnatural posture may result in increased errors,

Dewar's data suggest that the taller people are at more risk as the angle of a ladder becomes steeper. Hepburn's suggestion of the adoption of the quarter-length rule, therefore, might need slight revision.

Communication Considerations

As has been stressed constantly, the operator's communication requirements consist of links in both the operator–machine communication and operator–operator directions. These may occur via any of the operator's sensory systems, although the visual (for example, for displays), auditory (for example, for speech) and tactile (for example, for controls) systems will most often be used. This means that the operator must be able to see his machines, be able to move around quickly to operate them, and should be in a position to hear and to talk to other operators. The anthropometric and biomechanical requirements for movement around machines have already been discussed, so this section will consider the operator's visibility and auditory requirements, and the necessity to arrange machines so that movement from one to another is reduced.

Movement considerations

The principle governing the arrangement of men and machines in the workplace so that movement time between the components is reduced, follow the same principles suggested for the arrangement of controls and displays on the operator's console. These are that the most important machines are placed within easy access of the operator as are those which are used most frequently. Machines or workplace areas should be grouped according to function and, whenever possible, the movement of the operator from machine to machine should follow some sequence of use.

Unfortunately little data exists with which one is able to evaluate these guidelines. However, an experiment carried out by Fowler *et al.* (1968) (discussed in McCormick, 1976) indicates that when controls and displays are arranged on the operator's console according to one of these principles, the mean time to carry out a task was lowest for the sequentially arranged components. The 'functional', 'importance' and finally the 'frequency' arrangements respectively took progressively longer. Translating these results to the gross body movement time of the operator moving between machines, however, is difficult and no comparative data exists. Nevertheless, without any evidence to suggest the contrary, it would appear sensible to arrange machines as far as possible according to these principles.

Visibility considerations

Having ensured that the operator is able to move safely and quickly between machines, a further consideration is to ensure that he is able to see both the machines for which he is responsible and the operators with whom he may need to communicate. These visibility requirements could be hindered in two ways. First, if the level of illumination is too low for him to be able to see accurately (this is discussed in more detail in Chapter 11). Second, if his lines of sight are obstructed by other equipment or by other operators, and this is a problem which concerns the placement of both machines and operators.

The problem may be overcome in the design process by very simple procedures which involve mapping the operator's lines of sight. Usually this takes the form of running pieces of cotton or string from where the operator's eyes would normally be to the machines (and to the important parts of the machines) which should be under his control. As long as the cotton is able to run in a straight line, the operator's visibility requirements will be unimpaired. Bends in the cotton indicate machines or men which will be in the way.

Sell (1977) used a very similar procedure to design the structure of a moving crane cab used to transport steel in a mill. The shadows thrown by a small illuminated electric light bulb, placed at the position of the operator's eye in a scaled-down model of the cab, provided a preliminary idea of the directions in which obstructions to vision (caused by, for example, girders) would occur. He was then able to obtain more detailed information by running a thread from the place where the man's head would be to the various points where he has to look. Using these threads, it was easy to measure the angles of his sight lines and thus to design the cab structure on this basis.

Finally, it is well to remember that rather than providing for maximum visibility to as many parts of the operator's environment as is feasible, in some cases or in some directions of sight it may be more sensible to ensure that visibility is obscured, perhaps by erecting screens. This is particularly so in two cases: First, when requirements for privacy override communication needs (this is discussed in more detail below when landscaped offices are considered). In the second case, machines might be sited to shield the operator from too high a glare source, for example from a furnace.

Auditory considerations

Whereas the operator's channel of communication from his machines will primarily be in the visual mode, apart from non-verbal communication his communication with other men usually occurs in the auditory mode. For this reason assessing the levels of environmental noise and attempting to reduce the noise level is important, as is discussed in Chapter 10.

Noise reduction is usually effected at source by some method of noise absorption, for example by padding around the noisy machinery or by using sound absorbing wall and floor materials. However, it is also possible to reduce the levels of noise by a few decibels simply by an appropriate arrangement of the equipment. For example, Corlett, Morcombe, and Chanda (1970) suggest that altering the positions of moveable pieces of equipment produces both screening and a reduction in the level of reflected sound.

They measured noise levels at different points in machine shops and demonstrated the noise-reducing effects of storage racks containing work-in-progress material. On the basis of their data, Corlet, Morcombe and Chanda argue that a drop in sound level of 5 to 10 db could be obtained by the judicious placement of work-in-progress storage. Although this appears to be a small amount, compared with much larger reductions which can be obtained using sound-deadening materials, Corlett *et al.* suggest that small reductions can be beneficial in permitting considerable increases in tolerance and speech communication. However, they also concede that the siting of such storage areas (or, in the context of the present discussion, other machines) should not override the other considerations discussed elsewhere in this chapter. They conclude by suggesting that

For plants in which noisy machines are sited together and are drawing or feeding material from or to other processes in large quantities, the storage of this material in high racks between the machines should be attempted as a practical measure to reduce the 'build up' of noise. Suitable planning of such intermediate storage points may also give additional gains in terms of reduced handling and internal transports.

SOCIAL REQUIREMENTS IN THE WORKPLACE

So far, the discussion has centred around the concept of physical aspects of the working environment which may interfere directly with the operator's ability to carry out his task. However, it is important to remember that the environment also contains other operators who are likely to interact with one another, and these interactions can also affect performance. The physical arrangement of both men and machines, therefore, may inhibit or facilitate these interactions and can have important implications.

The social use of space is an extremely important aspect of man's interaction with his environment, but it is a topic which appears to have been neglected by many designers. Despite the fact that the operator's environment includes other men and women, little attention has been given to the influence of social environmental parameters on performance, safety or comfort. Two important aspects of the social space requirements are personal space and territoriality and these are discussed below.

Personal Space

Personal space has been defined as an area with invisible boundaries surrounding a person's body into which intruders may not enter (Sommer, 1969). It may be considered as a series of concentric globes of space, each defining a region in which certain types of interaction may occur. The regions, however, are not necessarily spherical in shape, sometimes extending unequally in different directions. Savinar (1975), for example, has shown that people have space requirements in both vertical and lateral directions. If the available space is limited in one dimension (for example, by reducing the ceiling height), then a person's spatial needs will increase in the other directions.

A. INTIMATE DISTANCE
B. PERSONAL DISTANCE

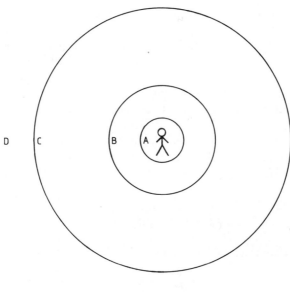

C. SOCIAL DISTANCE
D. PUBLIC DISTANCE

Figure 8.1   Personal space zones (to scale)

Hall (1976) broadly divides the social space areas surrounding a person into four distance zones from the centre: intimate, personal, social and public distance (Figure 8.1), each zone having both a close and a far phase. However, as is discussed below, the boundaries of each zone are not necessarily constant—they sometimes fluctuate under different conditions. The importance of these distances to the present discussion lies in the observation that only certain classes of people are allowed to enter each space area. The behaviour of a person whose space is violated may change considerably if the 'wrong' person infringes the 'wrong' zone.

At intimate distance (close phase 0 to 15 cm—far phase 15 to 45 cm) the presence of the other person is unmistakable and may sometimes be overwhelming because of increased sensory stimulation. The close phase is typified by actual physical contact which, in many cultures, is considered taboo between strangers. Within the far phase, the degree of physical contact is slightly reduced but increased visual awareness of the other person is maintained. As Hall describes it 'The head is seen as enlarged in size, and its features are distorted. . . . The iris of the other person's eye seen at only 6–9 in. is enlarged to more than life size. Small blood vessels in the sclera are clearly perceived, pores are enlarged. . . . '

Although, as its name suggests, personal distance is concerned with friendly contact, fairly large differences may exist between behaviours associated with the close and far phases (45 to 76 cm and 76 to 120 cm, respectively). The close phase is reserved for 'well-known' friends, since the movement of a stranger into this distance may imply a threatening situation. The far phase begins at about arm's length and is the area in which normal social contact is made. The personal distance, therefore, may be conceived as a buffer zone between the area reserved for intimate acquaintances and that in which less personal contact takes place. Perhaps because physical violence may be perpetrated, only well-known friends are allowed to enter.

Less personal business is normally conducted within social distance (close phase 1 to 2 m; far phase 2 to 3.5 m). People who work together tend to use close social distance. It is also a common distance for people who are attending a casual social gathering. If business is conducted at the far phase it has a more formal nature. Hall suggests that a feature of the far phase of social distance is that it may be used to insulate or screen people from each other: 'The distance makes it possible for them to continue to work in the presence of another person without appearing to be rude'. Public distance (close phase 3.5 to 7.5 m; far phase 7.5 + m) is well outside the circle of social involvement and, at this distance, other communication problems may occur similar to those discussed in Chapter 4.

When a stranger violates a person's space (that is, enters a particular zone not normally reserved for him) tension, discomfort and flight may result. Thus Sommer (1969) intruded mental patients' space by sitting close to them on benches. The typical reaction of the patients was to face away, become rigid, and pull in their shoulders. Within 2 minutes, 33 per cent of the victims departed while none of the control patients, whose space had not been violated, did so. Within 9 minutes, 50 per cent of the experimental subjects and only 8 per cent of the control subjects departed. In another study reported by Sommer (1969), female students working alone at a table in a library departed much faster than control subjects when another female student occupied the adjacent chair and moved it closer to the subject. Interestingly, the 'defence behaviour' is typified by a lack of verbal responses to the invasion. Only two of the 69 mental patients and only one of the 80 students asked the invader to move over. As Sommer points out, this provides support for Hall's (1959) contention that 'we treat space somewhat as we treat sex. It is there but we don't talk about it.'

Flight is an extreme reaction to invasion of personal space. McBride, King, and James (1965), however, showed that subjects' arousal (measured by skin resistance, GSR) rose more sharply when they were approached from the front in comparison with a lateral approach. Both of these directions of approach produced greater GSR changes than invasion from the rear. Other less extreme reactions were recorded by Patterson, Mullans, and Romano (1971) who described both 'leaning away' and 'blocking out' responses, both of which were designed to 'exclude' the intruder. These responses increased in frequency as the intruder seated himself closer to his victim. Mahoney (1974), however was unable to replicate the increased 'leaning' and 'blocking' responses when using control subjects whose space was not violated. This was primarily due to their subjects 'tensing up' and 'freezing' rather than activity moving to block out the violator. Consistent with this interpretation are the findings of Mehrabian and Diamond (1971) who showed that relaxation decreased as chairs were located in such a manner as to increase immediacy.

Another study which demonstrated the arousal that can result from an invasion of personal space was reported by Lundberg (1976). He measured the level of catecholamine (a pharmacological stress indicator) in the urine of selected passengers both before and after crowded or empty train journeys. Significant increases were obtained under crowded conditions and his finding were supported by subjective responses of increased feelings of discomfort as the number of passengers increased.

As an example of how the discomfort and stress which might result from personal space invasion may affect performance, Middlemist, Knowles, and Matter (1976) recorded performance reduction in a men's lavatory. In a three-urinal lavatory, a confederate stood either immediately adjacent to a subject, or one urinal removed, or was absent. They showed that both the onset and persistence of micturation was delayed significantly in the adjacent condition as compared to the other conditions. Leaving aside the ethical questions concerning privacy which such a study raises, the results do indicate that personal space invasion may result in a performance decrement, probably because the person's arousal is increased to a greater level than optimum. It is well known that overarousal is likely to lead to reduced performance.

It is clear, therefore, that the invasion of a personal space zone by someone who should not normally enter causes a complex series of behavioural responses. The purpose of these responses would appear to be to 'distance' the intruder from the victim by his leaning away, turning, or more simply, 'withdrawing' into himself. In this way he may, temporarily, shrink his personal space zone so that invasion is not perceived. For example, Sommer (1969) observed that people in a very crowded situation (for example in a crowded train) seemed to control their potential discomfort from overcrowding by behaviours such as staring at the floor or into space, thus relating to those close by as if they were non-persons. The importance of such invasion, therefore, lies in the increased feelings of unease, discomfort and stress with a possible loss of performance.

Many variables affect the distance of these different space zones and include personality, sex, age, culture and the status of the individuals involved.

Personality

Many studies suggest that subjects with personality abnormalities need more personal space, although the findings are not totally conclusive (Evans and Howard, 1973). Several studies have also been carried out to examine the relationship between personal space and personality types. Patterson and Sechrest (1970), for example, found that extraverts have smaller personal space zones than introverts, but Meisels and Canter (1970) found no such relationship.

Sex

An examination of sex differences in the way we use space indicates that females have smaller zones of personal space, and hence can tolerate closer interpersonal contacts than males (Liebman, 1970). When opposite sex pairs are mutually attracted, it is not surprising to find that the magnitude of such buffer zones decreases considerably (Allgeier and Byrne, 1973). One striking difference between the sexes is in the positions which each adopts; research on attraction indicates that males prefer to position themselves *across from* liked others, whereas females prefer to position themselves *adjacent to* liked others (Byrne, Baskett, and Hodges 1971).

Age

Very little research has been reported which explores the developmental aspects of personal space. Willis (1966) studied three age groups: older, younger and peer (age within 10 years). Peers approached one another more closely than they approached those who were older. Age differences, however, are often confounded with status differences and the effects of these are discussed below.

Culture

Clear differences may be observed in the spatial behaviour of members of different cultures. For example, Hall (1976) observed that Germans have a larger area of personal space and are less flexible than Americans in their spatial behaviour. Latin Americans, French and particularly Arabs, on the other hand, were found to have smaller personal space zones than Americans. Sommer (1968) found no significant differences in personal space between English, Americans and Swedes, but found that the Dutch had slightly larger, and Pakistanis slightly smaller, personal space areas. Sommer does suggest, however, that such studies are difficult to interpret because of language barriers which make it difficult for subjects to understand the purpose of the experiment.

Status and familiarity

In general, studies support the contention that external sources of threat (by meeting a higher status individual or one who is not known) lead to increased personal distance. Correspondingly, those who wish to convey a friendly impression or a positive attitude, choose smaller interpersonal distances than neutral or unfriendly communicators (Patterson and Sechrest, 1970).

Territoriality

Like personal space, territoriality is a concept which invokes social, unwritten rules of space behaviour, with infringement of these rules causing discomfort and/or other behavioural reactions. It is a concept which is widely understood in the animal world but has only been considered comparatively recently in relation to humans. Territoriality differs from personal space, however, in that territories have fixed locations and do not move around with the person himself. Furthermore, the boundaries are often quite visible, being marked by recognized stimuli (animals, for example, often use scent).

Although different societies and political systems have different rules governing territoriality, there appears to exist a commonly shared distinction between territory (or property) which is private and territory which is public. A private territory (for example, a house) may be occupied, or owned *in absentia*, by a single person who has the authority to decide who may and who may not physically enter it. Public territory (for example, stairs, streets, workplaces, etc.) is accessible to many diverse persons and cannot be owned by a single person (Fried and DeFazio, 1974).

Of interest to designers are the cases when a semi-private territory is set up in what is otherwise public territory. For example, the reservation of seats in a canteen, the marking of seat boundaries in transport, or the 'ownership' of equipment.

Fried and DeFazio (1974) observed the territorial behaviour of travellers in a New York City underground train. The seating accommodation in each car consisted of four long benches (two on each side of the car), each of which seated up to eight passengers, and two, two-passenger seats. No armrests or other means of delineating territory were present. The authors' interest centred on the use of the two-passenger seats. Under low-density conditions (up to 15 passengers) it was observed that only one passenger ever sat on a two-passenger seat, and the other seats were used in such a way as to provide a large degree of separation between passengers. It was clear, therefore, that the two-passenger seat was regarded as 'territory' which remained sacresanct even under medium-density conditions (16 to 40 passengers). Indeed many passengers were observed to prefer to stand rather than to sit in the 'free' space on the two-passenger seats. Territorial boundaries broke down under high densities (40 +) of passengers, however, with all seats being occupied and distances between passengers, both sitting and standing, being minimal.

From their study Fried and DeFazio concluded that certain implicit 'byelaws' exist with respect to territory which can be applied to other situations:

1. Seated passengers may not be challenged or deprived of their territory except under very special circumstances. Such circumstances include very high-density or other extreme conditions when a person may be asked to 'move over'.
2. There shall be little verbal interaction between passengers. This explains the success of the many territory markers which may be observed; coats or bags on the adjacent seat, and books on a desk all indicate territorial control. If another passenger wishes to sit at the seat, then he must either ask for the marker to be moved (thus breaking rule 2) or remove it himself without asking, which contravenes other social rules.

No matter how well the physical environment has been designed to match man's behaviour, therefore, social constraints may intervene to reduce the efficiency of the system. Thus the value of the perfect seat (in terms of comfort and reduced backache) may be reduced if it is placed so close to another that the two occupants' personal space requirements are infringed. A well-designed seminar or boardroom (in terms of noise level, temperature, ventilation, etc.) may be rendered ineffective if the seating arrangements are such that (a) the occupants are unable to mark their territories (for example, not enough desk space is provided to 'spread out' their papers), or (b) no account is taken of personal space requirements, with the result that the participants may tend to 'withdraw' into themselves to maintain 'distance'.

At this point it would be pertinent to ask how these social space requirements may be accommodated in the confines of a modern work environment. The obvious answer, clearly, would be to increase the space available to individuals in a potentially crowded condition. This would result in less density of people, and territorial and personal space requirements would be met.

In many cases, however, such propositions are not feasible. Indeed, sometimes the space available may not be increased by a single inch. In such cases territorial requirements may be accommodated by providing fixed markers, for example individual armrests to delineate seats, or small, individual tables. Furthermore, structures such as armrests might help to separate sitters enough to avoid actual physical contact.

Finally, although physical space may not be increased, it may be possible to increase apparent space by use of appropriate interior designs. Thus it is well known that broad horizontal stripes make an object appear wider than is actually the case, whereas vertical stripes have the reverse effect. Interior designers know of other perceptual 'tricks' for altering the appearance of shape, depth, etc. For example, Wyburn, Pickford, and Hirst (1964) suggest that colours and shadows may be used to produce apparent depth, distance and solidity. Little (1975) has demonstrated that people take account of their background surroundings when considering their personal space requirements. Although the surroundings used in his study were the interior of a living room, an office and a street corner, his results suggest that a background which makes the distance between two people appear to be greater than it actually is would help to reduced perceived space infringements.

The Landscaped Offices Concept

A further practical solution to the individual's need for space is to allow the user freedom to arrange his workplace as he wishes, thus permitting him to set up individual territories and ensuring that infringements do not occur unwittingly. The arrangement of both men and equipment in this freer way is embodied in the 'landscaped office' concept, said to have been proposed originally by West German furniture manufacturers, Ebehard and Wolfgang Schnelle, in the early 1960s (Brookes, 1972).

The important feature of the landscaped office lies in its lack of boundaries. Whereas a conventional office system might take the form of one floor of a building which has been further divided into smaller offices by fixed walls and doors, a landscaped office will use the same floor space but the different work groups are scattered around without being restricted by walls. The geometry of the workgroups is supposed to reflect the pattern of the work process, and is arranged by the individuals rather than being a superimposed rectilinear plan. The spatial needs of privacy and territory are meant to be accommodated by providing low, moveable screens and allowing the user, within limits, to arrange his desks (boundaries) as he feels fit. Furthermore it allows workgroups to adjust their work areas to the changing needs of their business. Brookes and Kaplan (1972) also point out that the concept insists that all staff participate—not just clerical workers and a few supervisors. Among the many advantages which were claimed for this type of spatial organization were that group cohesiveness would be enhanced by the mixture of executives, management, supervisors and clerical staff, and that productivity would be enhanced.

The few controlled studies which have been carried out to substantiate or refute these claims do not lead to any firm conclusions. Brookes (1972) has reviewed the literature which suggests that information flow between workgroups, and a perception of group cohesiveness, might increase in a landscaped office, and that employees prefer the brighter, more colourful and friendlier design that a landscaped office provides. On the other hand, this type of office arrangement produces a greater loss of privacy, an increase in distractions and interruptions and, paradoxically, a perceived loss of control of the space around workplaces. This last problem is extremely interesting in so far as it suggests that allowing an employee to have physical control over the size and arrangement of his immediate space is not enough to overcome territorial requirements—he must actually *feel* that he has control. Whereas the landscaped office concept allows him physical control, because of a lack of privacy it does not allow him to have subjective control.

Brookes (1972), therefore, concludes that there is no strong body of evidence to support the claims of landscapers. However, he also points out that evaluations so far carried out have been of a survey nature on existing designs rather than a systematic attempt to investigate variables which could be beneficial to the concept. He poses questions such as 'what type of staff can learn to live in what style of office?'; 'What changes in work, policy, goals or supervision can be made to increase productivity?'; 'can staff be taught to utilize their office spaces more efficiently?', etc. In other words his questions may be summed up by arguing that only when needs of the user are fully determined may the environment be designed to fit them.

SUMMARY

When considering how the operator's workplace should be arranged around him, two factors need to be assessed. The first concerns his communication requirements—communication with his colleagues and machines (his mobility, and visual and autitory needs). The second relates to his feelings of ease and comfort with respect to the position of other people in his immediate environment. In this case concepts such as personal space and territoriality have been borrowed from the behavioural analyses of animals to explain how these factors could help with performance.

CHAPTER 9

Seating and Posture

When discussing the dimensions and arrangement of the workplace, a crucial factor to be decided early in the design stage is whether it is better for the operator to perform his tasks sitting down. Seating has many advantages, indeed Grandjean (1973) describes sitting down as being 'a natural human posture'. Allowing the operator to sit relieves him of the need to maintain an upright posture, which reduces the overall static muscular workload required to 'lock' the joints of the foot, knee, hip, and spine, and reduces his energy consumption. Grandjean also points out that sitting is better than standing for the circulation. When a person is standing the blood and tissue fluids tend to accumlate in the legs—a tendency which is reduced when seated since the relaxed musculature and the lowered hydrostatic pressure in the veins of the legs offer less resistance to the return of the blood to the heart. Seating also helps the operator to adopt a more stable posture, which might help him to carry out tasks requiring fine or precise movements, and it produces a better posture for the operation of foot controls.

Despite these advantages, however, the seated operator is possibly at a disadvantage in some aspects. Perhaps the most important is that his mobility is severely restricted. As will be discussed later, a 'good' seat is one which helps the sitter to stabilize his body joints so that he can maintain a comfortable posture. If he needs to move around his workplace, however, he will have to upset this stability in order to move from the seat into a standing posture and, if this is repeated too frequently, fatigue will all most certainly occur.

Although the seat helps the operator to maintain a posture which is suitable for fine manipulation, this type of posture is unlikely to be useful if he need his hands or arms to operate controls with large forces or torques. Under such circumstances the operator's normal behaviour will be to rise from the seat to enable him to take up the necessary posture—an activity which again induces fatigue if repeated too frequently.

A third disadvantage of being seated arises if the workplace happens to be vibrating sufficiently for the vibrations to be transmitted through the seat. As is discussed in Chapter 10, if the operator also uses his backrest the vibration which may be transmitted through it to his back is often enough to reduce his manipulative performance.

Finally, prolonged sitting may itself cause health problems. For example, Grandjean (1973) points out that a sitting posture causes the abdominal muscles to slacken and curves the spine, in addition to impairing the function of some internal organs—particularly those of digestion and respiration. Furthermore Pottier, Dubreuil, and Mond (1969) have demonstrated that prolonged sitting (over 60 minutes) produces swelling in the lower legs of all sitters, which is caused by an increase in hydrostatic pressure in the veins and by compression of the thighs causing an obstruction in the returned blood flow.

Clearly, therefore, many factors need to be taken into account before a decision is made wheather an operator needs to sit to perform his task.

This chapter considers three primary aspects of maintaining a seated posture. First, the orthopeadic considerations, which relate to a possible damage to the sitter's health caused by poor design. Second, muscular aspects, and third, the behavioural and biomechanical aspects of the sitter who may be viewed as a 'comfort seeker' or as a 'discomfort avoider'. Finally the optimum seat dimensions which arise as a result of these considerations are discussed.

ORTHOPAEDIC ASPECTS OF SITTING

When seated, the primary support structures of the body are the spine, the pelvis, and the legs and feet.

The spine consists of 33 vertebrae joined together by multiple ligaments and intervening cartilages, the functions of which were discussed in Chapter 3. For convenience of description the vertebrae are dividied into four areas which

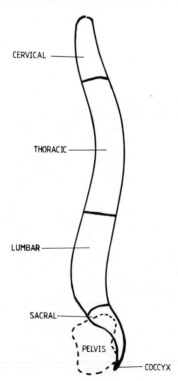

CERVICAL

THORACIC

LUMBAR

SACRAL

PELVIS

COCCYX

Figure 9.1 The shape of the normal spine viewed from the left side

167

correspond roughly to the changes in the shape of the spine. These areas are the topmost seven cervical, then twelve thoracic and five lumbar vertebrae, followed by five fused sacral and four fused coccygeal vertebrae. From the point of view of seating design, the orientation of the lumbar and sacral vertebrae are important, since it is these vertebrae and their respective discs and muscles which take most of the spinal load of a seated person.

Although when viewed from the front or back the normal, relaxed spine appears vertical, when viewed from the side its curved nature can be seen (Figure 9.1). The top, cervical, curve bends forwards leading into a convex backward bend throughout the thoracic region. The lumber region bends forward again, ending in the sacrum which is positioned on the pelvis.

Since the spine has evolved to this shape, it seems reasonable to suggest that

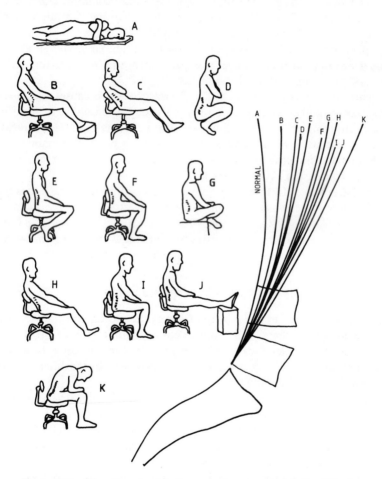

Figure 9.2 The degree of spine curvature produced in different postures (from Keegan and Radke, 1964, reprinted by permission © 1964 Society of Automotive Engineers, Inc.)

this 'natural' shape is one which produces both the optimum pressure distribution over the cervical discs, and the optimum level of static load on the intervertebral muscles. It follows, therefore, that a seat in which the sitter has to adopt a different spinal posture is likely to cause maldistributions in disc pressures and will result, over time, in lumber complaints.

Using X-rays to study the shape of the spine during different postures, Keegan and Radke (1964) suggest that the normal, relaxed spinal shape is produced when a person is lying comfortably on his side, with the thighs and legs moderately flexed. Comparing the lumbar curve produced by this position with the curves produced during ten other seated positions, wide variations in curve shapes can be seen (Figure 9.2). Keegan and Radke's data suggest that the sitting posture which produces the nearest approximation to the 'normal' lumber shape is one in which the trunk-thigh angle is about 115 degrees and the lumbar position of the spine is supported. Their figures also suggest that a 'sitting-up-straight' position produces a great deal of spinal distortion. In this position the compressive weight of the upper part of the body is harmful to the lower lumbar vertebrae, and this is the cause of discomfort and sometimes pain when a person is sitting in chairs which have this 90 degree backrest angle.

As may be seen from Figure 9.2, a forward posture causes the normally forwards-bent lumbar area to be straightened and, eventually, to bend backwards. Continuing up the spine this also affects the angles of the thoracic and cervical areas causing a 'hunchback' posture. Such a posture, if maintained for long periods, will increase the load on the musculature supporting the head, and produce fatigue in the neck and back.

As the back moves from an upright to a hunched position the pelvis makes a corresponding backward rotation, because it is attached to the sacral area. Grandjean (1973) suggests, therefore, that seats should be designed so that in both the forward and backward sitting postures they provide support to the upper edge of the pelvis, in an attempt to arrest this trend to rotate.

MUSCULAR ASPECTS OF SITTING

Because vertebrae are kept in position by muscles and tendons, any alteration to a 'natural' spinal shape will produce corresponding stresses on the spinal musculature. These increases in muscle activity can be domonstrated and measured by recording the electrical potentials produced by the different muscles (electromyography or e.m.g.)

A study of this nature was carried out by Floyd and Ward (1969) who recorded the activity of four groups of muscles (in the neck, on the collarbone, on the back just below the shoulder blades, and in the lumbar region), when the sitter adopted various positions. Their results (Figure 9.3) fully support the conclusions from orthopaedic studies. Thus, again, 'sitting-up-straight' without any backrest produced a fair degree of activity in the lumbar region—presumably because this posture attempts to straighten the lumbar curve (see Figure 9.1). However, the activity ceased as soon as a backrest was provided. A forward, hunchback

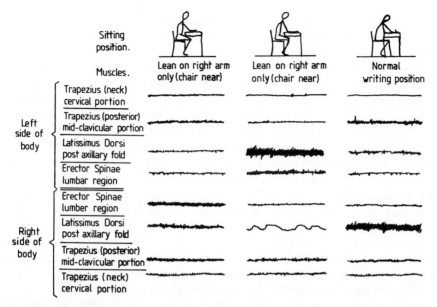

Figure 9.3 Muscle activity during different seating postures (from Floyd and Ward, 1969, *Ergonomics*, **12**, 132–139, and reproduced by permission of Taylor and Francis, Ltd.)

posture caused most activity to occur in the upper back and shoulder regions, again substantiating the orthopaedic observations. Finally, of interest to all who write for a living, Floyd and Ward's data indicate that providing a table support on which the author is able to rest his arms, does not reduce muscle activity. Indeed shoulder muscle activity on the side of the writer's preferred hand increased dramatically. Similar findings were obtained by Andersson *et al.* (1975) from both e.m.g. recordings and from recordings of the pressures produced in the lumbar discs.

Thus both the orthopaedic and muscular evidence suggest that:

1. an upright and a forward leaning posture will cause fatigue,
2. the provision of a backrest will reduce some of the lumbar fatigue, as will
3. an obtuse-angled backrest which helps to stabilize the pelvis rotation.

BEHAVIOURAL ASPECTS OF SITTING

The Concept of Comfort

The discussion so far has concerned the effects of different seating positions on the body's structure, but it will now turn to postural evidence gained from the sitter himself. At this point the subtle change in criterion which has occurred when discussing seats and seat design should be noted. In previous chapters the criteria for successful design have been discussed in terms of the speed and accuracy of the operator's response, and his ability to cope with the incoming information. Fatigue and poor anthropometric fit have been emphasized as contributing factors which help to reduce performance effectiveness. In seating, however, the criterion is in terms more of a reduction in fatigue and excessive posture on parts of the spinal column. Muscular fatigue and spinal deformation reduces comfort and increases the stress of the operators which, in turn, will reduce performance. The emphasis, therefore, is on designing to fit the user's physical requirements rather than his cognitive capacities.

Since it is comfort, rather than cognitive ability, which should be measured in respect to sitting behaviour, it follows that the term needs to be defined. (An accurate measuring tool can only be developed when the properties of the subject being measured are understood.) Unfortunately, however, it is extremely difficult to define comfort, primarily because it is such an entirely subjective concept; the features of a seat which one individual looks for in helping him to adopt a comfortable posture may be different from those chosen by another. As Branton (1972) has suggested 'the problem is. . . to relate measurements of a physical nature to subjective, and essentially private, experiences of feeling'.

The problem is increased further by the possibility that one cannot define or measure comfort. Indeed, the definition might be better couched in terms of discomfort. As an analogy Branton (1972) uses the definition of health; it is only possible to declare that a person is healthy when he does not have any illnesses. Branton further suggests that the absence of discomfort does not mean the presence of a positive feeling but merely the presence of no feeling at all. 'There appears to be no continuum of feelings, from maximum pleasure to maximum pain, along which any momentary state of feelings might be placed, but there appears to be a continuum from a point of indifference, or absence of discomfort, to another point of intolerance, or unbearable pain'. This argument suggests, therefore, that the ideal seat is one in which the person loses all awareness of his seat and of his posture. When in this state a person is able to give his undivided attention to whatever activities he may wish to pursue.

This concept of comfort leads to another aspect which deserves consideration—that of the reason why the sitter is sitting in the seat. In other words, his motivation to sit. If an optimum seat is one which allows the operator to pursue his activities as he wishes, it is reasonable to suggest that a soft, low easy chair, for example, will be unsuitable for a task which requires fine, manipulative work at a console. Branton (1969), therefore, has suggested strongly that a seat may be 'inefficient' to the degree to which it interferes with the primary activity. The assessment of 'comfort', therefore, needs sometimes to give way to a consideration of the operator's 'efficiency' since it is unlikely that one can exist without the other.

Although they are probably related, the two concepts of comfort and efficiency also include antagonistic principles: an efficient chair may not have all of the features of a comfortable chair, and vice versa. For this reason, the behavioural repertoire of the sitter is likely to include activities such as fidgeting, which alter the pressure distribution on parts of his body caused by the 'uncomfortable' aspects of the 'efficient' seat. Indeed the degree of fidgeting could act as an indicator of seat discomfort, as was demonstrated by Branton and Grayson (1967). They recorded the changes in sitting posture of 18 subjects during a 5 hour train journey while sitting in one of two types of seats. On the basis of the significant increase in the number of fidgets produced in one seat over the other, the authors were able to recommend which seat should be adopted.

A factor closely related to the motivation for sitting is the duration for which the sitter is sitting. If sitting comfort changes over time, a chair which is designed for ease, comfort and long-term sitting needs to be assessed according to different criteria than one which is to be used for only a few minutes at a time.

The few data which are available on this point suggest that chair comfort is not affected significantly by the length of time for which the sitter is seated. Grandjean (1973), for example, cites an experiment in which 38 subjects were asked to assess the degree of discomfort in different parts of their body after both 5 and 60 minutes of sitting. No significant difference was observed between the responses. This is not to say, however, that the duration of sitting has no effect at all on sitting comfort—it is simply that for the durations studied it has no effect.

The Behavioural Dynamics of Sitting

Sitting behaviour, therefore, can be characterized by regular movements or fidgeting which helps to relieve pressure maldistributions on parts of the spine. However, the spine is not the only important body structure when considering seat design. Branton (1966), for example, has emphasized the importance of the legs and pelvis, since they take the form of a simple mechanical lever and suspension system which helps to stabilize the body.

The hip portion of the sitting body—the pelvis—can be likened to an inverted pyramid. In fact, contact with the seat is made by only two rounded bones, the ischial tuberosities, covered with very little flesh. Indeed Dempsey (1963) has pointed out that the human body supports approximately 75 per cent of the total body weight on 25 square cm of the ischial tuberosities and the underlying flesh (this is illustrated in Figure 9.4). He also suggests that this load is sufficient to produce 'compression fatigue', which varies according to the compressive load of the body and the duration of loading. 'In physiological terms, compression fatigue is the reduction of blood circulation through the capillaries, which affects the local nerve endings and results in sensations of ache, numbness, and pain.' Branton (1966) further adds that although the findings illustrated in Figure 9.4 relate to hard seats, they are not altered drastically when sitting on a cushion: 'If the cushion is soft, it gives as little support to the flabby muscle or fatty tissue; if it is compressed to the point of being solid, i.e. if 'bottoming' occurs, the cushion is no different from a hard seat'.

The use of modern upholstery, however, may alter Branton's position somewhat. Thus Diebschlag and Muller-Limroth (1980) recorded the pressure distributions of subjects sitting on either a hard seat or on one which was padded using polyurethane foam. As a result of the foam the pressure values under the

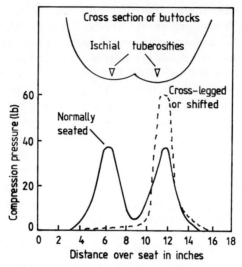

Figure 9.4 Pressure distributions produced over the seat during normal and cross-legged positions (from Dempsey, 1963, reproduced by permission of McGraw Hill)

buttocks were reduced by nearly 400 per cent and the supportive area of contact between buttocks and seat were increased from approximately 900 to 1050 sq cm (Figure 9.5). Similar conclusions were arrived at when the backrest was also changed.

Over longer periods blood circulation to the capillaries in the buttocks is also

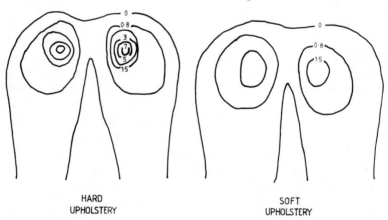

Figure 9.5 Pressure distributions under the buttocks whilst sitting on hard and soft upholstery (pressure is expressed in N/cm^2); reproduced with permission from Diebschlag and Muller-Limroth (1980). Physiological requirements on car seats: Some results of experimental studies. In D. J. Oborne and J. A. Levis (eds.) *Human Factors in Transport Research, Vol. II.* Copyright by Acadamic Press Inc. (London) Ltd.

reduced if the body is unable to move at regular intervals. If the body is confined in a relatively fixed seating position for more than 4 hours, the physiological functions that control the flow of body fluids slow down. This action, coupled with continuous pressure loading on the flesh, accelerates the rate of compression fatigue. Fatigue can be delayed, however, by periodic movements of all of the major body segments. This results in changes of the loading conditions and allows muscular expansion and contraction for adaptation to new weight conditions.

With the mass of the body suspended over what is virtually a two-point suspension on the seat pan, the system is mechanically unstable. This instability is increased as a result of the body's centre of gravity not being situated vertically over the tuberosities when sitting upright, but about $2\frac{1}{2}$ cm in front of the navel. The system is only made stable by the addition of the levers provided by the legs and feet. Crossing the legs, for example, is a normal means of locking the system as is resting the arms on a table or the head on the arms. The whole forms a most complex mechanical system which allows movement in both the horizontal and the vertical planes.

All of these observations lead to the antagonistic requirements discussed earlier: On the one hand the sitter needs to vary his posture to relieve pressure maldistribution; on the other hand he needs to maintain, and actively seeks, stability. To accommodate these two possibilities, Branton (1966) has formulated a theory of postural homeostasis. Homeostasis is a concept which is widely understood in physiology and concerns the self-regulation of body functions. A common example is body temperature regulation: if the body temperature rises, sweat is produced which, when it evaporates, has a cooling effect. If the body becomes cold, however, the blood is routed away from the skin to warmer, central, parts of the body (thus giving cold skin its characteristic bluish appearance). One of the characteristics of such homeostatic activities is that they are autonomic, in other words, they are not under deliberate, conscious control. Awareness only comes with drastic changes of conditions and then only after the event has occurred.

Branton argues that postural activity falls into the same category of autonomic regulation, and that postural homeostasis is a process by which the sitter strikes a compromise between his needs for both stability and variety. Thus sitting behaviour will be characterized by cycles of both inactivity and activity representing the changing needs for stability and variety. An efficient and comfortable chair, therefore, needs to be able to accommodate these homeostatic requirements and allow the sitter both stability and flexibility.

SEAT DESIGN

A number of principles for seat design can be extracted from the above discussions. These include:

1. the type and dimensions of the seat are related to the reason for sitting,

2. the dimensions of the seat should fit the appropriate anthropometric dimensions of the sitter,
3. the chair should be designed to provide support and stability for the sitter,
4. the chair should be designed to allow the sitter to vary his posture, but the fabric needs to resist slipping when there is fidgeting,
5. backrests, particularly prominent in the lumbar region, will reduce the stresses on this part of the spinal column, and
6. the seat pan needs sufficient padding and firmness to help to distribute the body weight pressures from the ischial tuberosities.

With regard to the motivation for sitting, seats may be divided simply into three groups. First easy, comfortable chairs for relaxation. In these the criterion for an effective chair should be, in Branton's terms, a loss of awareness of the seat and minimal discomfort of any part of the body's supporting structure. The second group includes chairs which are used for work. In these cases stability is an important consideration, requiring adequate support of the lumbar area and distribution of the body weight over the seat pan. The final group includes those chairs which Grandjean (1973) describes as 'multipurpose' chairs. These may be needed for a variety of dfferent purposes, for example they could be used at a table, occasionally for working, or as spare chairs, and frequently need to be stacked.

Anthropometric Considerations

Regardless of the function of the seat, its linear dimensions must fit those of the likely user population. This by now is axiomatic, and appropriate anthropometric data exist. Normally these figures all relate to an unclothed sitter, so the presence of clothing and footwear will increase the dimensions by a proportional amount. The dimensions indicated below take account of this factor.

Seat height: easy chair 38 to 45 cm; workchair 43 to 50 cm

The seat height is adjusted correctly when the thighs of the sitter are horizontal, the lower legs are vertical and the feet are flat on the floor. This is because the soft undersides of the thighs are not suitable for sustained compression, and pressure from the front edge of the seat pan can become uncomfortable. For this reason, the limiting case for the seat height is that of the short-legged person who would be prevented from resting his feet on the floor if the seat is higher than his leg length.

The reason for the differing recommended seat heights between easy and working or multipurpose chairs is the way in which they are likely to be used. The height of an easy chair should allow the legs to be stretched well forward since this is one of the preferred relaxing postures for the feet, in addition to helping to stabilize the body. For a working chair, however, the sitter is likely to be in a more upright position with his feet flat on the floor.

Many authors recommend that working chairs should be made to allow the height to be adjusted, to accommodate the wide range of workers who may be called upon to use them. If the chair needs to be higher than the recommended dimensions (perhaps because of a tall machine or high workbench), an adjustable footrest is also recommended.

Finally Grandjean (1973) argues that the optimal height for a workseat can only be decided in relation to the height of the working surface, which he suggests should be between 24 and 30 cm below the working surface.

Seat width: 43 to 45 cm

In this case, the largest person needs to be accommodated. Since the appropriate dimension is the hip width, and since major sex differences are apparent in this dimension, the limiting case should be the upper range of a female sitter.

Seat depth: easy chair 40 to 43 cm; workchair 35 to 40 cm

The importance of an appropriate seat depth is to ensure that all potential sitters find support in the lumbar area from the backrest. If the seat is any deeper than the thigh length of the shortest person, the front edge of the seat will restrain him, causing his lumbar area to curve to reach the backrest. In addition the pressure-sensitive areas at the back of the knee will become pressed against the seat. For a workchair, which will be used by a larger proportion of the population, the recommendation is to make the seat depth accommodate even shorter people, since the only consequence of a taller member sitting in such a seat will be that his knees protrude slightly at the front. Providing the seat height is adequate and the feet are able to be placed flat on the floor, this is unlikely to induce compression fatigue in the thighs.

Seat angle: easy chair 19 to 20 degrees; workchair less than 3 degrees

This refers to the angle of the seat pan to the horizontal. A seat pan which is tilted backwards will produce two effects. First, by the force of gravity the sitters' back is moved towards the backrest, so supporting the back and reducing static load on the back muscles. Secondly the slight inclination of the seat pan at the front helps to prevent the gradual slippage out of the seat which was observed to occur over long periods in the sitting study by Branton and Grayson (1967). The 20 degree optimal inclination is supported by Andersson (1980) who measured the amount of activity in the back muscles (e.m.g.) at different backrest angles. His results, shown in Figure 9.6, illustrate well the value of the sloping backrest.

There seem, however, to be major divergencies in the recommendations proposed for easy chairs and those suggested for workchairs, which relate to the chairs' function and to the motivation for sitting. In an easy chair the sitter wishes to relax, the most relaxing position, of course, being horizontal, and a

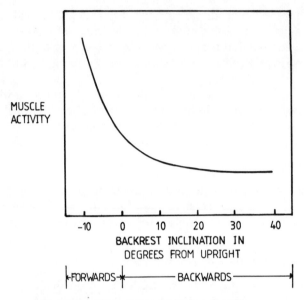

Figure 9.6 The relationship between back muscle activity and backrest inclination; reproduced with permission from Andersson (1980). The load on the lumbar spine in sitting postures. In D. J. Oborne and J. A. Levis (eds.), *Human Factors in Transport Research Vol. II.* Copyright by Academic Press Inc. (London) Ltd.

backward sloping seat pan helps to attain this. In a working chair, however, his requirements are to be positioned for easy access to his work area in front of him. A backward tilting seat would cause him to have to bend forward and would curve his spine unnecessarily.

Mandal (1976) has taken this further to argue that since most work is carried out in a forward bent posture, a forward-sloping seat is most appropriate. He suggests that a backward-sloping seated workchair of even 5 degrees only will cause a straightening of the lumbar area and thus discomfort. For this reason he suggests that a sitter will gradually tend to sit on the front edge of a seat, pivoted on his thighs. 'That this position really is one of the most frequently used can quite clearly be seen as only the front part of the seat covers of old office chairs is worn; the rear part being almost untouched'. (This is a good example of an unobtrusive investigation, as discussed in Chapter 14).

Mandal measured the pressure distribution obtained over the seat and the extent to which the dorsal muscles were stretched, using five combinations of seat pan angle and sitting position. His results indicated less muscle elongation and a more even pressure distribution using a forward-tilting (15 degrees) than a backward-tilting seat (Figure 9.7). However, the suggestion that workchairs ought to be forward tilting should be treated with some caution since this would tend to destabilize the body and increase its tendency to slip forward. In addition, the

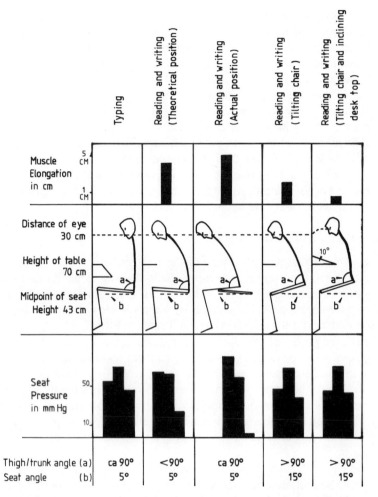

Figure 9.7 Elongation of the dorsal muscles and pressure distributions over the seat produced by different seat angles (from Mandal, 1976, *Ergonomics*, 19, 157–164, reproduced by permission of Taylor and Francis Ltd.

supporting advantages of the backrest will be less apparent; other muscles, therefore, could be overworked to compensate for the reduced dorsal muscle load.

Backrest height and width: up to 48 to 63 cm high; 35 to 48 cm wide

The proposed dimensions of the backrest relate, quite simply, to the distance from the shoulder to the underside of the buttock (height) and to the shoulder width (width). The height dimension, of course, extends from the compressed seat if padding is present.

As has become apparent, however, the linear dimensions of the backrest are only part of the question. Since its function is as much concerned with maintaining a relaxed (that is, non-fatiguing) spinal posture, the shape and the angle of the backrest are extremely important. In addition, since the curvature of the spine varies greatly from one person to another, a complex relation between height and shape arises.

In order that the sacrum and fleshy parts of the buttocks which protrude behind a sitter can be accommodated, while at the same time allowing the lumbar region to fit firmly into the backrest, many authors suggest that the backrest should have an open area or should recede just above the seat pan. A space of at least 12.5 to 20 cm is required to accommodate the buttocks in this way.

Finally, a high backrest might prevent full mobility of the arms and shoulders in certain tasks, for example, typing. In such cases small backrests which support the lumbar region only are suggested by many authors.

Backrest angle: 103 to 112 degrees

Like an angled seat pan, the angle of the backrest to the seat pan serves two purposes. First, it prevents the occupant from slipping forwards, and second, it causes him to lean against the backrest with the lower (lumbar) part of his back and his sacrum supported. From an orthopaedic viewpoint, the appropriate angle would be about 115 degrees which Kegan and Radke (1964), for example, demonstrated produced the nearest to a 'natural' lumbar shape. However, when sitting comfort responses have been elicited from seated laboratory subjects, a less obtuse angle has consistently been found to be more 'comfortable'.

Jones (1969) studied posture and feelings of comfort in a highly adjustable car seat in many different positions. Subjects were trained to recognize the sensations of 'no sensation', 'conscious of contact with the seat', 'numbness', 'ache' and 'pain' after varying intervals. From his data he suggested a backrest angle of 108 degrees.

Grandjean (1973) discusses work which he carried out with Burandt to determine the optimum backrest angle for easy chairs when used for different reasons. Their data suggests that an angle of 101 to 104 degrees is optimum when reading, whereas 105 to 108 degrees is an optimum relaxed angle.

Armrest height: easy chair 21 to 22 cm above the compressed seat

The primary function of armrests is to rest the arm in order to lock the body in a stable position. In an easy chair this is often accomplished by the arm being used to support the head. Armrests may also be useful in helping to change position or as an aid to getting up from the chair. However, it should be remembered that armrests may prove to be restrictive to the free movement of the arms and shoulders if they are incorporated in a working chair.

Cushioning and Upholstery

The importance of cushioning was demonstrated by Branton and Grayson (1967) in the observational study of sitters in two types of train seats. Although the dimensions of the two seats were approximately similar, the type and strength of seat spring and padding differed. One gave a relatively 'soft' subjective feel whereas the other appeared subjectively 'firm'. After analysing the number of 'fidgets' observed in the sitters and the length of time for which stable postures were maintained, the authors were able to state that 'by almost all counts II (the firmer seat) is much the better'. Furthermore, not only was the number of different postures greater in II, but more were 'healthy'.

Cushioning performs two important functions. First, it helps to distribute the pressures on the ischial tuberosities and on the buttocks caused by the sitter's weight; as was shown earlier, if not relieved this pressure will cause discomfort and fatigue. The second function is to allow the body to adopt a stable posture. To this end the body will be able to 'sink' into the cushioning which then supports it. However in this respect Branton (1966) raises a warning against the cushioning being too soft.

A state can easily be reached when cushioning, while relieving pressure, deprives the body structure of support altogether and greatly increases instability. The body then 'flounders about' in the soft mass of the easy chair and only the feet rest on firm ground. Too springy a seat would therefore not allow proper rest, but may indeed be tiring because increased internal work is needed to maintain any posture.

Kroemer and Robinette (1968) agree with Branton's position and also caution that soft upholstery will allow the buttocks and thighs to sink deeply into the cushioning. If this occurs all areas of the body that come into contact with the seat will be fully compressed, offering little chance for the sitter to adjust his position to gain relief from the pressure. In addition, the body often 'floats' on soft upholstery, again causing the posture to be stabilized by muscle contraction.

With regard to the seat covering, the important aspects are its ability to dissipate the heat and moisture generated from the sitting body (which will, in turn, be related to the type of environment in which the sitter is sitting), and its ability to resist the natural forward slipping movement of the body over time. For both of these criteria, adequate thermal and mechanical techniques exist to allow the designer to make appropriate measurements.

SUMMARY

Sitting is often considered to be a natural posture, relieving the sitter of the need to maintain an upright position. However, as this chapter has demonstrated, a seated posture can sometimes cause more problems than are solved. A seat which requires the sitter to adopt certain postures can create at the least muscular fatigue due to the static loads placed on the spinal and other muscles. At worst, permanent orthopaedic damage can result from spinal pressure maldistributions. The second part of this chapter considered chair dimensions for alleviating some of these problems—particularly in relation to increased sitter comfort.

CHAPTER 10

The Physical Environment:
I   Vibration and Noise

In addition to discussing specific details, the previous chapters have also introduced the concept of the environment in the ergonomic design of the total work system. The operator's immediate workplace, his machines, their arrangement within the environment, his interaction with other operators and the manner in which he operates his system were all shown to affect significantly his performance and sensations of ease. These environmental considerations, however, are 'visible'; he can see them and they affect the operator in so far as they restrict his actions, his judgement and his immediate perceptions. The present and following chapter will consider the less tangible (perhaps the 'invisible') aspects of his environment; the ubiquitous sensations to which he is exposed from different pieces of machinery or from other components in his workplace. In addition to others, then, these 'pollutants' include the vibration, the noise, the temperature, and the illumination, and these aspects will be discussed in this and the next chapter.

The order in which the factors are discussed in these two chapters should not be taken as any suggestion of their relative importance to the operator, however. This depends entirely on the work situation which is being considered and the individual who is being stimulated. Take, for example, two seamen standing on a small ship in cold, rough weather: One, being a 'poor' sailor will be concerned more about the motion of the ship than any other factor; the other, a 'good' sailor will probably worry more about the cold.

Each environmental factor can affect the operator in one or more of three ways: first his health, second his performance and third his comfort. These effects, however, are not independent. For example, poor health can lead to poor performance; poor performance can lead to reduced work satisfaction; and, as was shown when discussing seating, a lack of comfort may be a precursor of a health hazard and can lead to poor performance; etc. Although these aspects will be considered independently, therefore, their affects are probably combined.

The purely physiological effects of environmental stressors on an operator's health occur mainly at the extreme levels of the stressor intensity. Even before the extremes are reached, however, the stressor (be it noise, vibration, temperature, illumination or any other aspects of the environment) can affect the operator detrimentally, although this may not be apparent to the observer.

Despite the fact that man is extremely adaptable, the optimum range of any environmental factor for comfort or for performance is very narrow. As has been stressed throughout this book, having to adapt to conditions outside this range can require more effort on the part of the operator which can lead to a breakdown in performance. It is well known, for example, that there is a limit to our ability to process information (see, for example, Miller, 1956). Brown (for example, Brown, 1965; Brown and Poulton, 1961) and others have taken this argument further to suggest that we have a limited store of 'mental capacity'. Carrying out a task under normal circumstances and within the narrow optimum range of environmental factors will take up a certain amount of this mental capacity. Having to adapt to more severe conditions (for example, an extremely difficult task or adverse environmental conditions) will reduce the store further, until a point is reached when the 'spare mental capacity' is exhausted. When this happens, the argument suggests, performance decrements will begin to occur.

VIBRATION

Vibration will be discussed first because it is the most fundamental environmental factor. Understanding the definitions and parameters of structure-borne vibration will help the reader to understand the processes of most other environmental parameters—particularly noise and, to a lesser extent, illumination.

Definitions

Vibration may be defined simply as any movement which a body makes about a fixed point. This movement can be regular, like the motion of a weight on the end of a spring, or it can be random in nature. The vibration which is experienced from machinery is usually a very complex, but regular, motion. Using appropriate analysing techniques, however, any complex motion can be defined in terms of a number of simple components. This type of analysis— Fourier analysis—is outside the bounds of this book but can be found in most texts dealing with vibration theory.

In simple terms, the movement of a vibrating body can normally be defined in terms of two parameters: the vibration frequency and intensity.

The frequency is essentially an indication of the speed of movement, and is measured in cycles per second or Hertz (1 c.p.s. = 1 Hz). Thus the vibrating body is said to have moved through one cycle when it has moved from its fixed point to its highest deviation from the point, back to the lowest deviation, and then returned to the position of the original fixed point (Figure 10.1). The number of times that it does this in a specified time (usually 1 second) is its frequency of movement (or its number of cycles per second). The type of motion, shown in Figure 10.1, is the most basic of all motions and is known as sinusoidal motion.

Vibration intensity can be measured in a number of ways, although amplitude or acceleration are the units normally used. The amplitude is expressed in the

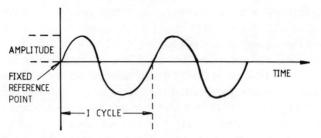

Figure 10.1 The frequency and amplitude of a body vibrating sinusoidally

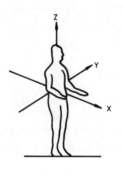

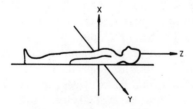

Figure 10.2 The three vibration coordinates for standing, seated and lying man

normal units of distance (inches, feet, cm or mm) and is the maximum distance that the body moves from its starting point. However it is now more common to express intensity in terms of the body's acceleration, the units being metres per second or 'g' units (1 g = 9.81 m.sec^{-2}, and is the acceleration which a body needs before it can overcome the force of gravity and lift from the earth's surface). The measurement is normally made using a small acceleration sensor (an accelerometer) placed on the vibrating body.

Each of these parameters is related — thus acceleration in 'g' units = $\left[\dfrac{4\pi^2 f^2 a}{981} \right]$

where f is the frequency of the vibration, and a is the vibration amplitude in cm.

By convention the direction of vibration is defined in terms of three coordinates: vertical (z), lateral (y) and fore–aft (x). For a human being these coordinates are deemed to pass through the chest in the heart region and are related to the back, chest, sides, feet and head. This means that the physical directions of motion for an operator who is lying down are different from those of a standing man (Figure 10.2).

The Limits of Vibration Perception

Before considering the different ways in which vibration can affect the operator, it is useful to have some idea of the upper and lower limits (of tolerance and perception) which the human body can accept. This should enable the vibration levels which will be discussed later to be put into perspective.

The lower threshold of perception

In the early investigations of how humans respond to vibration, many experimenters attempted to obtain definitive data concerning the lowest level of vibration needed for perception. However when all of these data are collated, little agreement can be seen between the various graphs purporting to illustrate how the threshold of the perception of acceleration varies over different frequencies. As can be seen from Figure 10.3a, these threshold curves spread over an intensity range of about 10:1. This may be due in part to the various investigators' use of different apparatus, experimental methods, instructions, subject populations and statistical analysis. Unfortunately, however, the sole conclusion which can be drawn from the curves is that it would be unjustifiable, at least on the basis of the work carried out so far, to define a lower limit of whole-body vibration perception.

The threshold of tolerance

The results which purport to describe the upper limit of vibration perception are as difficult to compare as are those for threshold values, and for many of

184

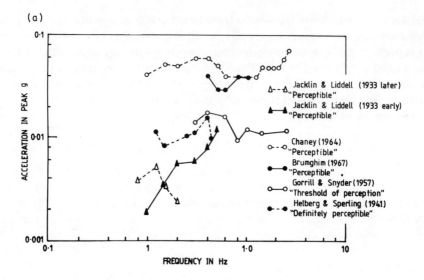

(a)

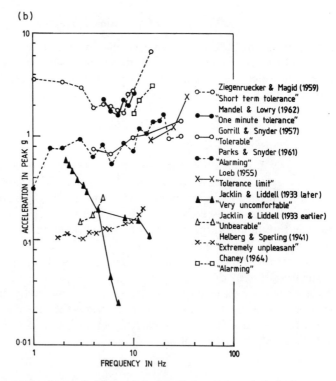

(b)

Figure 10.3(a) Experimentally derived threshold curves for whole body vibration; (**b**) experimentally derived tolerance curves for whole-body vibration (from Hanes, 1970, reproduced by permission of the Johns Hopkins University Applied Physics Laboratory).

the same reasons. Differences in equipment, methodology and adequacy of reporting make comparisons between the various studies at best difficult and at worst useless. To take extreme examples: one group of subjects tested by Jacklin and Liddell (1933) considered 0.024 g at 7 Hz to be 'Very uncomfortable, painful or unbearable', but this level coincides with one which Brumaghim's (1967) subjects described as 'perceptible'.

An indication of the spread in levels of different tolerance curves is also shown in Figure 10.3b. Again it is unfortunately clear that previous investigations of threshold or tolerance limits add very little to our knowledge of how people react to vibration.

The Effect of Vibration on Health

When excited by vibration, any physical structure will amplify the intensity of the input motion if it is at certain frequencies (the structure's resonant frequency) and attenuate it at others. Since the human body is a very complex structure (being composed of different organs, bones, joints, muscles, etc.),

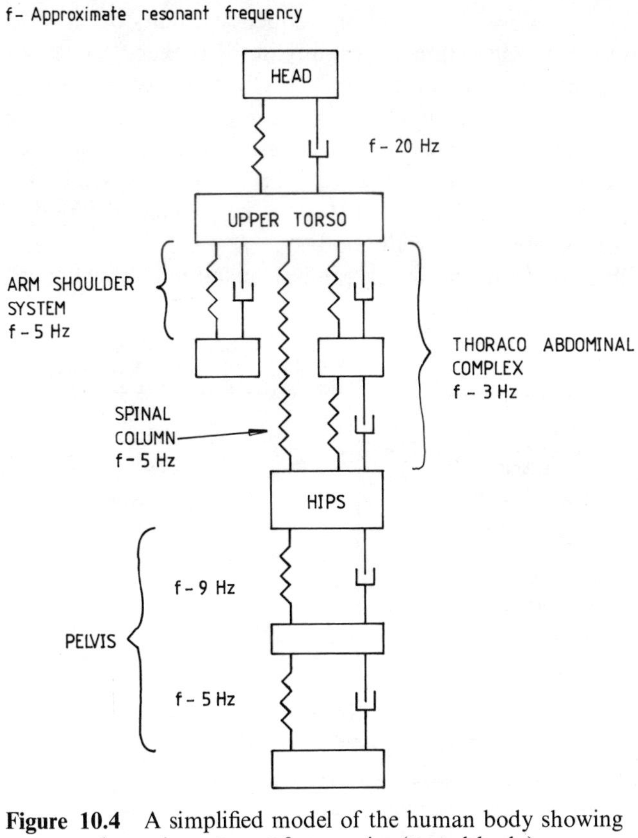

Figure 10.4 A simplified model of the human body showing the main resonant frequencies (seated body)

186

different parts have different resonant frequencies. Structural damage due to vibration amplification may soon occur, therefore, if the body is vibrated by strong vibration stimuli with frequencies close to the resonant frequencies. As an illustration of these frequencies, Figure 10.4 provides a crude indication of the primary resonance characteristics of different parts of the human body. These data, however, should only be taken as a rough guide since body resonances are likely to be affected by many factors such as the muscle suppleness, the dimensions of the bones, the amount of fatty tissue, etc.

The damage to health which can be caused by mechanical vibration tends to fall into two broad categories. The first contains changes which can be directly attributed to the vibration frequency, and occurs as a result of different body structures being excited at or near their resonant frequencies. Effects in the second category show less obvious dependence on frequency and are related more to the 'impact' of the stimulus on the body—that is, the vibration intensity and duration.

Frequency-dependent health effects

Injuries which are caused by vibration frequency usually occur after prolonged exposure to the vibrating stimulus, mainly in the higher frequency ranges. The type of appliance which could cause such effects includes many of the kinds of powered tools used in industry, for example chipping hammers, scalers, screwdrivers, rivet-drivers, road-drills, stone-breakers, and the forester's chain-saw. Examples of some of these effects are described in Figure 10.5. Most injury reports indicate that the damage occurs to the peripheral blood and nervous systems in the exposed part of the body.

Intense vibration from hand tools can be transmitted to the fingers, hands and

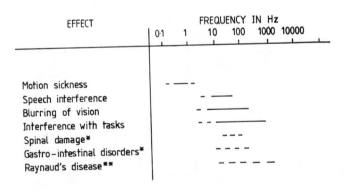

Figure 10.5 Some frequency-dependent effects of vibration

arms of the operator both from the handles of the machine and from structures which are held or steadied by the hand and vibrated by the appliance. Complaints usually include many of the symptoms of intermittent numbness and clumsiness of the fingers, intermittent blanching of either all or part of the extremities, and a temporary loss of muscular control of the exposed parts of the body (Agate, 1949). These symptoms are commonly referred to as 'white finger disease' or Raynaud's disease. Relief from all or some of these symptoms usually occurs after prolonged rest from exposure to the vibration stimulus, although they are likely to reappear promptly with renewed exposure.

It is difficult to estimate the extent of such damage in industry. Basing their assessment on studies of large groups of different types of workers using vibrating tools over periods of time, however, Guillemin and Wechsberg (1953) report that the incidence of neurovascular damage due to vibration exposure varied between 89.5, 66.2, and 11.6 per cent and as low as 0.2 per cent from different groups. This wide range is probably due to differences between actual daily exposure times, to large variations in the frequency and intensity of the vibration stimuli, to the use of various types of materials, and to varying individual susceptibility to the vibration.

From their review of the available information Guillemin and Wechsberg concluded:

Mechanical vibration endured repeatedly over long periods of time by human subjects produce disabilities that differ in nature and extent in three broad frequency ranges. Those below 1,000 per minute (approximately 16 Hz), occuring with large amplitudes (up to an inch or more), typically produce lesions in the bones, joints and tendons. In the range 2,000 to 10,000 cycles per minute (33–166 Hz), with amplitudes of tenths or hundredths of an inch, the symptoms are mainly cardiovascular .... Above 20,000 cycles per minute (over 300 Hz), where amplitudes are measured in thousandths of an inch, neurovascular disturbances are accompanied by continuous burning pain which may be the predominant symptom.

These conclusions have not been altered radically in the light of any new knowledge.

Intensity and duration physiological effects

Intensity-dependent effects of vibration occur mainly as a result of body parts moving against one another; in addition large organs pull on supporting ligaments and it is also possible for soft tissues to be crushed. Examples of these injuries can be seen in normal working life. For example, Rosegger and Rosegger (1960) carried out a comprehensive survey of the health complaints of 371 tractor drivers and demonstrated two forms of damage resulting from long periods of tractor operation: stomach complaints and disorders of the spine (particularly in the lumbar and thoracic regions). Both the number and severity of such cases increased proportionately with the length of service as a tractor driver. Traces of blood in the urine have also been found in the drivers of these vehicles. This is probably due to kidney damage, although this breakdown could be caused as much by resonating the kidneys as by damage through high intensity vibration *per se*.

The effects of these types of stimulation have been investigated at a descriptive level by Magid *et al.* (1962), who asked ten subjects to describe their feelings about different intense vertical stimuli in the frequency range 1–10 Hz. Many of their subjects reported 'chest pain' at between 4 and 8 Hz stimulation (at about 0.5 g) with all subjects reporting these symptoms at 7 Hz. Some reports of 'abdominal pain' were produced at lower frequencies (3–6 Hz) but the number of reports was fewer than for 'chest pain'. At 10 Hz (1.2 g) skeletomuscular and abdominal pain was reported.

The Effects of Vibration on Performance

Because body parts will tend to vibrate in sympathy with any vibrating machinery on which they may rest, the effects of vibration on performance lie mainly in the degredation of motor control. This might be the control of a limb (causing, for example, reduced hand steadiness) or of the eyeballs (causing fixation difficulties and blurring). Little evidence exists to suggest that vibration can detrimentally affect central, intellectual processes.

The Effects of Vibration on Vision

A clearly formed image of an object will only be perceived in the visual cortex of the brain if a stable image falls on the retina. If the image is moving it will fall on differing sets of receptors in the retina, so producing a signal of overlapping and confused images, which makes it difficult to detect any of the object's detail. This is particularly so if the image falling on the retina oscillates with a relatively large amplitude.

Three situations can exist to cause a moving image to appear on the retina: First, if the object alone is vibrating, with the observer stationary; second, if the observer is being vibrated while the object is stationary; and finally, if both the object and the observer are being vibrated. In this last case the resultant degree of blurring will be determined by the extent to which the man and object are vibrating in phase (in other words, whether they go 'up' and 'down' at the same time), and the frequency of the whole body vibration.

Vibrating the object

The detrimental effects of vibrating the object alone appear to be related to the frequency at which the object is vibrating. At low enough frequencies (say less than 1 Hz) the eyes are able to compensate for the movement by tracking the object and are thus able to produce relatively stable images on the retina. (Over time, of course, this is likely to cause fatigue in the muscles which control the eye movements). As the frequency of the vibrating object increases, however, performance (usually measured in terms of reading or of tracking errors) is likely to deteriorate because, although the eye muscles attempt to do so, they are unable to maintain this tracking adequately. This occurs up to some critical frequency, after which performance appears to improve slightly. This apparent improvement in performance is thought to be due to the eye being unable to track the movement of the object efficiently, so producing a slightly blurred image. Unless the object movement is too great, however, the blurring is unlikely to cause performance decrements. The precise critical frequency at which this occurs is uncertain, but various authors have suggested between 3 and 4 Hz.

At higher frequencies and intensities any performance reduction is due solely to a blurred image being produced on the retina. For example Griffin and Lewis (1978) report a study which indicated that the number of reading errors, the reading time and the subjective report of reading difficulty increased as the object frequency increased from 5 to 30 Hz (details of the vibration intensity were not reported).

Since the degree of blurring is related to the amount of image movement on the retina, it could reasonably be expected that the amplitude of the object vibration will also have a direct effect on performance. Experimental results suggest, however, that this is only the case in particularly unfavourable circumstances. For example Griffin and Lewis (1978) report a series of experiments carried out by Crook *et al.* who vibrated their objects at different amplitudes but with a constant frequency. In one experiment, when vibrating an object at 17.5 Hz, with amplitudes of between 0.15 and 0.75 mm, the authors found no significant change in either reading errors or in reading time over the different amplitudes. In later experiments, however, they showed the amplitude of the vibrating object to be an important determinant of performance when either the type or the level of room lighting was poor. They concluded that, providing other environmental conditions are optimal, vibration amplitudes of up to 0.5 mm will not impair performance.

The extent to which the amplitude of the object vibration affects performance will be mitigated somewhat by the distance of the observer from the object. Because the image of an object which is nearer to the observer appears larger than an object further away, if the vibration amplitude remains constant as the object moves further from the eye the amount of blurring is likely also to be reduced. For this reason the effect of vibration amplitude is often discussed in terms of the angle which the object makes at the eye (which takes account of both the object's apparent size and its distance from the observer). In this respect Griffin (1976) suggests that a vibration intensity which produces a blur of plus or minus 1 min of arc at the eye represents the 'threshold of blur'. Since this level would probably be over-restrictive in all but the most critical conditions, he suggests that displacements of greater than plus or minus 2 min of arc may, depending on the difficulty of the task, start to affect reading ability.

Vibrating the observer

The visual effects produced when only the observer is vibrated are often similar to those resulting from vibrating the object alone—a blurred image which, in

190

this case, is due to the movement of the eyeball itself relative to the object. At low frequencies (less than about 2–3 Hz) the whole body moves in sympathy with the motion, so the head and eyes move in a similar fashion. At higher frequencies, however, the head and eyes are likely to resonate and so produce a blurring effect which is stronger than would have been predicated from the intensity of the vibration input alone. For this reason, therefore, it is important to understand the resonance characteristics of the head and eye relative to the rest of the body.

Resonance effects were briefly described at the beginning of this section. Depending on the frequency, any part of the body may treat the incoming vibration in one of three ways: first, the internal structure can be such that it reduces the level of vibration as it is transmitted through it (as, for example, would a soft cushion); second, it can transmit the vibration in a 1:1 ratio (in other words, the input level to the body is not affected); and third, it can accentuate the vibration (this occurs at the resonant frequencies of the structure). The resonance characteristics of the body can be measured, therefore, by placing two vibration sensing devices at each 'end' of the object (one measuring the level of vibration input (I) and the other the level of vibration which has been transmitted through the body (T)). The ratio T/I provides an index of the object's transmissibility. Thus a ratio of greater than 1 indicates resonance, while a ratio of less than 1 indicates attenuation.

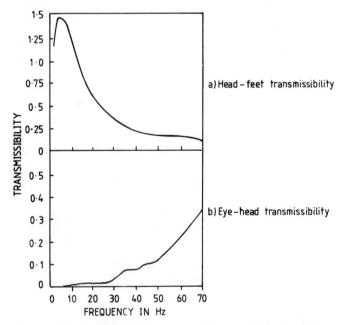

Figure 10.6 Transmissibility characteristics of the body and the eyes (from Lee and King, 1971, reproduced by permission of the American Physiological Society)

When this procedure is carried out on a human body, by placing a vibration sensor at the feet (input) and on the head (output) of a standing man, a transmissibility curve can be drawn over different frequencies which indicates peak head resonance at about 8–10 Hz (see Figure 10.6a, from Lee and King 1971). Figure 10.6b illustrates eye-to-head transmissibility. From these two sets of results, a clear eyeball resonance can be seen which starts at about 20 Hz, although Thomas (1965) has demonstrated a slightly higher eyeball resonance at about 30 Hz. Taken together these data suggest that the eyeball will begin to accentuate the input vibration if the vibration frequency is above about 20–25 Hz. At lower frequencies, between about 8 and 10 Hz, the head and neck resonance is also likely to cause the eyes to move at higher intensities than those which enter the body. As long as the movement is not too great, however, this may still be compensated by the eye muscles to allow a relatively stable image to fall on the retina.

The question of whether it is 'better' or 'worse' to vibrate the object rather than the observer might appear at this stage to have practical consequences. However it is one which is difficult to answer, primarily because of the effects of the frequency of the stimulation. Dennis (1965) for example, investigated the effects of two vibration conditions (frequency range 5–37 Hz) on a task which involved reading different numbers on a display. In the first condition the numbers were vibrating and the subject remained stationary; in the second the subject was vibrating with the display stationary. Dennis's results indicated

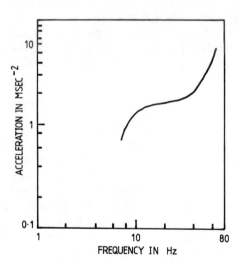

Figure 10.7 The levels of vibration likely to cause decrements in visual performance (reproduced with permission from Griffin, 1976, Vibration and visual acuity. In W. Tempest (ed.) *Infrasound and Low Frequency Vibration.* Copyright by Academic Press Inc. (London) Ltd)

that below 6 Hz vibrating the display produced higher visual impairment than when the subject was vibrated. Above 14 Hz, however, the converse was found to be the case. This implies, therefore, that above approximately 14 Hz the decrement in performance is primarily subject-based (probably resulting from eyeball resonance), whereas below this level any decrement is likely to be display-based.

The levels of vibration likely to cause decrements in visual performance are shown in Figure 10.7 (from Griffin, 1976).

Combined object and observer vibration

In many situations, for example in transport, the vibration will be applied to both the observer and to the object which he is required to view. In such cases if the eyes remain in a fixed position in the head and the vibration frequency is low enough, the motions of the object would be accompanied by similar motions of the man. In this case the image of the vehicle interior, for example, would maintain a fixed position on the retina.

Unfortunately, however, vibrations are not always of low frequency. Because of different resonance chararcteristics of the man and parts of his machine there are differences in both the amplitude, phase and possibly direction of the observer and object motions which cause complications. For example, even if the observer and object are vibrating with the same frequency and amplitude, if they are vibrating out of phase (that is, if one 'goes up' while the other 'goes down') their relative amplitude will be twice that of the individual amplitude.

Because of these problems, very few well-controlled studies have been carried out to investigate combined object and subject vibration effects, but the few studies which are available have been reviewed by Griffin and Lewis (1978).

The Effects of Vibration on Motor Performance

Whereas the operator can usually compensate for small amounts of vibration when carrying out a visual task, decrements in motor performance are less easily overcome. Because the vibration tends physically to move the operator's limbs, possibly out of phase with the rest of his body, any compensating has to be done by tensing different muscles to steady the limb (in the same way that the seated body seeks stability). Even if the operator is able to compensate for the motion, therefore, it is likely to cause fatigue.

The task most commonly used to investigate the effects of vibration on motor performance has been some type of tracking task. The subject is given a control which operates part of a display such as a spot on an oscilloscope, and he is required either to keep the otherwise moving spot stationary or to make it follow some predetermined pattern. Clearly, this is a type of activity which is used constantly in different working situations, particularly by such operators as drivers, pilots and astronauts. It is also a task which has been consistently shown to be detrimentally affected by vibration.

For vibration stimuli below 20 Hz, there appears to be fairly good agreement that any decrease in performance is related to the amount of vibration which is transmitted through the body. For example Buckhout (1964) has shown that, for vertical vibration at 5, 7 and 11 Hz, tracking error was highly and positively correlated with the amount of vibration transmitted to the sternum. This observation is particularly important since it implies that when in vibrating conditions the presence of a backrest or armrests may reduce, rather than increase, body stability. Vibrations transmitted through the seat back to the shoulder and the chest areas are likely to play a part in reducing an operator's performance on a motor tracking task. This effect will be accentuated if the operator is using a safety harness or seat belt, which tends to force the body back into the seat and into the backrest (see, for example, Lovesey, 1975).

With regard to specific intensities there is good agreement amongst investigators that, for a given vibration spectrum performance is progressively degraded as the level of vibration is increased above a threshold. The threshold level, however, is extremely variable and depends, amongst other factors, on the type of task (primarily its difficulty) and on the operator's motivation and workload. These factors affect his ability to compensate for the effects of the vibration on his limb control.

The frequencies which affect motor performance lie between 4 and 5 Hz, with decrements in performance becoming progressively smaller as the frequency becomes higher or lower (see, for example, Huddleston, 1970).

In addition to the frequency, the direction of the vibration stimulus will also determine the direction of the control task which is affected. For example, although vertical vibration has clear effects on a vertical tracking task, it has little effect if the operator is required to track an object in the horizontal axis. This has obvious implications for the design of a control system in a vibrating environment.

Finally, it could be questioned whether these decrements in tracking performance under vibration could not be due, in some small part, to the reduced visual perfomance as already discussed. After all, the value of visual feedback to tracking performance is obvious. Lewis and Griffin (1978), however, discount this suggestion. They argue that the visual performance decrements due to vibration will have little effect since the tracking displays which are used do not generally present such a finely detailed task as some of the reading tasks used to measure the visual performance effects. They also suggest that vision is not the only feedback loop involved in performing a tracking task and that other neuromuscular control loops, including proprioceptive feedback, may also be affected by vibration.

The Effects of Vibration on Cognitive Performance

A number of studies have been carried out to investigate the effects of vibration on the information-processing capabilities of the operator, in which the tasks used have mainly been either the operator's speed of reaction (reaction time), or his ability to do some complex reasoning task. Most studies, however, have demonstrated either no effect at all or no consistent effect due to vibration (Grether, 1971). Any reduction in performance which might occur was shown by Shoenberger (1974) to be due to difficulties in perceiving the stimulus due to visual interference by the vibration.

Despite this, Poulton (1978) has argued that, given the correct conditions, vibration can positively affect vigilance. He suggests that the reason why experiments have shown no effects of vibration on central processing is that in a laboratory the type of environment and the tasks which a subject has to carry out are likely to maintain arousal or alertness. In real life, however, this is not necessarily the case and the additional stimulation which the vibration provides might be enough to raise the operator's level of alertness when doing tasks which involve a lot of monitoring but little movement. This could happen despite the accompanying detrimental effects of the vibration on vision and motor control. He supports his argument with the results of experiments by Shoenberger (1967) and by Wilkinson and Gray (1975) which demonstrate that vigilance improved with vertical vibration at 5 Hz, but deteriorated reliably at 7 Hz and 11 Hz.

Combining the evidence from all studies, therefore, it would appear that vibration has detrimental effects on most peripheral and sensory processes (for example, vision or motor control). It has little effect, however, on central information processing other than perhaps on vigilance.

The Effect of Vibration on Comfort

The concept of comfort has already been discussed, but two points should be re-emphasized: first, comfort is extremely difficult to define, and second, discomfort may both affect and be affected by performance decrements. However a detailed review of the relationship between vibration and comfort will not be considered here, except to point out that Oborne (1978a) suggests that levels of vibration above about 0.06–0.09 g at frequencies between about 4 and 20 Hz will be judged, by transport passengers, to cause discomfort.

One aspect of vibration which certainly causes discomfort and performance loss, however, is motion sickness. This is the distressing effect experienced by a large proportion of travellers, caused by the low frequency, high amplitude vibration conditions in rough seas or air turbulence. The most commonly reported symptom is nausea, often leading to vomiting. For many the act of vomiting leads to a rapid recovery of wellbeing, but in some people the nausea persists even after vomiting—leading to further bouts of vomiting and retching, and a continued decline in wellbeing.

The part of the body which triggers the sensation of motion sickness very clearly lies in the vestibular system in the ear, since an intact system is needed before the symptoms can occur. However Reason (see, for example, Reason, 1978; Reason and Brand, 1975) has suggested that its cause also lies in the fact that in a moving environment discrepant information is received by the brain from either the eyes and the vestibular apparatus (visual–inertial conflict) on the one hand, or from the semicircular canals and the otoliths within the vestibular apparatus (canal–otolith conflicts) on the other hand (see Chapter 2). The suggestion, therefore, is that the sensations received from the body's position sensors (the vestibular apparatus providing balance information, and the eyes providing visual information) are at variance not only with one another, but also—and this is the crucial factor—with what is expected on the basis of previous dealings with the environment, that is, from past experience.

Reason (1978) makes various practical suggestions to alleviate such conflicts, all of which attempt to reduce the discrepancies of the incoming information:

... canal–otolith conflicts can be minimised by keeping the head as still as possible, or by adopting a supine position where that is possible. In a sickness-provoking land vehicle, such as a car or a coach, visual–inertial conflicts can be minimised by maintaining a clear view of the road ahead. Aboard ship, visual–inertial conflicts can be reduced by fixating on the horizon or upon any visible landfall. Almost all other directions of looking will result in visual information that is discordant with the prevailing inertial inputs. Where a visual–inertial conflict is inevitable, as in a backwards or sideways facing seat, or below decks, or in any totally enclosed vehicle, it often helps to close the eyes. Should this prove impracticable, it is likely that some benefit can be gained from wearing an optical device that occludes the peripheral visual field leaving the central area unobscured. Even thick spectacle frames may confer some advantage. (This material appeared in Vol. 9, pp. 163–167, of *Applied Ergonomics*, published by IPC Science and Technology Press Ltd., Guildford, Surrey, UK)

Another way of alleviating motion sickness is to use drugs, but before an appropriate drug can be prescribed the true site of the cause of motion sickness needs to be determined. In this respect, controversy has raged for a number of years over whether the site of the important vestibular sensations lie in the otoliths (which sense linear motion) or in the semicircular canals (which sense angular acceleration). A great deal of this controversy has been caused by the difficulty of delivering a stimulus to one part of the vestibular apparatus (for example, the semicircular canals by rotation) without also stimulating the other part (for example, the otoliths by centrifugal force). Recent evidence using high amplitude lifts, however, tends to suggest that the otoliths represent the prime motion sickness centres.

With this information it should be possible to prescribe effective antimotion sickness drugs, and Reason (1978) suggests a number drugs each having different qualities:

British investigators favour l-hyoscine hydrobromide (trade name 'Kwells'), while Americans tend to recommend antihistamines. There is evidence, however, to indicate that both classes of drugs are effective but in different circumstances. Where the traveller seeks protection against a short but relatively severe exposure such as a Channel crossing in a ship or hovercraft, then hyoscine given orally in doses of 0.3 to 0.6 mg one hour before departure is recommended. But for longer journeys in which the motion is likely to be less provocative, the antihistamines are more satisfactory since they are longer acting and have less severe side effects. Cyclizine 50 mg ('Marzine', 'Valoid'), Dimenhydrinate 50–100 mg ('Dramamine', 'Gravol'), Promethazine 25 mg ('Phenergan') and

meclozine 50 mg ('Ancolan') will provide useful protection against long duration motion where repeated medication is needed.
(This material appeared in Vol. 9, pp. 163–167, of *Applied Ergonomics*, published by IPC Science and Technology Press Ltd., Guildford, Surrey, UK)

A Vibration Standard

Since 1964, attempts have been made to combine all the preceding data on the effects of vibration on health, performance and comfort, to compose an accepted standard for human exposure to whole-body vibration. Such a standard was finally produced in 1974—ISO 2631: *Guide for the Evaluation of Human Exposure to Whole Body Vibration*.

The guide lays down limits of exposure to both vertical and lateral vibration under three criteria: the preservation of health (exposure limit), working efficiency (fatigue-decreased proficiency boundary, FDP) and comfort (reduced comfort boundary, RCB). As an innovation the levels for each criterion, within the frequency range 1–80 Hz, are defined in terms of the maximum time for which an operator should be exposed to the vibration–from 1 minute to 8 hours. For any exposure time, exposure limit $= 2 \times$ FDP, and RCB $=$ FDP$/3.15$. Figure 10.8 illustrates some of the levels produced for vertical vibration.

Before commenting on the standard, it should be remembered that the limits are not the result of an experimental investigation specifically designed to produce a standard. They have been produced by a Technical Committee set up by the International Standards Organization (TC 108) to review previous work. This leads to a number of confusions, not the least being which experimental data was used to produce the standard, and how much weight was put on the evidence of any one experiment. As Oborne (1976) comments on the final draft (ISO/DIS 2631): 'It is not clear from which sources the ISO working party drew their information. The reference list at the end of the document covers most of the major experimental and review papers published prior to 1972 and does not distinguish between those of general interest and those used in producing the standard.'

The major innovation of the standard is to relate vibration to exposure time. Thus an operator who is likely to be working in an environment for 8 hours in which, for example, 20 Hz vibration predominates, should only receive an average of $0.08 g$ vibration before his performance deteriorates. The same man experiencing the same frequency vibration, however, but for a short, 16-minute period would be able to accept about $0.65 g$.

Unfortunately, however, like the rest of the standard, the experimental evidence for this time dependency is sadly lacking. Indeed Allen (1975) points to recent evidence indicating that for casual exposures any performance decrement which is due to vibration does not get worse with time—at least for exposures up to 3 hours (see for example, Wilkinson and Gray, 1975).

How much trust can be laid on many aspects of the guide, therefore, is open to debate. As a guide the ISO document provides a useful starting point from

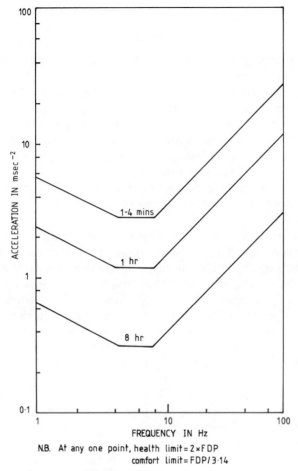

Figure 10.8   ISO vertical vibration exposure limits for
the preservation of performance (fatigue decreased
proficiency, FDP)

which designers may begin to decide appropriate vibration levels, but these
decisions will be more valid if its restrictions are understood. Thus the document
refers, primarily, to whole-body vibration applied to seated or standing man.
It provisionally applies to recumbent or reclining man but not to local (for
example, hand–arm) vibration. Further, it covers only people in 'normal health',
and most of the evidence for the limits appears to be based on average laboratory
data obtained from fit young men. (Jones and Saunders, 1972, for example,
have demonstrated some significant differences between men and women's
reactions to higher frequency (greater than 30 Hz) vibrations stimuli). Finally,
and this restriction applies to all composite curves, different individuals react
differently to vibration, and curves such as the ISO standards which depict
average reactions, may mask these differences (see, for example, Oborne and
Humphreys, 1976).

NOISE

Definitions

Noise is an aspect of the working environment which has received much attention for many decades. Indeed audiometric surveys were being carried out before the effects of vibration on man began to be considered. This is probably because the sites of the body which receive the noise stimuli (the ears) are obvious and are able to be stimulated directly, as long as adequate earphones are used. Furthermore the equipment used to produce acoustic stimuli is fairly easy to obtain. Vibration reception, on the other hand, occurs over the whole body making it more difficult for the experimenter to be sure that the stimulus is presented to the appropiate receptors. Furthermore, vibration researchers also need some fairly sophisticated equipment to produce their stimuli accurately and safely.

Noise is conveniently and frequently defined as 'unwanted sound', a definition which in its looseness enables a sound source to be considered as 'noise' or 'not noise' solely on the basis of the listener's reaction to it. Unlike the concept of noise in electronics, for example, no specific characteristic of the noise source (frequency, groups of frequencies or intensity) is involved. Acoustic noise is simply sound which is unwanted by the listener—presumably because it is unpleasant or bothersome, it interferes with the perception of wanted sound, or it is physiologically harmful. Furthermore this definition of noise implies that sounds which are labelled as being 'noise' by an individual on one occasion may not be so labelled on other occasions or in a different environment.

Because sound is vibration which is normally experienced through the air, the parameters of a single tone are precisely those of a single vibration stimulus—frequency and intensity.

In terms of the frequency, for the human listener sound is defined as acoustic energy between 2 and 20 000 Hz (2 Hz to 20 KHz) (Kryter, 1970), the typical frequency limits of the ear. The ear is still able to separate the wave changes of the air below about 16 Hz, but the sensations are perceived as 'beats', whereas above about 16 Hz, the beats begin to fuse to produce a tone-like quality.

Noise below about 16 Hz is normally described as infrasound—a stimulus which is gradually receiving more attention. This can be produced by any pulsating or throbbing piece of equipment normally encountered in the workplace. Leventhall and Kyriakides (1976), for example, have measured significant levels of infrasound in transport, close to blast furnaces and diesel engines in industry, and from ventilation systems in offices.

The effects of the levels of infrasound encountered in normal working conditions are not clear. At high intensities, such as those produced near rockets and extremely large jet engines, detrimental performance effects have certainly been described. However, the levels of infrasound measured in industry and in transport, for example, do not reach these high intensities and the evidence of reduced performance caused by levels of infrasound less than about 120 db is very weak. This has led authors such as Harris, Sommer, and Johnson (1976) to conclude: 'Regardless of whether performance, nystagmus (loss of balance), or subjective measures are considered, it seems certain that the adverse effects of infrasound reported at low-intensity levels either do not exist or have been exaggerated'.

Whereas the basic intensity measure of a vibration stimulus is its acceleration, the intensity of a pure tone is defined in terms of pressure changes associated with the compression and refraction in the air caused by the sound source. The description of sound intensity, therefore, is the sound pressure level (SPL) and is measured in the logarithmic units of decibels (db).

Because the decibel scale is logarithmic in nature a simple, linear relationship between decibel level and sound intensity does not exist. Thus 100 db is not twice as intense as 50 db. Being logarithmic, a 10–fold increase in power (intensity) occurs with each 10 db increase; 50 db, therefore, represents a 100 000 times intensity increase (10^5) whereas 100 db represents 10 000 000 000 (10^{10}) increase (thus 100 db is 100 000 times as intense as 50 db). The starting point

Table 10.1 Some commonly encountered noise levels

Sound 'power' increase	Noise level in db A	
100 000 000 000 000	140	
		'Threshold' of pain
10 000 000 000 000	130	
		Pneumatic chipper
100 000 0000 000	120	Loud automobile horn (from 1 metre)
100 000 000 000	110	
10 000 000 000	100	
		Inside subway train
1000 000 000	90	
100 000 000	80	
		Average traffic on street corner
10 000 000	70	
		Conversational speech
1000 000	60	
		Typical business office
100 000	50	
		Living room
10 000	40	
		Library
1000	30	
		Bedroom at night
100	20	
		Broadcasting studio
10	10	
1	0	Threshold of hearing

of the logarithmic scales (0 db) is arbitrarily set to be at a sound pressure level of 0.0002 dynes/square cm. For this reason a negative decibel value indicates an extremely quiet sound, with a SPL of less than 0.0002 dynes/square cm. Examples of different sound sources on a decibel scale are shown in Table 10.1.

Just as the ear converts acoustic energy into nerve impulses to be decoded and 'measured' by the auditory cortex, electrical sound measuring machines, sound level meters, also convert the sound received from the microphone into electrical energy which is then measured. The ear, however, does not respond equally to all frequencies. Just as the human body is differentially sensitive to different vibration stimuli, the ear is more sensitive to some frequencies than to others. Unless modified, however, the microphone and amplifier would treat all frequencies in the same way and so, for this reason, weighting networks are often included in sound level meters—the three internationally recongnized scales being the A, B, and C scale.

As can be seen from Figure 10.9, the A scale tends to attenuate the lower frequencies more than do the B and C scales respectively. These weightings have been produced after measuring the response of the human ear under normal (40 db), moderate (70 db) and intense noise conditions (the A, B, and C scales respectively) and thus would be used in these conditions. Because they affect any readings which are made, however, the scale used should always be quoted in any noise level report, the normal convention being to describe the sound pressure level in either db(A), db(B) or db(C) terms.

The Effects of Noise on Health

Perhaps the most obvious effect of continuous exposure to intense noise is damage to hearing which results in deafness. As will be discussed later, however, this is not the sole effect of noise exposure.

A reduction in the ability to hear can have two causes. The first, conduction deafness, results from the airborne vibration being unable to make the eardrum

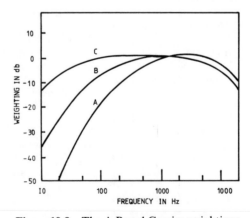

Figure 10.9 The A B and C noise weightings

vibrate adequately, and it might be caused by such factors as wax in the ear canal, infection or a scarred eardrum. Although conduction deafness does not occur as a result of a noisy environment, it still has consequences for the individual's social life and for his safety.

The second type of deafness, nerve deafness, is due to a reduction in sensitivity of the nerve cells in the inner ear. This is likely to be caused by noise, with the operator's hearing loss occurring at or near to the frequency range of the environmental noise which is experienced. However it can also occur as a result of the normal process of ageing: the normal threshold of noise detectability declines more rapidly at higher frequencies than at the lower frequencies. This ageing factor is termed presbycusis and, because it does not occur consistently in all individuals, it complicates the task of determining the extent to which long-term noise exposure contributes to deafness overall.

Kryter (1970) has suggested that noise-induced deafness is a significant health problem in most modern countries. It is an insidious complaint since an operator whose hearing is being damaged (in other words, who is becoming deaf) is unlikely to demonstrate any decrease in performance as a result of the deafness when he is in a noisy environment. Without continuous, objective audiometric testing the gradual hearing loss which may be caused by noise will not become apparent.

Noise-induced deafness can be temporary (up to 16 hours) or permanent, and these effects are commonly described as temporary threshold shift (TTS) or permanent threshold shift (PTS) respectively. Although TTS is not damaging to health, Kryter (1970) considers that the two types of deafness have so many factors in common that it is likely that the continual experience of TTS will lead eventually to PTS. Perhaps the most important of these similarities is that the same areas of the ear are affected in workers suffering from both TTS and PTS.

Variables which affect susceptibility to hearing loss

Duration of exposure

Since becoming progressively deaf is a relatively long process, as he is continually exposed to a noise stimulus the operator will also become progressively older. This brings all of age's attendant problems of presbycusis. Any assessment of the effects of long-term noise exposure, therefore, will be complicated by the need to account for the normal ageing effects.

As was disussed earlier, the symptoms of presbycusis are a more rapid decline in the hearing threshold of high frequencies than of low. In addition it appears to occur more severely for men than for women. Whether or not it is due to physiological ageing (perhaps a hardening of the eardrum or a degradation of the auditory nerves), or to wear and tear on the auditory system by the intense noises or sounds of every day living, is an open question. However Corso (1963) has shown that presbycusis occurs even in subjects from a

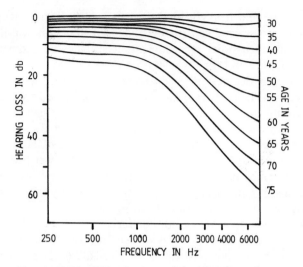

Figure 10.10 The amount of hearing loss due to presbycusis with increasing age (male subjects) (from Spoor, 1967, reproduced by permission of the International Society of Andiology)

non-industrial, relatively noise-free environment. These variations in hearing threshold at differing ages are illustrated in Figure 10.10 (from Spoor, 1967).

Nixon and Glorig (1961) have considered in detail the relationship between presbycusis and hearing loss. Their study involved measuring the hearing threshold (at 2 and 4 kHz) of a number of workers in three different industries. Working on the assumption that any change in hearing threshold is due to an additive effect of presbycusis and PTS, they compared the threshold data from workers in these three industries with a sample taken from 'non-noise-exposed' individuals. Using this information Nixon and Glorig were able to demonstrate that a significant increase in hearing loss with exposure was produced in only one of the industries, that is the industry in which the noise level exceeded 94 db(A). Furthermore, only workers with more than 6 years' exposure to the noise in this industry had significant hearing losses at 2 kHz. Other authors have also demonstrated similar effects in other industries (see, for example, Baughn, 1966).

Intensity of exposure

In addition to duration, the noise intensity is an obvious variable to affect the extent to which an operator may become deaf (PTS). As Nixon and Glorig's results show, for example, workers in the industry which had an average noise level greater than 94 db(A) were liable to PTS whereas the incidence of hearing loss was less in the other industries with lower noise levels.

Because high noise levels can damage hearing permanently, and lower levels can temporarily interfere with hearing, most modern industrialized countries

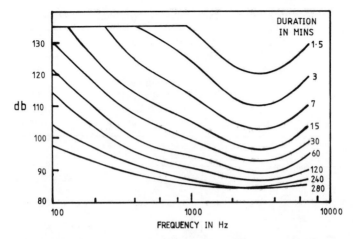

Figure 10.11   Damage risk contours for continuous noise (from
Kryter *et al.*, 1966, reproduced by permission of the *Journal of the
Acoustical Society of America*)

have produced legally enforcable maximum noise levels for workers. Such levels
are often derived from 'damage risk contours' and Figure 10.11 illustrates one
such set of contours produced by Kryter *et al.* (1966). Like the ISO standard
for exposure to whole-body vibration, the damage risk contours also relate
exposure risk to exposure duration.

Type of noise

In addition to being defined in terms of the total length of time for which the
operator is exposed to the noise, the duration of exposure can also be regarded
in relation to the intermittency of the stimulation, in other words, whether the
noise is continuous or whether the continuity is broken either by bursts of louder
noise or periods of quiet.

   Although most of the investigations of hearing loss have been carried out
using continuous noise, the effects of intermittent noise on hearing are gradually
but slowly being documented. The few results available appear to suggest that
the important characteristic of the stimulus, at least for recovery from
TTS, is the average energy level of the noise to which the operator is exposed.
Johnson, Nixon, and Stephenson (1976), for example, exposed subjects to
different types of noise for 24 hours, as shown in Table 10.2 (pink noise is a
special type of random noise which is composed of frequencies whithin a
specified bandwidth and with constant energy levels). It is important to note
that the levels of intermittent noises were set to produce equivalent energies to
the 85 db(A) continuous noise. The authors measured their subjects' hearing
levels at different times during the 24-hour exposure period and for 24 hours after
(recovery). Their results are shown in Figure 10.12, from which it is clear that
the continuous noise produced slightly more TTS than did the intermittent

Table 10.2 Types of noise used by Johnson *et al.*, (1976)

Condition	Noise	Time pattern	Exposure level
A	Pink	continuous	85 db(A)
B	Pink	3 min on/6 min off	90 db(A)
C	Pink	20 sec on/40 sec off	90 db(A)
D	Pink	3 min on/87 min off	100 db(A)
E	Pink	20 sec on/580 sec off	100 db(A)

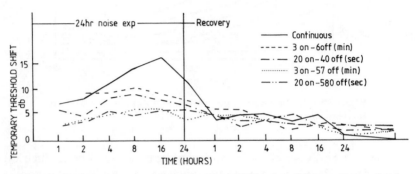

Figure 10.12 Growth of and recovery from TTS using different types of noise (from Johnson, Nixon, and Stephenson 1976, reproduced by permission of the Aerospace Medical Association)

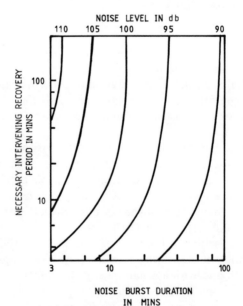

Figure 10.13 Damage risk contours for intermittent noise containing frequencies 1200–2400 Hz (from Kryter *et al.*, 1966, reproduced by permission of the *Journal of the Acoustical Society of America*)

noises. However the trends in recovery for each type of noise were very similar. Kryter *et al.* (1966) discuss in more detail the damaging effects of interrupted noise, and have produced a series of damage risk contours for such noises (see Figure 10.13).

Intermittent noise of another type can occur as sudden large 'bangs' or impulses arising from any percussion type of machine such as a gun, steam hammer or road hammer. Impulsive sound is characterized by an initial surge of high energy which reaches its peak extremely quickly (a near instantaneous rise time), but which decays gradually over about 1 msec. The damaging effects of the stimulus, therefore, are the intensity of the initial surge and the decay duration.

Coles *et al.* (1968) have reviewed many of the studies concerning the TTS and PTS resulting from such noises. On the basis of these studies they produced a damage risk contour for impulsive sound (shown in Figure 10.14) which relates the maximum sound pressure level acceptable to the ear (in other words, the noise from both the machine itself and reflected from the surroundings) to the impulse duration. They point out that such criteria will protect 75 per cent of the population at risk; to protect 90 per cent the contour should be reduced by 5 db at each duration. They also emphasize that the criterion is based on repetition rates in the order of 6 to 30 impulses per minute, 'with the total number of impulses limited to around 100 per exposure'.

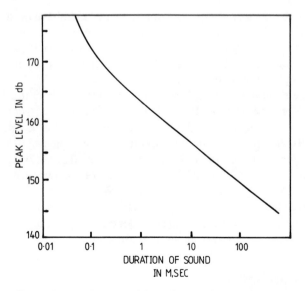

Figure 10.14 Damage risk contour for impulse sounds having near instantaneous rise times (from Coles *et al.*, 1968, reproduced by permission of the *Journal of the Acoustical Society of America*)

The Effects of Noise on Performance

Communication

Effective verbal communication depends on both the ability of the speaker to produce the correct speech sounds, and on the ability of the listener to receive and to decode these sounds. A noisy environment is likely to interfere with this last stage in the speech transmission, due to an effect which is described as 'masking'.

The American Standards Association (1960) defines auditory masking as 'the process by which the threshold of audibility for one sound is raised by the presence of another sound'. Deatherage and Evans (1969), however, have redefined the concept as 'The process by which the detectability of one sound, the signal, is impaired by the presence of another sound, the masker'. This reformulation, then, recognizes the fact that masking occurs at signal intensities which are higher than threshold (minimum audible) levels.

Since the effect of masking is to impair the perception of a signal (in the present context the 'signal' is either speech or the output of an auditory display), it is important to realize under what conditions masking occurs so that its effects can be reduced. Unfortunately, however, masking is dependent on almost any aspect of the signal and of the masker (noise) that could be considered: their respective intensities, frequencies, phases, durations, etc. The first three of these will now be considered in more detail.

Intensity relationships

The effects of the masker intensity on auditory masking have been investigated by a number of experimenters, and the results are fairly consistent: more masking occurs as the masker (noise) intensity increases when the signal intensity remains constant; less masking occurs as the signal intensity increases if the masker intensity remains constant.

Perhaps the earliest experimenters to demonstrate this effect in any consistent way were Wegel and Lane in 1924. They asked subjects to listen to different signal tones which were masked with a 1200 Hz tone which was set at 160, 1000, and 10 000 times 'its minimum audible value'. Their classic results are shown in Figure 10.15, from which it can be seen that masking (defined as the difference between the threshold of the signal on its own and the threshold of the signal in the presence of the noise) increased as the noise intensity increased. Fletcher and Munson (1937) later showed that the level of this masking increases in a nearly linear fashion with masker intensity (the relationship became less linear at lower masker intensities).

Frequency relationships

Wegel and Lane's graphs in Figure 10.15 also demonstrate the effect of varying the frequency relationships of the signal and the masker, showing that masking

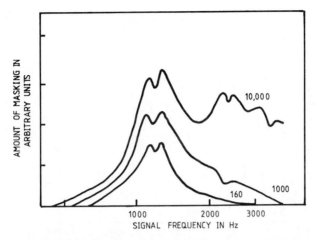

Figure 10.15 The amount of masking of signal frequencies by a 1200 Hz masking tone at 160, 1000, and 10 000 times its 'minimum audible value' (from Wegel and Lane, 1924, reproduced by permission of *Physics Review*)

increases as the frequency of the signal approaches that of the masker. The 'dips' in the curves are due to 'beats' which occurred when the signal and masker were at the same frequency (1200 Hz). These fluctuations in the intensity of the tone which is heard, caused by the interaction of the signal and the masker, have the effect of making the signal become more apparent and thus less easily masked.

Masking increases, therefore, as the masker frequency approaches that of the signal. Fletcher (1940), however, took this observation further when he produced his concept of the 'critical band' (of masking). This arose from experimental results which indicated that a pure tone is really masked only by a certain narrow band of noises which are centred at the frequency of the tone. Frequency components outside of this critical band are relatively ineffective in their masking ability. Fletcher also demonstrated that the breadth of this critical band increases as the central frequency of the masker (noise) increases above 1000 Hz (Figure 10.16).

These data, therefore, have practical implications since they might imply that it is important to know the frequency, in addition to the intensity, characteristics of the ambient noise. If the noise is composed of frequencies which are outside the critical band needed to mask, for example, a tone produced by an auditory display, then interference will not occur. If the noise frequencies lie inside the critical band, however, performance decrements due to masking will become increasingly important.

Although the notion of a critical band is adequate for most circumstances, if the noise intensity is 'sufficiently' high it can mask tones which lie outside its critical band. In Wegel and Lane's results shown in Figure 10.15, for example, signal tones as high as 3000+ Hz were able to be masked. Bilger and Hirsh

208

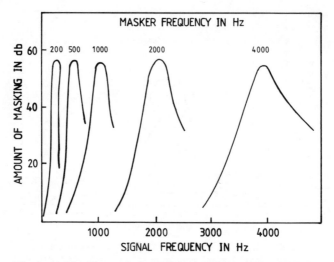

Figure 10.16 The variation of the width of the critical band with increasing masker frequency (from Fletcher, 1940, reproduced by permission of *Review* of *Modern Physics*)

(1956) describe these effects as 'remote masking', and suggest that they occur as a result of the high noise intensity 'scattering' acoustic energy in the cochlea causing distortion, the products of which may act as masking stimuli.

Phase relationships

A pure tone is produced by the action of an earphone diaphragm (A) vibrating sinusoidally. If a second diaphragm (B) is made to vibrate with the same frequency but when A is at its peak position B is at its lowest position, then the two diaphragms are said to be out of phase. They are in phase when both move in the same direction at the same time. Different phase relationships occur between these two extremes of 0 degrees (in phase) to 180 degrees (out of phase).

During the Second World War, workers at the Harvard psycho-acoustic laboratory conducted many experiments on masking—particularly the masking of speech by noise. It was discovered that, when the noise and the signal (speech) are presented to both ears, a substantial improvement in hearing could occur if either was reversed in phase at one ear relative to the other ear.

The initial research into these phase relations and their effect on masking was first described by Licklider in 1948, who considered the six conditions shown in Table 10.3. Thus Licklider describes the case where either the noise or the signal is reversed in phase between the two ears as antiphasic; the case where neither is reversed or where both are reversed as homophasic; and where the masker is random in nature as heterophasic. Using speech as his signal, Licklider showed that intelligibility was highest with the antiphasic relations

Table 10.3   The classes of phase relationships used by Licklider (1948)

Arrangement	Signal	Masker	Class
1	In phase	In phase	Homophasic
2	In phase	Out of phase	Antiphasic
3	In phase	Random	Heterophasic
4	Out of phase	In phase	Homophasic
5	Out of phase	Out of phase	Antiphasic
6	Out of phase	Random	Heterophasic

and lowest with the homophasic arrangements—with the heterophasic conditions allowing intermediate intelligibility.

The reason for such differential phase effects probably lies in the fact that as a tone changes in phase between the two ears, the position of the resultant tone appears to 'rotate' around the head. If both the signal and masker are either in or out of phase, therefore, the two would appear to be in the same position in space. As one moves out of phase between the ears, however, it appears to 'separate' from the other and thus becomes more discriminable. This is simply a further example of the general observation that masking decreases as the signal and masker become less 'similar'.

Implications for communication

All the results described above have clear implications for speech communication. For example, if the environmental noise has a similar bandwidth to that of normal speech, then masking will occur.

Kryter (1970) indicates that speech (for both males and females) tends to predominate at around 400–500 Hz, but includes frequencies up to 5000 Hz. Clearly, therefore, significant environmental noise with these frequencies should be avoided. However, Licklider's data suggest that if the speech can be presented to the observer through earphones, adjusted so that one diaphragm is out of phase with the other, then masking may be reduced.

The normal behavioural response to increased ambient noise levels will be for the speaker to raise his voice. However as Kryter (1970) points out, the increase in effort is not sufficient to completely override the increases in noise level. Indeed Korn (1954) showed that speech intensity needs to be increased by about $3\frac{1}{2}$ db for every 10 db increase in room noise.

If the signal is masked by noise which is interrupted by periods of quiet, then the degree of masking changes in quite a complex manner which is determined by the rate of interruption. Miller and Licklider (1950), for example, explain the relationship as follows:

At interruptions of less than 2/sec, whole words or syllables within a word tend to be masked; at interruption rates of between 2 and 30/sec, noise detection is so brief that the listener is able to hear a portion of each syllable or phoneme of the speech signal, thereby tending to reduce the amount of masking; when the interruption rate is more frequent than 30/sec the spread of masking in time around the moment of occurrence of a burst of noise results in increased masking until by 100 interruptions per sec there is, effectively, continuous masking.

(Reprinted from the *Journal of the Acoustical Society of America*, with permission from the publishers.)

Cognitive performance

There is presently a great deal of controversy raging over the question of whether environmental noise affects anything other than auditory-based performance. Basing his conclusions on work carried out at the Harvard psycho-acoustic laboratory during the Second World War, for example, Stevens (1972) insisted that noise has no direct detrimental effects on man, apart from producing deafness and annoyance. After including masking as an effect, this suggestion has also been supported by both Kryter (1970) and Poulton (1977).

Other authors, however, most notably Broadbent, claim that continuous broad band noise at levels greater than 100 db(A) has a detrimental effect on work which is distinct from the effects of noise due to the masking of any auditory feedback cues which the operator is able to derive from his task.

The controversy was sparked off by experiments similar to those reported by Broadbent (1954). In his experiment (now known as the '20 dials test') subjects were asked to monitor 20 steam pressure gauges for $1\frac{1}{2}$ hours; if they saw any of the pointers reading above a danger mark they were required to turn a knob below the gauge to bring the pointer back to the mark. This was carried out under two conditions: 'noise' (100 db) and 'quiet' (70 db). His results indicated that the subject's performance was impaired in the noisy relative to the quiet condition.

Other experimenters have also indicated that high levels (greater than 100 db) of noise can have an effect on monotonous vigilance tasks over relatively long periods of time. Jerison (1959), for example, demonstrated that environmental noise at 114 db(A) produced significantly more errors on both a 'clock-watching' and a 'complex mental counting' task after at least 2 hours work.

Poulton (1976), however, claims that many of these effects can be placed at the door of equipment deficiencies. For example the equipment used by Broadbent contained microswitches mounted directly behind the knob which recorded the subject's response. He argues that this feedback cue could have helped the subjects enough in the quiet condition to be able to produce a faster and a more accurate response than if they had not been present. The 'noise' conditions, however, probably masked these cues. In his 1977 review, Poulton lists 32 experiments indicating a performance decrease in the presence of noise; in each case he suggests aspects of the equipment which could, if masked, have reduced the number or quality of cues given to the operator.

It is questionable, however, whether acoustic masking can account for all of the measured performance decrements. For example in his 'quiet' condition Jerrison set the level of environmental noise to 'mask the sounds of equipment' and it ranged from 77.5 to 83 db(A). These levels certainly should have overcome Poulton's (1977) objections to this experiment—that 'masking of the tap of the spring-loaded switch, which indicates that it has been depressed far enough, occurs in the "noisy" but not in the "quiet" condition'.

An extension of the 'masking' hypothesis has been proposed by Poulton (1976, 1977). He suggests that 'inner speech' is also masked by the noise: 'you can't hear yourself think in noise'. Many of the tasks which demonstrated a detrimental effect of noise had a large short-term memory component: subtracting a four-digit number from a memorized six-digit number; counting, and holding in separate cumulative totals, the number of flashes from each of three sources; searching for a series of two-digit numbers; etc. In such tasks Poulton argues that the noise masks the man's internal verbal rehearsal loop, causing him to work slower and to make more errors.

A separate hypothesis to account for some of the possibly observed detrimental effects due to noise has been advanced by Jerison (1959). He suggests that the noise affects the operator's judgement of time. While performing the 'counting' task described above, Jerison's subjects were also asked to press a key at what they considered to be 10-minute intervals. His results show that throughout the experimental period his subjects progressively contracted their internal time-scale when in the noisy condition but not in the 'quiet' one. Whereas in the first 15 minutes of noise the key was pressed after an average period of $8\frac{3}{4}$ minutes (to signal the end of a 10-minute period), after $2\frac{3}{4}$ hours '10 minutes' was contracted to about 7 minutes. It is not entirely clear, however, how such distortions in time judgement can influence all cognitive or motor work performance.

From laboratory-based studies at least, therefore, a detrimental effect of noise on cognitive performance alone has not been conclusively shown to occur. Noise clearly has an effect on overall performance, but this could be due as much to the masking of acoustic cues as to any deficiency in central cognitive processing.

The conclusion, therefore, must be that the relationship between noise and congnitive performace is similar to that between any environmental stressor and cognitive performance as discussed at the beginning of this chapter. That is, the stressor is unlikely to affect cognitive performance as long as it does not demand more mental capacity than the task leaves spare.

Support for this contention is given by Wohlwill et al. (1976) who suggest that individuals are able to cope with noise through increased concentration and effort. They make the observation that subjects in such experiments sometimes experience considerable release of tension after the experience—even to the extent of breaking down to cry. Glass and Singer (1972), for example, report an experiment in which subjects were given arithmetic problems to solve in the presence of an unpredictable burst of noise. Both physiological and performance indices indicated adaptation to the noise, but the subjects showed a lowering of resistance to frustration on a subsequent task given after exposure to the noise. Noise, therefore, can be even more insidious than at first thought.

The increased concentration and effort required to overcome its effects can produce performance decrements even after it has stopped.

The Effects of Background Music on Productivity

In many respects this topic should not be considered in the section which discusses the effect of noise on performance since noise, by definition, is unwanted sound. Background music, however, is often wanted and is enjoyed by many workers. In cases in which it is not wanted (and thus become noise), the sound pressure levels produced are not high enough to cause the direct performance decrements discussed above. The effect is more likely then to become one of annoyance as will be discussed later. The effects of background music on performance will be discussed in this section, however, since music is an acoustic stimulus which, it has been argued, may possibly affect performance. However, as with the effects of noise on cognitive performance, the evidence regarding music and productivity is also controversial.

The theoretical basis for suggesting that music might aid performance lies in the alleviation of boredom and fatigue which occurs with repetitive work. The normal stimulation which is received by the operator from his task is used not only to give information about the job, but to provide stimulation for the part of the brain, known as the reticular activating system, which determines how much attention, alertness or vigilance the operator will bring to his job. This in turn is related to the irregularity, rate of occurrence and variability of these signals, but not by their intensity or frequency. Repetitive work with little stimulation, therefore, can lead to underarousal and a loss of efficiency.

The suggestion is made, therefore, that varying, secondary stimulation might provide the stimuli needed to 'reactivate' the reticular activating system, and that background music is an obvious source of stimulation.

With music, however, other influencers of performance might also be present. Thus, music may influence not only attention and vigilance but also feelings of wellbeing and job satisfaction, and these effects could be reflected by drops in absenteeism, bad timekeeping and labour turnover. It is for this reason that controversy surrounds the claims for background music—are any performance increases due to increased arousal or are they due to increased operator happiness? At a pragmatic level, however, although this question is debatable and, given the difficulties of performing properly controlled industrial studies, unlikely ever to be fully resolved, an increase in performance remains an increase in performance.

Studies investigating the effects of background music on productivity have been reviewed by Fox (1971). Although many were poorly controlled he concluded, both from laboratory and industry-based studies, that under the right conditions music was beneficial. Thus subjects increased their performance in the laboratory while different industrial studies showed reductions in errors, poor timekeeping, staff turnover and accidents, and increases in output and production quality.

It must be pointed out, however, that some studies have shown no effect of background music on productivity. For example Gladstones (1969) compared keypunch operators' productivity with and without background music, for approximately $5\frac{1}{2}$ months. He concluded that 'background music had no significant sustained effect on either work rates or error rates in this situation'. However, the music did appear to be appreciated by the workers.

The type and presentation of the music in such situations are clearly important considerations—for example, in Gladstones' study the music periods appeared to be almost continuous during the experimental periods. Fox (1971), however, argues strongly that continuous music is *not* desirable (and this may be one reason why Gladstones failed to find an effect). He suggests that 'with continuous music playing, the music becomes part of the fixtures and loses its stimulating value. Equally obvious is that music should be timed to counteract the peaks of fatigue during the day'. The studies which are available, however, do not suggest for how long or at what times of the day the music should be played.

A further consideration is the type of music which should be played and, in this respect a series of experiments carried out by Fox and Embrey (1972) suggest that the workers themselves should be able to choose their music. In a laboratory experiment they tested six subjects on a detection task using four conditions:

1. No music,
2. Music played during the 15th to 20th minute of the test session using a programme of randomly selected music,
3. As for (2), but using a commercially prepared lively programme,
4. As for (2), but allowing the subjects to select a programme from the tapes used in (3).

Fox and Embrey's results demonstrated that average detection efficiency increased significantly from the 'no music' to the 'commercial music' condition (3). Furthermore the detection rate was higher in (4) than in (3), but the authors offer no statistical evidence to suggest that the increase is significant.

The studies suggest, therefore, that selected music does have some effect on operator morale or performance. However all of the important parameters have not yet been fully investigated.

The Effects of Noise on Annoyance

Annoyance is a common subjective response experienced by us all when we are exposed to any stimulation which we do not want. It can be caused by something said, seen, heard, smelt, etc., or any combination of these sensations. By definition noise is unwanted sound, and so in whatever form it presents itself (speech, music or random noise) it is likely to cause annoyance. The important consideration, therefore, is the degree of annoyance which is caused by a given

214

noise, but since this depends on the extent to which the sound is unwanted it is clear that what is annoying to one person may not be so to another. Thus, whereas it can be reliably predicted that a given noise will cause either temporary or permanent deafness, or will mask an important signal, with such a subjective concept as annoyance, like the concept of comfort, no such certainties exist.

Physical aspects of noise annoyance

Contrary to popular opinion the mere physical intensity of noise is not a sufficient criterion on which to predict the degree to which the noise is likely to be annoying. Two noises may have the same intensity but cause different degrees of annoyance due, perhaps, to the frequencies which they contain, their respective durations, or even their meanings to the listener.

Kryter (1970) suggests that five aspects of a noise stimulus can be identified as affecting its annoyance level: (a) the spectrum content and level; (b) the spectrum complexity; (c) the sound duration; (d) the sound rise time; and (e) the maximum level reached (for impulsive sounds).

With regard to the content of the spectrum, by asking subjects to adjust tones of different frequencies to make them equally 'noisy', Kryter and Pearsons (1963) were able to produce bands of equal noisiness over a frequency range of 40–10 000 Hz (a distinction was made between 'noisiness' and 'loudness'). These bands indicated that the higher frequencies (above about 2000 Hz) tend to be more annoying than the lower frequencies, even though they were equally loud. This relationship can be seen in Figure 10.17. As the noise frequency increases above about 1000 Hz it appears to become more 'noisy' (or, interpreting the graph literally, a lower intensity is needed to keep the noise equally noisy). Although the same relationship is apparent for equal loudness, it is not so marked.

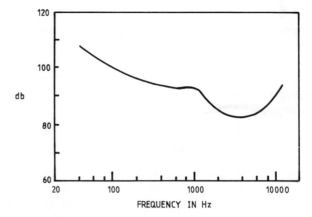

Figure 10.17 An equal noisiness contour (from Kryter and Pearsons, 1963, reproduced by permission of the *Journal of the Acoustical Society of America*)

In addition to the specific frequencies which make up the noise, Kryter also suggests that the complexity of the noise spectrum (or the ways in which the frequencies are distributed) is also an important determinant of the degree to which the noise will be judged to be 'noisy' or 'annoying'. In this respect Kryter and Pearsons (1963) again demonstrated that the higher the proportion of high frequency tones contained in the noise, the more unacceptable the noise became.

Finally, regarding the duration of the total sound experienced, Kryter and Pearsons (1963) demonstrated that, over a range of durations from $1\frac{1}{2}$ to 12 sec, for every doubling of the duration of the sound, its intensity needed to be reduced by $4\frac{1}{2}$ db(A) for it to be judged to be equally 'acceptable'.

Subjective aspects of noise annoyance

When discussing the physical aspects of noise the concept of 'noisiness' is used synonymously with 'annoyance'. In many respects, however, what is conveyed by the word 'annoyance' is more than just 'noisiness'. Annoyance commonly signifies one's reaction to sound that is based not only on perceived noisiness but also on the emotional content and novelty which the sound may have for the particular individual (both of which are excluded from the concept of noisiness). In addition, as Wilson (1963) suggests,

the annoyance may be ascribed to the 'information' which sounds may carry from the source to the recipient. The physical energy in the noise of a creaking door, a crying baby, or a distant party may be very small, and if distributed in the form of random noise probably would be quite unnoticed. But it may convey manifold suggestions of alarm, neglect, sadness, loneliness, and so in some people it has an emotional effect out of all proportion to its physical intensity.

Reviewing the many studies and surveys which have been carried out to investigate the types and levels of noise which cause disturbance, it is clear that annoyance generally occurs when the noise interferes with a person's ability to carry out some activity which he wants to carry out. Foremost in these activities is the interference with speech due to masking by environmental noise.

Nemecek and Grandjean (1973) surveyed the noise requirements of employees working in landscaped offices. Of those employees who considered that they were disturbed by noise, 46 per cent felt that the noise produced by conversations was most annoying; 25 per cent disliked office machinary noise, and 19 per cent disliked the telephones. Interestingly, however, further questioning revealed that the majority who indicated that conversation noise disturbed them most, felt that it was the content of the conversation rather than its loudness which was most disturbing.

This problem of overhearing conversations was also advanced as a cause of annoyance by Cavanaugh et al. (1962). Proposing the concept of 'speech privacy' they argued that the disturbance might be caused by the worry that if one can hear other people talking then one can also be heard by other people. They also suggest that it is the degree to which the intruding speech can be understood, rather than its loudness, which destroys the feeling of office privacy.

In addition to the direct interference of tasks by noise, its indirect consequences can also be annoying. For example Griffiths and Langdon (1968) investigated community responses to road traffic noise. In addition to causing problems such as headaches, a further annoying aspect of the noise was due to the necessity of having to keep windows shut during the summer months. In addition their respondents also complained that living next to a noisy road resulted in the value of their house being reduced. Similar points were also raised by Stockbridge and Lee (1973).

The effects of aircraft noise on schools around London Airport were considered by Crook and Langdon (1974). In addition to interfering with the lessons, the authors also reported changes in the style of teaching on the noisier days (lessons abandoned and more pauses in the flow of the teacher), increased pupil fidgeting, and a reduction in the teacher's satisfaction with the class as a whole (teachers often felt that the noise caused the whole atmosphere of the proceedings to deteriorate, that they and their pupils became irritable and tired, that they developed headaches and that their pupils became noisier and less inclined to work).

Annoyance, therefore, appears to be a subjective reaction to the inability to carry out a preferred task (for example, talking or sleeping). It can also arise as a result of changes in the listener's physiological state, perhaps causing headaches and high blood pressure. In these cases, however, the effects of noise are due not to the specific attributes of the noise but to the (dis)stress which it causes.

SUMMARY

This chapter has considered in detail the effects which two environmental parameters, vibration and noise, have on man. In both cases their effects can be threefold. At extreme intensities they can be dangerous, causing either a breakdown of body parts (in the case of vibration) or deafness (from noise). At lower intensities, however, they are more likely to affect performance, although these effects are more diverse. Thus vibration interferes primarily with the ability to control the affected body part (either a limb or the eyes). The performance effects of noise, however, are likely to occur in interference with hearing. Some evidence exists to suggest that both environmental parameters affect cognitive performance, although it is likely that they act in the manner of general stressors rather than affecting the brain directly. The final reaction to both vibration and noise is one of reduced comfort and increased annoyance. In both cases, however, although it is possible to relate the degree of annoyance or comfort reduction to physical aspect of the stimuli, the reaction is likely to be determined by the meaning of the stimulus to the individual and its effects on his ability to carry out a preferred task.

CHAPTER 11

The Physical Environment:
II Temperature and Illumination

The previous chapter looked at two aspects of the worker's environment (vibration and noise) which could play a part in affecting his health, performance and comfort. With these two environmental attributes, however, it was apparent that for any particular frequency the importance of the vibration or noise increased as its intensity increased. In other words the intensity effects fell on a single dimension of no (or relatively little) effect to increased interference. For both temperature and illumination, however, the position is slightly different. In both cases there is a narrow range of intensities under which a worker can, and should, operate. Departure from this optimum, either by increasing or by reducing the intensity, is likely to affect performance, comfort and, in extreme cases, health. For example a man's ability to see fine detail is impaired in both dark and extremely bright conditions; his ability to perform complex mental operations is affected in conditions of both extreme cold and extreme heat.

In the same way that vibration and noise were treated separately in Chapter 10, so will temperature and illumination in this chapter. Again, however, this is not to imply that they are independent factors or that temperature, for example, is more important than illumination. As was stressed in Chapter 10, the importance of any environmental factor can change with changing circumstances, and their interactive effects can be unpredictable.

TEMPERATURE

Man's response to the thermal environment depends primarily on a very complex balance between his level of heat production and heat loss. The heat which results from the normal body metabolism, particularly during work, maintains the body at a temperature well above that of the usual surrounding environment. At the same time heat is constantly being lost from the body by radiation, convection and evaporation so that under ordinary resting conditions the body temperature is maintained within its normal narrow range of between 36.1 and 37.2 degrees C (97–99 degrees F). These temperatures refer to the conditions deep in the body and not simply to the surface temperature of the skin. Whereas the skin temperatures may fluctuate over quite a wide range without serious damage to performance or health, the deep body temperature must be maintained within its narrow range of operating temperatures.

217

The body's thermal balance is maintained by a very complex self-regulating system controlled by the hypothalamus in the brain. Whenever the body needs to lose some heat this area causes the blood vessels to dilate, the sweat glands to produce cooling sweat, the respiration rate to increase and the body's metabolic rate to be lowered. Under cold conditions, however, when the body needs to conserve and even generate heat, the hypothalamus causes the blood vessels to constrict and to move the blood away from the extremities (causing a 'blue' appearance), and increases the metabolic rate by inducing the often uncontrollable muscle activity described as shivering. By these processes, therefore, the optimum body temperature is maintained through quite adverse external environmental conditions.

In some cases, however, this self-regulating system proves to be inadequate and the body either gains or loses too much heat. When this happens, depending on the amount of heat gain or loss it can lead, progressively, to reduced performance, to damage to health and, eventually, to death. Any discussion of the effects of the thermal environments, therefore, is complicated by the need to assess the effects of heat and cold separately. Thus the first part of this section deals with health and performance effects of hot conditions. In the second part these aspects are discussed in relation to cold conditions. The final part deals with less extreme departures from optimal balance and their effects on comfort.

Hot Environmental Conditions

The effects of heat on health (hyperthermia)

A worker exposed to high levels of radiant or convected heat may have his health damaged in one or both of two ways. First the elevated temperature on the skin can cause tissue damage from burning, particularly at skin temperatures of over 45 degrees C (113 degrees F). These effects, however, are immediately observable and, as long as he has no neurological diseases, under normal circumstances the pain will cause the operator to remove the exposed part of the body from danger.

The more insidious effects of an elevated body temperature occurs if the deep body temperature increases to a level of about 42 degrees C (108 degrees F) (that is, a core temperature increase of 5 degrees C (10 degrees F)) or more. When this occurs the onset of heat stroke (hyperthermia) can be very sudden with the collapse and (unless treated promptly with cooling agents) the imminent death of the individual. In some cases the loss of conciousness is preceded by a short period of general weakness or confusion and irrational behaviour.

Since hyperthermia is likely to occur if the body is unable to rid itself of the excess heat that it has generated, there would appear to be three related ways by which this could happen. First, by exposure to environmental conditions which are so humid that the body is unable to reduce its heat by evaporating sweat. A small microclimate is produced around the body, perhaps inside the protective clothing, which becomes supersaturated with water and thus impairs adequate evaporation. Second, hyperthermia can occur when the environmental conditions are too hot (but not dangerously so) and interfere with the ability of the sweat produced to cool the body. Finally, it can be caused by the insulating effects of the protective clothing. In this case the stress is the result of an impairment of evaporation due, perhaps, to the impermeability of the clothing and to its heat-retaining properties. It is imperative, therefore, that protective clothing is able to dissipate adequate amounts of both heat and water if hyperthermia is to be avoided in very hot conditions.

Since heat stress is a function of the temperature of the body core rather than the external conditions, it may still occur even if the surrounding temperature is less than the critical 42 degrees C (108 degrees F). Thus strenuous physical exercise can, by itself, cause heat stroke if the level of metabolic heat released by the effort is greater than the body's ability to rid itself of the excess. Indeed after reviewing many studies of the affliction, Shibolet, Lancaster, and Danon (1976) argue that heat stroke is most likely to strike at highly motivated young individuals who are engaged in hard work, military training and sporting endeavours. Under other circumstances these individuals would have rested when tired, taken liquid when dry, or remained at home when ill. The authors suggest, therefore, that the prevention of heat stroke requires adequate rest and liquid consumption (to compensate for the increased liquid loss during sweating) before physical exertion, as well as periods of rest during work when the individual can cool off and drink adequately. When such seemingly obvious precautions were implemented in the South African mining industry, Wyndham (1966) demonstrated a significant fall in the incidence and mortality of heat stroke.

A major factor in the likelihood of collapse from heat stroke is the duration for which the worker is exposed to the heat source. In this respect Bell, Crowder, and Walters (1971) carried out an experiment using eight fit, unacclimatized young men (18–29 years) who were asked to perform a stool-stepping task under different thermal conditions (the task being to step on and off a 22.9 cm high stool in time with a signal light flashing at a rate of 12 times per minute). Each subject was encouraged to push himself to his individual limit of exhaustion, knowing that he would be protected from hazard by the presence of two competent and experienced observers in the hot chamber with him. From their data Bell, Crowder, and Walters were able to produce a series of graphs of predicted safe exposure times for different combinations of dry and wet bulb temperature conditions; two of these curves are shown in Figure 11.1. As can be seen, the figures match very closely similar limits produced by the American Society of Heating and Refrigeration Engineers (ASHRAE) (1965) based on physiological data (the average time taken to raise the pulse rate from 75 to 125 beats per minute, and the core temperature from 37 to 38.3 degrees C (98 to 101 degrees F)), although the ASHRAE guidelines are more conservative in their levels.

Bell, Crowder, and Walters do caution against the overuse of these figures,

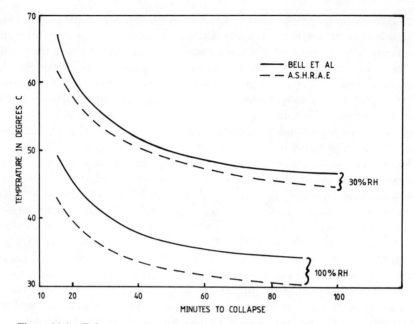

Figure 11.1 Tolerance to heat at different relative humidities, produced by Bell, Crowder, and Walters (1971) and the American Society of Heating and Refrigeration Engineers (ASHRAE) (1965)

however. Thus they point out that their data were obtained from fit young men and it would be dangerous to extrapolate the findings to other populations of workers with different work rates, ages and physical fitnesses, etc. Second, the safe exposure times relate to subjects who were required to work continuously, whereas workers who normally carry out lighter or more intermittent work could be expected to endure longer exposures. Third, their recommended limits apply to workers with very little acclimatization to heat, whereas it has been shown that acclimatized men may be able to endure the heat for increased durations. For example Wyndham *et al.* (1970) have shown that, for durations. of 400 minutes or longer, resting, acclimatized men can tolerate nearly 2 degrees C more than can unacclimatized men. However, the value of acclimatization appears to fall as the environmental conditions become more severe.

The effects of heat on performance

Some of the earliest and most comprehensive experiments of the effects of various stressors (including temperature) on performance were carried out by Mackworth over a period of 7 years at the Applied Psychology Unit at Cambridge. These were published in 1950 and his experiments on the performance effects of heat have, perhaps, formed the basis for a number of subsequent investigations. It is useful, therefore, to consider his series of studies in more detail.

Mackworth used five different performance tasks in his experiments. First the 'clock test' which tested the subject's vigilance. He was asked to watch a clockface whose hand moved around the face with a jerk every second. Occasionaly, however, after long and irregular intervals, the pointer moved through double the usual distance and the subject was asked to respond (by pressing a key) to this 'stimulus'. This task, therefore, tested the subjects' ability to sustain attention. *Vigilance Tasks* .

A second group of two tasks tested the subjects' cognitive abilities. In the first, the 'wireless telegraphy reception test', 11 experienced wireless operators were required to perform their normal job of rapidly writing down morse code messages heard over their headphones. In the second test, the 'coding test', 12 subjects were each presented with a set of small, flat, squared shapes and a formboard. Their work involved putting each shape into its corresponding hole on the board.

The final group of two tasks tested more physical work performance. First arm strength was measured in the 'pull test' in which the subject was asked to raise and lower a 15 lb weight attached to pulleys, by bending and straightening his arm in time with a metronome. This was continued until the men could not lift the weight again. In the second task, the 'heavy pursuit meter' task, the subject had to keep a pointer in a specific position on a display using both hands to operate a very heavy control. This experiment lasted for 3 hours.

Each test was carried out under different combinations of wet and dry bulb temperatures (see p. 229), and the results have been combined in Figure 11.2. Because each task produced different types of data (rate of failure; number correct; etc) the results from the five experiments have been reanalysed to show the groups' performances at each temperature relative to their worst performance, in other words, the vertical axis indicates the percentage increase in performance at different temperatures.

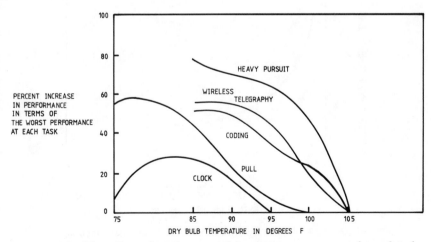

Figure 11.2 The effect of environmental temperature on a number of tasks (Adapted from Mackworth, 1950)

From Figure 11.2 it can readily be seen that performance on all tasks remains fairly constant until a dry/ wet bulb temperature of about 30/24 to 32/27 degrees (86/75 to 90/80 degrees F) is reached, after which the performance on all tasks decreases dramatically.

In addition to considering the critical temperature levels for decreased performance, Mackworth also investigated the effects of such variables as the subjects' experience and their motivation to complete the task.

As a possible moderator of the effects of the heat, Mackworth considered the experience of the subjects during both the clock test and the wireless telegraphy test. In both, his results suggest that over all conditions the increased temperature did not affect the various skilled groups differently. When the results were analysed in terms of the different hours of keeping watch on the task, however, the higher temperatures appeared to affect the unskilled operators in the final hour more than they did the skilled operators.

To test the effects of incentives, each subject who undertook the pull test completed the task twice. In one condition the subjects were given no idea of how they were performing. In the second condition, however, they knew their results and were urged and encouraged to perform on each trial 25 per cent better than previously.

As would be expected, Mackworth's results showed that the encouragement had an effect on overall performance so that the higher incentive group performed significantly better than when they did not know how they were doing. However, his data also demonstrated that the temperature conditions did not affect the two groups differently. Although the now predictable reduction

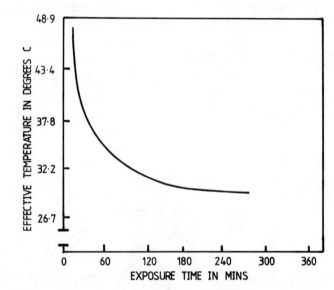

Figure 11.3 Upper tolerance limit for impaired mental performance (from Wing, 1965), reproduced by permission of the Aeorospace Medical Association)

in performance occurred after about 29 to 32 degrees C (85–90 degrees F), it occurred in both groups to the same extent.

A number of studies subsequent to those of Mackworth have been carried out and demonstrate similar trends for a variety of tasks (for example, grip strength, Clarke, Hellon, and Lind, 1958; manual dexterity, Weiner and Hutchinson, 1945; a tracking task, Azer, McNall, and Leurig, 1972; and cognitive tasks, Pepler, 1958).

Wing (1965) has combined the data from some of these studies in an attempt to indicate, in terms of the duration of exposure, the temperature levels able to be endured before cognitive performance decrements become apparent. As can be seen from Figure 11.3, Wing's upper limit for impaired performance follows a very similar course to the predicted curves of the physiological limit as illustrated in Figure 11.1.

Cold Environmental Conditions

The effects on health (hypothermia)

In a similar fashion to hyperthermia, any significant cooling of the body from its optimum core temperature is likely to produce a severe risk to health.

Clinically, a state of hypothermia can be said to exist when the body's core temperature falls to about 35 degrees C (95 degrees F). Below this temperature the risk of fatality increases, until at temperatures below 30 degrees C (86 degrees F) the imminent death of an individual is likely due to cardiac arrest. Once again, therefore, it requires the core temperature to depart only slightly from its optimum temperature (in this case 2 to 6 degrees C) for fatalities to occur.

Following exposure to cold, the body's regulation system attempts to produce heat rapidly by increased muscular activity, manifested by an increase in muscle tone and shivering. Violent shivering appears not to begin until the core temperature begins to fall but, once started, its effect can be marked causing the falling trend to be reversed and the core temperature to rise. During the shivering stage the cardiovascular system responds to the cold by constricting the peripheral blood vessels and this increases the blood pressure. If the core temperature falls further, however, the heart rate decreases due to a direct effect on the heart pacemaker. When the heat loss cannot be compensated further, the core temperature falls until at temperatures around 30–33 degrees C (86–91 degrees F) shivering gradually ceases and is replaced by muscular rigidity.

The risk of hypothermia at work is probably less than hyperthermia. Very few environments are likely to be cold enough to induce hypothermia in active, working individuals as long as dry, warm, protective clothing is used. With protective clothing, however, as was pointed out earlier, if the clothing does not dissipate the excess heat produced by the body fast enough the reverse problem, hyperthermia, may occur.

Tolerance to cold exposure, and thus susceptibility to hypothermia, varies greatly between individuals. For example Timbal, Loncle, and Boutelier (1976) immersed subjects in cold water at 30 degrees C for 15 minutes. After this period some subjects' core temperatures had dropped by 2 degrees C, whereas others had dropped by only one-tenth of this amount. These differences in susceptibility are due, primarily, to morphological factors of which the amount of subcutaneous fat around the body is probably the most important. This acts as good insulating material, particularly when the blood vessels constrict in response to the cold and move the blood away from the body surface. In addition, the size and weight of the individual are important factors in cold susceptibility. This is because the degree of heat loss is proportional to the body's surface area, and the amount of heat which can be generated (perhaps by shivering) depends on the mass of active muscular tissue in the body.

The effects of cold on performance

Cold environmental conditions which are not severe enough to induce hypothermia (or in which the worker is adequately protected) may, nevertheless, still create performance decrements.

After reviewing much of the work of the effects of environmental cold on human performance, Fox (1967) concluded that cold can effect performance in five areas: (a) tactile sensitivity, (b) manual performance, (c) tracking, (d) reaction time, and (e) 'complex behaviours'. These five behaviours, however, can be further subdivided into two main categories: motor performance and cognitive performance.

Motor performance

As far as motor performance is concerned, two factors appear to be important. First, the temperature of the limb which is being used, and second, the rate of cooling.

The limb temperature, rather than the overall body temperature, affects motor ability because of the effects which cold has on muscular control. It causes a loss of cutaneous sensitivity, changes in the characteristics of synovial fluid in the joints, and a loss of muscular strength.

It is, perhaps, obvious to point out that the important consideration in relation to limb sensitivity is not the level of environmental temperature but ultimately the temperature of the limb itself. As Fox (1967) suggests, therefore, it is more appropriate to consider the effects of factors such as hand skin, or finger skin temperatures. In this respect Morton and Provins (1960) have demonstrated a significant reduction in dexterity whenever the skin temperature fell below 20–25 degrees C (68–97 degrees F), although there was large individual variability. They suggest that the true relationship for any one person may be such that most of the change in performance is spread over only a few degrees. Thus skin sensitivity remains fairly normal for small drops in skin temperature, but deteriorates considerably after the individual's critical hand skin temperature has been reached.

Data from Clark (1961) tends to support this contention of a critical temperature. He showed that as the length of time for which his subjects were exposed to cold increased, their performance decreased as the skin temperature was lowered from 15–10.5 degrees C (51–59 degrees F). At a higher overall temperature, however (13–18 degrees C; 56–64 degrees F) this performance — exposure-duration relationship was not apparent.

With regard to the rate of cooling, this was investigated by Clark and Cohen (1960) on a knot-tying task. These experimenters found that slow cooling to a 7.2 degrees C finger temperature (45 degrees F) resulted in greater knot-tying errors than did fast cooling to the same temperature. The authors suggested that this was because the slow-cooling procedure allowed relatively lower subsurface temperatures to occur. Although the rate of cooling results could not be replicated, their data do indicate significant decrements in performance on a number of dexterity tasks when the temperature was reduced from 18–9 degrees C (65–48 degrees F).

To investigate the effects of long-term exposure to cold conditions on manual performance, Teichener and Kobrick (1955) asked their subjects to perform a tracking task for approximately 5 minutes each day over a 41-day period, during which time they lived in a constant temperature chamber. For the first 16 days the room temperature was maintained at 24 degrees C (65 degrees F); during the next, cold, period lasting 12 days the temperature was held at 13 degrees C (55 degrees F); and finally, during the 13-day recovery period, the temperature was again set to 24 degrees C.

Teichener and Kobrick's results are shown in Figure 11.4, and a clear decrease in performance under the cold conditions can be seen. Interestingly, however,

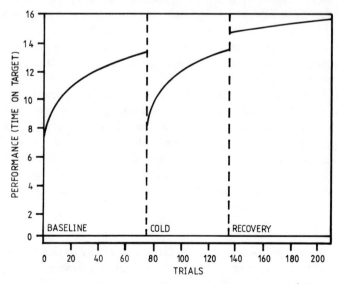

Figure 11.4 Performance changes in a cold environment (from Teichener and Kobrick, 1955)

the results also suggest that the subject became adapted to the cold conditions fairly quickly. Whereas even at the end of the cold period the subjects' performance was not as good as their performance the next day in the warm conditions, within the cold period their performance had increased by approximately 50 per cent.

Because the main factor affecting manual performance in the cold is related to the skin temperature of the affected limb, when it is impractical for gloves to be worn it would appear sensible to attempt some means of warming the skin locally. Lockhart and Keiss (1971), for example, showed that for most tasks the impaired manual performance which was experienced during the cold conditions (when the average skin temperature of the little finger was 12 degrees C (54 degrees F)) was greatly alleviated by applying radiant heat to the hands only (raising the temperature of the little finger to 19.5 degrees C (67 degrees F)). Indeed performance using the radiated heat in very cold conditions (− 18 degrees C; 0 degrees F) was no worse than working in 15.5 degrees C (60 degrees F) conditions (when the average temperature of the little finger was 30 degrees C (80 degrees F)).

Artificially raising the temperature of the affected limb, however, cannot fully overcome the effects of the cold if the rest of the body is not similarly heated. For example Lockhart (1968) maintained the temperature of his subject's hands at the 'normal' temperature, while cooling the body from 25.5 to 19 degrees C (78–66 degrees F). His results suggest that cooling the whole body affects the operator's ability to carry out some manual tasks even though his hands were kept 'warm'. The tasks affected involved fine dexterity, and Lockhart suggests that part of the reason for the performance reduction could have been due to shivering. Support for this contention has been provided by Peacock (1956) who examined the effects of cold exposure on rifle-aiming steadiness after vigorous exercise. Although steadiness was reduced, Peacock's results showed that if shivering is excluded as a cause of unsteadiness, rifle steadiness was not seriously affected in the standing position when compared with results taken during normal temperatures.

The effects of cold on manual performance, therefore, appear to be twofold. First if the cold is applied locally to the operating limb, it can have a direct effect on the muscular control of that limb, reducing such abilities as dexterity and strength. This may be overcome somewhat by locally warming the limb. Cold applied to the whole body, however, can reduce performance due to shivering. If the body can be stabilized during this shivering, generalized cold would appear not to affect manual performance to any great extent.

Cognitive performance

As far as cognitive performance is concerned, the ability to think, to judge and to reason under conditions of low environmental temperatures have been investigated primarily from the point of view of the diver. It is frequently necessary for a diver to work in cold water and his life depends on his ability to think clearly and efficiently. For example, typical Continental Shelf temperatures, where a great deal of diving takes place, are in the range 7–13 degrees C (45–55 degrees F) or less. Although diving suits have been developed to combat the effects of these temperatures, they are not entirely reliable and pose problems for the diver, particularly in relation to his manoeuverability.

The experimental work which has been carried out in this area, however, does not lead to any firm conclusions. In an investigation of the effects of cold on various tasks Bowen (1968) required his subjects to carry out their tasks whilst immersed in cold water at 7 degrees C (45 degrees F) for up to 30 minutes. Both manual performance (for example tactile sensitivity, grip strength, dexterity) and cognitive aspects (for example mental arithmetic, complex coding tasks) were tested. On all tasks Bowen's results indicated an impairment in performance in the cold compared with the ability in 'normal' temperature (22 degrees C; 72 degrees F) water. The problems involved in interpreting these results, however, stem from the fact that the cognitive tasks were scored not in terms of accuracy but in terms of the number of calculations attempted. Even Bowen himself realized that the reduction in performance in the cold could have been attributed to the problem of writing in the cold water, in which the dexterity of gripping a grease pencil was reduced.

Although the cold appeared to have no effect on the mental arithmetic task, accuracy on the complex coding task did show a significant decrease in the cold. During this task the subjects were given a set of symbols and colours which they were required to process in the following way: by entering the number and colour into a matrix they found a value for each colour; they were then given a problem number which they used to read, from another array, four numbers; each of these numbers was paired with the previously ascertained number of the four colours; each pair was multiplied together and the products summed; the subjects then marked their answers on the scoreboard using a grease pencil. This task, therefore, required accurate translation of symbols and colours into numbers, short-term storage of the different numbers, and mental arithmetic. Bowen concludes that his data indicate that 'mental performance is impaired providing the tasks are sufficiently demanding in terms of concentration and short-term memory requirements'. Hence the reason for an impairment in the cognitive task but not the mental arithmetic task.

Bowen's conclusions, however, have not been substantiated by other experimenters. For example Baddeley *et al.* (1975) asked their divers to perform different cognitive tasks (vigilance, memory, reasoning) in colder water and for longer than usually accepted (4 degrees C (40 degrees F) for 50 minutes). However, a reduction in performance was not demonstrated on either the reasoning test (judging a series of statements as being either 'true' or 'false') or the vigilance test (detecting the occasional onset of a dim light mounted in their helmet).

To test the diver's memory, the experimenters read to them part of a passage containing a number of 'facts'. The subjects were subsequently asked first to recall as many of the facts as possible, and were then given a list of features and asked to identify any of them which they recognized as being in the original list. No significant impairment in recall was obtained in the cold, although recognition was significantly affected.

On the basis of their experiments Baddeley *et al.* concluded that cognitive efficiency may be surprisingly resistant to the effects of cold, given good conditions and well-motivated subjects.

The results of these different studies, therefore, are inconclusive. Both were well-controlled but each seems to lead to contrary conclusions. However, if the 'riders' introduced by both authors are noted, some conclusions may be drawn. Thus Bowen suggests that only the very complex tasks will be affected by the cold. Certainly the vigilance and reasoning tasks used by Baddeley *et al.* were not as complex as Bowen's complex reasoning task. Secondly Baddeley *et al.* suggest that the cold will not affect the cognitive efficiency of well-motivated subjects since the cold 'tends to reduce his motivation by making him uncomfortable, and his discomfort may in turn distract him from the task in hand'. This is less likely to happen to a well-motivated subject. Unfortunately, neither experimenter really considered the distracting effects of the cold on a complex task.

Some evidence is also available to suggest that cold interferes with an individual's ability to accurately estimate time. Baddeley (1966) asked 20 amateur scuba divers to count up to 60 at what they considered to be a 1 sec rate. His data indicate that after the divers had entered the 4 degrees C (39 degrees F) water their estimation of time increased—in other words 1 sec was judged to be a longer period than it in fact is. It may be remembered that a similar effect was obtained by Jerison (1959) from subjects working under noisy conditions (Chapter 10). The similarity of these results with those of noise suggests that the cold is acting in the way of a general stressor when dealing with cognitive efficiency, rather than on particular physiological functions such as muscle control or skin sensitivity.

In summary, therefore, it appears that the cold is likely to affect the efficiency of less well-motivated individuals carrying out complex cognitive tasks which contain an element of time estimation. The reason for this is likely to be due to distraction caused by the stressing effects of the cold.

The Effects of Temperature on Comfort

From the foregoing discussion it is quite clear that the body's core temperature must be kept to within quite narrow limits. A temperature increase of about 5 degrees C (9 degrees F) is likely to lead to death, as is a reduction by only 3–4 degrees C (5.5–7 degrees F). Furthermore, small departures from this critical band can lead to decrements in both motor and cognitive performance, as can large changes in the general environmental conditions.

In addition to these effects, changes in the thermal environment can affect an individual's comfort. As with any interaction between man and his environment, the resultant sensations of comfort experienced by an individual depend both on the conditions of the environment and on the individual factors which the man brings to the situation. For most purposes the physical environmental factors can be distilled to a consideration of the air temperature, the air humidity and the amount of air movement. Each of these is likely to become stressful when it interferes with the body's ability to maintain an adequate thermal balance.

The effect of the air temperature, measured with any type of dry bulb thermometer, is likely to be to raise or lower the overall body temperature, requiring the balancing mechanisms to produce sweat in hot conditions or to constrict the peripheral blood vessels when cold. However, in very hot conditions, for example near a furnace, the amount of heat which is radiated from the hot source (the furnace) is also an important determinant of the degree to which the operator is likely to feel 'comfortable'. Radiation is emitted by the heat source as electromagnetic waves, which are absorbed and converted back into heat when they fall on to solid objects. If the radiation is severe enough, the skin tissue can be burnt—as most people who lay for too long on a Mediterranean beach will know.

Radiant temperature is usually measured using a globe thermometer, which takes the form of a hollow metal globe about 15 cm in diameter and painted matt black (the original equipment used by Bedford and Warner in the early 1930s was a cistern ballcock). A normal mercury thermometer is placed in an opening in the bulb. In this apparatus, the thermometer receives radiant heat from all directions and an indication of the 'mean radiant temperature' can be obtained after also noting the air temperature and the speed of air movement.

The humidity is an index of the amount of water vapour in the air and is normally measured using two mercury thermometers. On one, the 'wet bulb' thermometer, the bulb is kept damp and is thus cooled by the evaporating water. The other, the 'dry bulb' thermometer indicates the 'normal temperature'. Since the amount of water which is able to evaporate from the wet bulb is dependent on the air humidity, and since the degree of evaporation is reflected by a lower wet bulb temperature, taken together the two temperatures (wet and dry bulb) indicate the amount of water vapour present in the air.

Excessive water vapour in the air, indicated by a high relative humidity, is likely to interfere with the efficiency by which sweat will evaporate from the skin for cooling purposes. At the other end of the scale, very low humidities are likely to cause discomfort by drying the normally moist membranes in the nose and throat, particularly if the air temperature is rather high.

When the air moves the body it has a cooling effect as a result of both helping to evaporate the sweat and by dissipating the heat from the body surface. Naturally, this is likely to result in increased comfort, but too large an air velocity is likely to lead to complaints of draughts. It should be remembered, however, that velocity is a relative term which in this case applies to the relative motions of the air and the observer. If the observer is stationary the relative velocity is equal to the air speed. The velocity of the air relative to a moving observer, however, is determined by both the speed of the observer and his direction of movement with respect to that of the air.

The two important variables that the observer is likely to bring to the situation are the nature of his activity (type and level) and his clothing. Other additional variables such as his age and physical condition are also likely to affect whether he perceives a particular thermal environment as being 'comfortable' or 'uncomfortable', but these act solely on his body's ability to maintain a thermal balance through sweating or vasoconstriction. Activity and clothing, on the other hand, will act independently of the observer's physical capacities—unless he has any physiological deficiency.

The effects of activity level on body temperature have already been considered when discussing heat stress. Thus when work is done by the muscles, heat is produced as a by product of the oxidation process, which must then be dissipated.

The transfer of dry heat between the skin and the outer surface of the clothed body, however, is quite complicated and involves internal convection and radiation processes in the intervening air spaces, and conduction through the cloth itself. These variables are accounted for by the now internationally accepted, dimensionless unit of thermal resistance from the skin to the outer surface of the body—the 'clo'.

Gagge, Stolwijk, and Nishi (1971) have defined 1 clo as being 'the amount of insulation necessary to maintain comfort and a mean skin temperature of 33 degrees C in a room at 21 degrees C with air movement not over 0.1 m/sec and humidity not over 50 per cent with a metabolism of 50 cals/sq m/hr'. This applies, therefore, to the resting subject only.

It is extremely difficult, however, to measure the clo value of any clothing systems. As Fanger (1970) points out, to obtain reasonably accurate measurements it is necessary to use a lifesized heated manikin which is clothed in the clothing ensemble. As heated manikins which include the necessary electronic equipment are expensive to produce, however, few are available outside the large (usually military) laboratories, which means that there exist in the literature results for relatively few clothing ensembles. The range of clothes investigated extends from nude (clo 0) to heavy wool suits for polar environments (clo 3–4), but within these extremes Fanger is able to list the clo values of only 12 other clothing ensembles.

Apart from the conductive resistance of the textile itself, the air between textile layers also acts as an important insulator, and so the quality of tailoring and fit will influence the clo value. For very loose hanging clothing such as the dress worn by, for example, the Arabs, the heat which rises from the bottom of the airspace can cause slight 'chimney' effects to occur which produce a forced ventilation over the skin.

A combined temperature scale

Since each of the thermal variables, air temperature, air velocity, radiation and humidity can, both separately and together, affect thermal comfort, it is useful to attempt to combine them to produce a single temperature scale. The most used such scale is the effective temperature scale, first proposed by Houghton and Yaglou in 1923. Unfortunately, however, since the scale only took account of air temperature, humidity and velocity it is inapplicable for environments which contain high levels of thermal radiation.

Vernon and Warner (1932) applied a radiation correction factor to the ET scale by using the globe thermometer to measure thermal radiation, rather than a simple mercury thermometer to measure dry bulb temperature. Their resultant scale, the corrected effective temperature scale (CET) has been used extensively in many conditions and is shown in Figure 11.5. Under 'normal' conditions, however, the ET and the CET scales can be taken as synonymous.

In more recent years attempts have been made to reassess the CET scale to bring it more up to date. For example Nevins and Gagge (1972) suggest that there is now substantial evidence that the temperature criteria for thermal comfort have risen steadily from a range of 18–21 degrees C (65–70 degrees F)

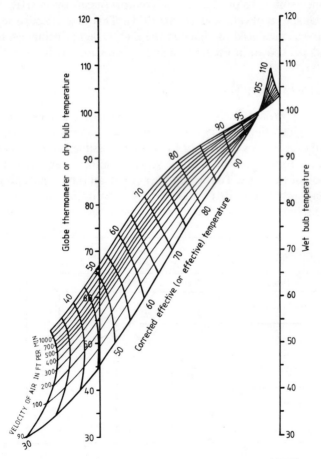

Figure 11.5 The corrected effective temperature (CET) scale

in 1900 to 24–26 degrees C (75–80 degrees F) in 1960. Although they do not support their contention with any written evidence, they suggest that this shift has probably resulted from changes in the clothing worn by both men and women and from changes in building construction.

A second reason for reassessing the scale has been the fear that both scales overemphasize the effects of humidity in cool conditions and underemphasize its effects in warm environments. Indeed 25 years after the introduction of his original scale, Yaglou (1947) himself recognized that the scale may perhaps have overemphasized the effects of humidity towards the lower temperatures. Finally it was felt that the old scales did not account sufficiently for the physiological processes which regulate the body's thermal response, including the 'insensible' heat lost from surfaces such as the lungs during breathing and vaporized water diffusing through the skin.

The American Society of Heating and Refrigeration Engineers, therefore, commissioned studies to produce a new comfort scale—the ASHRAE comfort scale (see Gagge, Stolwijk, and Nishi, 1971). This new effective temperature scale which was produced (designated the ET* scale to distinguish it from the old ET scale) is shown in Figure 11.6 and is compared with the old ET scale.

Variables which affect thermal sensations

The number of variables which may contribute to the assessment of thermal comfort is legion. Besides the obvious physical aspects of the environment discussed above, Rohles (1967) has listed such possible variables as the colour and size of the room, the season of the year, the subject's age, activity, clothing and duration of exposure. To these Fanger (1970) adds national and geographical location, body build, the position of a woman during her menstrual cycle, circadian rhythms and ethnic differences. Some of these will be discussed below.

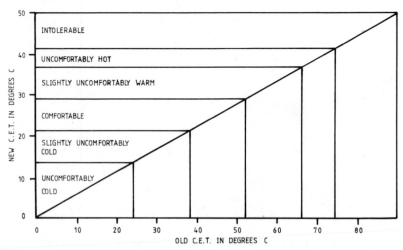

Figure 11.6 The old and new corrected effective temperature (CET) scales

Age

It would be reasonable to expect the elderly to prefer higher temperatures than younger people, since it is well known that metabolic rate decreases slightly with age. Furthermore, the elderly are less likely to lead an active life which would help to maintain a higher metabolic rate. The reduced body heat, then, should be compensated by an increased need for warmth.

The research which has been carried out in this area, however, suggests that, providing their physiological regulatory mechanisms remain effective, the thermal preference of the elderly is no higher than for younger subjects. For example Rohles (1969) sent to 64 elderly subjects (mean age 75 years) questionnaires which asked the respondent to indicate whether they thought that each of 41 temperatures ranging from 32 to 110 degrees F (0–43 degrees C) would be 'too warm', 'too cold', or 'comfortable'. Despite the obvious experimental problems of this study, the distributions of responses for each category compared well with the distributions of more empirically derived responses from younger groups in other experiments. It is unfortunate, however, that the author did not send the same questionnaire to subjects of different age groups—then at least direct comparisons could have been made.

Fanger (1970) conducted a more controlled series of experiments to investigate the same problem and found similar results. He ascribes the failure to show differences in preference to the body's bioregulation system. His measurements indicated that 'insensible' sweating (in other words, the evaporation of imperceptible amounts of sweat from the body which help to maintain thermal balance) in the elderly decreased in proportion to their reduced metabolic rate. Thus, although less heat is produced by the body, less is lost.

Sex

In an experiment which measured the metabolic rate of subjects carrying out different levels of activity, McNall *et al.* (1968) demonstrated that males have a higher rate than females. As with the elderly subjects, however, the insensible sweat rate of females was slightly lower than that of the males. Under these conditions, therefore, it would seem likely that any need for warmer conditions by women (due to their lower metabolic rate) would be offset by their reduction in heat loss—in other words, there should be no significant difference in thermal preference between the two sexes. This contention is supported by much of the experimental data (for example, Fanger, 1970). Furthermore after recording the time of his experiments in relation to the menstrual cycle of his subjects, Fanger also concluded that the position of a woman's cycle has no effect on her thermal comfort.

Circadian rhythm

The concept of the circadian rhythm is discussed in more detail in the next chapter. However, it is well known that the internal body temperature fluctuates on an hourly basis over a 24-hour period, so some rhythm in comfort preferences may also be feasible. However Nevins *et al.* (1966) found no difference in preferred temperature between experiments performed in the afternoon and those carried out in the evening, an observation which was confirmed by Fanger (1970). This is not to say, of course, that finer experimental analysis would not show a rhythm in thermal preferences. Both Nevins *et al.* and Fanger only divided the rhythm into two or three periods.

Colour

A considerable body of anecdotal evidence has been built up to suggest that the colour of the surroundings can influence man's thermal preferences. For example, an individual is presumed to feel warmer in a space which is lighted, painted or furnished in a colour scheme in which red predominates, as compared with one in which blue is the prevailing colour.

Perhaps the first experimental study which bears on this question was conducted by Morgensen and English in 1926, who asked subjects to judge the temperature of heating coils wrapped in different colours. Apparently, the green ones were judged to be the hottest.

Unfortunately, however, experimental results relating to thermal preference show no such conclusive results. For example, Berry (1961) exposed each of 25 subjects to a temperature chamber illuminated by one of five colours: blue, green, white, yellow or amber. His subjects, however, showed no change in the upper comfort limit as a function of the colour of the surroundings. Similar conclusions were also reached by Fanger, Breum, and Jerking (1977) who concluded that 'the effect of colour on man's comfort is so small that it hardly has any practical significance'.

In conclusion, however, Fanger (1970) rightly points out that any influence which colour has on the thermal sensation must be of a 'psychological nature'. The possibility cannot be denied, therefore, that an individual when away from the oppressive environment of a laboratory and controlled temperature chamber may feel more at ease, more comfortable and, perhaps, experience more 'warmth' under different lighting conditions.

ILLUMINATION

When discussing vibration and noise, the two defining characteristics of a stimulus which were used were the signal's intensity and frequency. With illumination, on the other hand, although light energy can be conceived of as fluctuating energy which reaches the eye, the defining parameters of a light stimulus are its intensity and wavelength. The concept of wavelength is similar to that of frequency, except that the wavelength is related more to the distance between two peaks of a sinusoidal stimulus (rather like the distance between two waves on the sea) rather than the frequency with which the peaks occur. Wavelength, therefore, is measured in terms of distance but, because the distances between two energy 'peaks' are so small, the wavelength of light is normally described in terms of nanometers ($1\,nm = 10^{-9}$, or one-billionth, of a metre). Visible light is simply a form of radiation with a wavelength between 380 and 780 nm, and the eye discriminates between different wavelengths in this range by the sensation of colour. The violets are around 400 nm, blending into the blues (around 450 nm), the greens (around 500 nm), the yellow–oranges (around 600 nm), and the reds (around 700 nm and above).

The intensity of a light source is expressed in terms of the amount of luminous flux (or energy) which it generates. Just as the power of a car is expressed in relation to the power likely to be produced by a 'standard' horse (horsepower), the luminous flux produced by a light source is measured by comparing it with the flux produced by burning a standard candle of a specified material and weight. This candle is said to produce one candle power (or 1 candela) of energy.

Although flux so far has been used effectively as a synonym for energy, strictly speaking it refers to the rate at which the energy is produced and is measured in 'lumens'. However as energy radiates from its source it spreads out and loses its intensity as it travels through a dense medium such as air, water or glass (the rate at which this energy is lost is inversely related to the square of the distance which is travelled (known as the inverse square law)). Thus the level of illumination which falls on a surface will be lower than that which originated from the light source and will be spread over the surface area. Illumination, therefore, is defined in terms of the rate of flux (lumens) produced and the surface area over which it spreads (that is, lumens per square foot). Sometimes, however, illumination is also expressed in terms of footcandles ($1\,fc = 1\,lumen/sqft$).

Although a particular level of light may fall on a body, this is not to say that the observer will 'see' the same level since different bodies absorb and reflect different amounts and qualities of light depending on their surface characteristics. For example a highly polished surface may reflect around 90 per cent of the light energy falling on it, whereas a dull, matt surface might reflect only about 10 per cent of the light. In this case the term used to define the amount of light reflected from the body is its luminance, and this is equal to the amount of light falling on the body multiplied by the proportion of light which the body reflects (its reflectance). Unfortunately for the novice illumination engineer, however, there are a multitude of terms used to define luminance. It can be expressed either in terms of luminous intensity per unit area (candela/sq m), or in terms of the 'equivalent illumination', expressed in equivalent lux, or apostilbs (asb), or in foot-lamberts (ft-L). The most modern unit chosen is the candela/sq metre but, because so much of the experimental work carried out in this area uses foot-lamberts, this unit will be used in the remainder of this chapter ($1\,cd/sq\,m = 0.29\,ft\text{-}L$).

The luminance qualities of an object, however, will only be perceived by an observer after the reflected light has stimulated his retinal cells and the information passed to the optic cortex of his brain. At this point the concept of the body's brightness is invoked, which is the subjective aspect of a body's luminance.

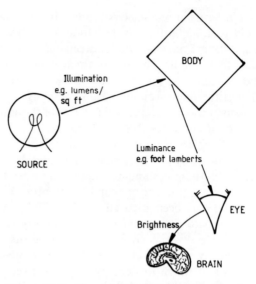

Figure 11.7 The relationship between illumination, luminance and brightness

The relationship between these definitions can perhaps be explained by reference to Figure 11.7. Thus a source (for example a light bulb) illuminates a body (the intensity being measured in lumens/sq ft). This body reflects the light to the eye of the observer, the intensity being in terms of the body's luminance and measured in many units, including foot-lamberts. The observer then reports the body's brightness and this, being subjective in nature, is likely to depend on many factors including his past experience and the brightness of other bodies in his visual field.

Illumination and Visual Performance

Before we can see any object we need to be able to separate its image from its background or surroundings, otherwise it becomes 'camouflaged'. The quality of an individual's visual performance, therefore, can be related simply to the extent to which the viewed object stands out from other stimuli in the observer's visual field. This is perhaps the fundamental principle to operate in helping to decide on the appropriate level and type of illumination to use. However, as will be demonstrated, it is often one which is violated.

Three factors affect the degree to which an object stands out and is able to be perceived; first the overall illumination, secondly its size, and thirdly the contrast between its luminance and the surround luminance. These factors, however, do not act independently; they are highly interrelated in their effects, as will be demonstrated.

Overall illumination levels

As was discussed in Chapter 2, the eye responds differently under different levels of illumination, allowing us to be able to see within a large range of luminance levels from about 10^{-6} to 10^7 ft-L. At low illumination levels, the rods in the retina are most responsive thus allowing scotopic vision, but as the luminance rises above about 1 ft-L the cones become more stimulated, allowing photopic vision.

Despite the fact that we are able to perceive objects over a wide range of luminances, however, it is clear that some illumination levels are more ideal at work than are others. To investigate the effects of illumination level Gilbert and Hopkinson (1949) asked subjects to read various letters on a Snellen chart (the chart used by opticians as an aid to assess eye deficiencies—the authors suggest that this constitutes a simple test for assessing visual acuity). The charts were lit with different levels of illumination ranging from 0.1 to 100 lumens/sq ft.

The results showed that the acuity of subjects with 'normal' vision increased with increasing illumination, although the increased advantage tended to level out above about 10 lumens/sq ft. Children with 'subnormal' vision, however, did not demonstrate this levelling off, even up to an illumination level of 100 lumens/sq ft. Hopkinson and Collins (1970) suggest that these data confirm the generally held opinion that people with poor eyesight benefit more from increased levels or lighting than do people with normal sight. However, they also point out that some pathological sight conditions may require great caution to be exercised when increasing the illumination level, to ensure that the light falling directly on the eye is not increased at the same time.

The size of the object being viewed

The level of illumination chosen for a particular work area will depend on the type of work which is being carried out. As Gilbert and Hopkinson's results demonstrated, as the objects (letters on the Snellen chart) became smaller, more light was required for them to be read accurately. Any suggestions for appropriate lighting levels in various situations, therefore, need to take account of the type of detail required in the task.

In its code for interior lighting, the Illuminating Engineering Society suggest illumination levels for many different types of work interiors which are related to the type of work normally carried out. Overall, seven illumination levels are suggested, as shown in Table 11.1 below. The code suggests, however, that before deciding on an appropriate illumination level for the task in hand, two important questions should be asked:

1. are the reflectances or contrasts unusually low (for example having to pick out dark objects from a dark, matt background), and
2. will errors have serious consequences?

Table 11.1 Suggested illumination levels for different types of work (From the Illuminating Engineering Society Code, 1973)

Type of work	Recommended Illumination level in Lumens/sq ft.
Storage areas with no continuous work	150
Rough work (rough machining and assembly)	300
Routine work (Offices, control rooms, medium machine and assembly)	500
Demanding work (deep plan, drawing or business machine offices, inspection of medium machinery)	750
Fine work (colour discrimination, textile processing, fine machinery and assembly)	1000
Very fine work (hand engraving, inspection of fine machinery or assembly)	1500
Minute work (inspection of very fine assembly)	3000

If the answer to either of these questions is 'yes' then the next higher level of illumination should be used.

Contrast

Contrast effects provide vivid examples of the need for the object to stand out from its surroundings before it can be perceived. Without contrast an object cannot be seen, however large it is, and this applies equally to such stimuli as words on the printed page, to a large machine in a dimly lit room, or to a well-camouflaged insect.

Definitions

The normal way to express contrast is in terms of a ratio between the luminance difference of the object and its surround to the surround luminance. So an object which is illuminated by 100 units of light placed on a background of 10 units will have a contrast ratio of $\frac{(100-10)}{10} = 9$, as will a 10-unit object placed on a 1-unit background.

Unfortunately, however, as Hopkinson and Collins (1970) rightly point out, defining the contrast simply in terms of the physical characteristics of the object and its surround represents only part of the calculation. Both the object and its surround are only perceived by the observer when the image falling on the retina has been passed to the optic cortex, so the observer's perceptual behaviour also needs to be taken into account in the calculation. For example, whereas two light meters may register 100 units and 10 units in one case, and 10 units and 1 unit in another (giving the same contrast ratio), the observer's perceptual system is likely to treat the two sets of stimuli differently. For this reason the authors suggest that contrast should be expressed not in terms of the actual luminance levels but in terms of the difference between the 'apparent brightnesses' of the object and surround.

The concept of apparent brightness rests on the assumption that everything that we see is evaluated, as far as its brightness is concerned, in terms of some reference level which Hopkinson and Collins suggest is associated with the state of adaptation of the eye at the time. So a luminance of, say, 100 units, will produce higher apparent brightness when presented to a dark-adapted eye than to one which is light adapted. Indeed the light will appear 'bright' to the dark adapted eye and 'dark' to the light-adapted eye. Apparent brightness, therefore is a function not only of the physical luminance or the object itself, but of the luminance of the surrounds which govern the adaptation of the eye at the time.

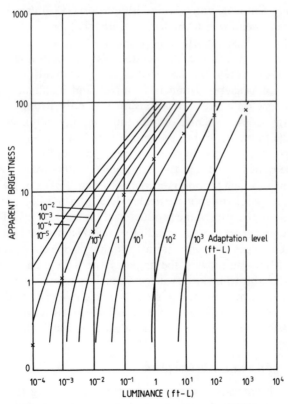

Figure 11.8 Curves of equal apparent brightness (from Hopkinson, Waldram, and Stevens, 1941)

Hopkinson, Waldram, and Stevens (1941) have produced a set of curves from which one can read the apparent brightness of the object or surround given their respective brightnesses (see Figure 11.8). Thus, for example, an object with a luminance or 1 ft-L placed on a background (adaptation level) having a luminance or 100 ft-L will have an apparent brightness of about 1 unit. The background or 100 ft-L itself has an apparent brightness of 80 units, so the contrast is said to be $(80-1)/1 = 79$ units.

The effects on performance

As an example of the importance of contrast to efficient visual performance, in their illumination study Gilbert and Hopkinson (1949) asked their subjects to read the letters on a Snellen chart under different levels of illumination, using different contrast ratios between the letters and the background. Their results indicated strongly that as the contrast was increased the subject's ability to read the letters accurately also increased. This effect was particularly marked at the lower levels of overall illumination, whereas the performance increase was not so marked when the overall illumination was reasonably high.

More recently Timmers, Van Nes, and Blommaert (1980) also obtained similar results. However, their experiment demonstrated that the place where the image falls on the retina has important implications for the extent to which this performance–contrast relationship occurs. Taking both the percentage of correct recognition scores and the speed of seeing Dutch three-letter words as their performance measures, Timmers, Van Nes, and Blommaert demonstrated that the performance–contrast relationship was only strong when the word images fell away from the centre of the retina (parafoveally). When they were projected to the fovea (centre) of the retina, very little increase in performance occurred despite the increase in contrast (see Figure 11.9).

It should be remembered that the term 'object' refers solely to the aspect(s) of his environment which the observer wishes to view. For example, an industrial inspector whose task is to spot particular defects in finished products will be interested solely in the area of the product likely to be defective. The remainder of the product represents the 'surround'. Fox (1977) illustrated this aspect when he used different types of illumination to highlight defects in coins produced at the Royal Mint. By moving the angle of the light illuminating the work surface, Fox was able to make any defects which were present 'stand out' from the rest of the coin because of the shadows which were caused. In this case, therefore, the defect stood out from its shadow—in other words, the contrast between the defect luminance and the shadow area was high.

The brightness of the surround and glare

Although a high contrast is clearly important in ensuring that the object is perceived accurately, it is also important that the direction of the contrast effect is considered, for two reasons.

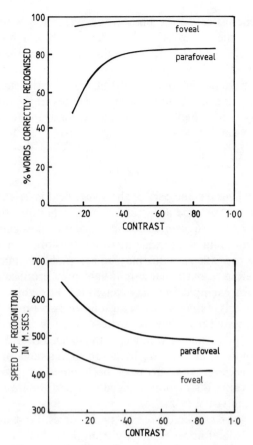

Figure 11.9 The effect of contrast on the accuracy and the time taken to recognize three-letter words when the words are presented to different areas of the retina (from Timmers, Van Nes, and Blommaert, 1980. In *Ergonomic Aspects of Visual Display Terminals*, reproduced by permission of Taylor and Francis Ltd.)

First, if the object is brighter than its surround, it is more likely to be perceived. As was discussed in Chapter 4 an object or symbol will be perceived and interpreted more quickly and more accurately if it is seen as a figure rather than as the ground. Various ways of ensuring that a symbol is seen as a figure were discussed in Chapter 4, and Hochberg (1972) adds that the figure should also be 'more impressive, more apt to suggest meaning, and better remembered'. These criteria will be met if the contrast direction is such that the object to be viewed is brighter than its surround. In addition to the brighter object standing out and being seen better because it presents itself as the figure, it is also likely to stand out because the eye tends to move to the brightest part of the visual field. This tendency is called the phototropic effect and is well known to car drivers at night as the 'compulsion' to look at the headlights of an oncoming car rather than the road in front, and to the shoppers attracted towards neon advertising signs.

The second consideration concerns the fact that, if the surround is brighter than its object, it will be likely to reduce the visibility of the object due to glare. Glare, as a reducer of visual efficiency and as an agent for visual fatigue, is discussed in more detail below.

Glare

Glare is caused whenever one part of the visual field is brighter than the level to which the eye has become accustomed. This definition, therefore, is much like that given for noise (unwanted sound) since it defines glare in terms of the circumstances prevailing at the time at which the 'bright' stimulus occurred. Just as the same acoustic stimulus may be acceptable at one time but 'noise' at another, therefore, a particular bright light may produce no adverse effect during the day, for example, but may cause glare at night.

Glare is commonly described as being one of two types. If there is direct interference with visual performance, the condition is referred to as disability glare. However if performance is not directly affected, but the bright stimulus still causes discomfort, annoyance, irritation or distraction, the condition is called discomfort glare. The effects of both types of glare, however, may eventually be to cause distraction and reduced performance and to draw the eyes away from the visual task.

Both types of glare can be caused by two sources. First direct glare occurs when the excessive light appears directly from the light source itself, such as a headlight of a car at night or from the sun during the day. Second, and perhaps more insidious, reflected or specular glare is caused by reflections of high brightness from polished or glossy surfaces that are reflected towards the observer. These effects can be more problematic since they may occur under various, and often unpredictable, circumstances—for example, when the object is placed at a certain angle to the light or when a number of lights are brought together in a particular way. Only then will the object be unable to be seen adequately.

Disability glare

The reduced ability to see accurately due to interference from a bright light source has probably been experienced by all at some time or other, so the potential disabling effects of glare are well known.

At a descriptive level, two of the earliest workers in the field, Luckiesh and Holladay (1925) considered disability glare to be of three types: first, veiling glare due, they felt, to light from the glare source being scattered in the fluids in the eye, so reducing contrast and hence visibility. An effect is caused, therefore, which is similar to illuminated fog or mist. The second type of disability glare, which they describe as 'dazzle glare', occurs as a short-term effect for the duration of the glare source. Finally 'blinding glare' lasts beyond the period of the glare stimulus due to the formation of 'blinding after-images'.

Whatever the form, disability glare may be due to two causes. First scattering of the light in the optic fluids of the eye is likely to cause a 'veiling' effect, and this argument has been advanced by Stiles and his co-workers (Stiles, 1929; Crawford and Stiles, 1937). For example, in one experiment the authors induced the glaring light source to fall on the blind spot of the retina, so that the light source itself was invisible. The veiling effect of the glaring source, however, was still present.

The second cause of disability glare lies in the inhibitory effect which cells in the retina have on adjacent cells. This was demonstrated in a famous experiment carried out by Hartline, Wagner, and Ratliffe (1956) who recorded the activity of single retinal cells in the horseshoe crab. This animal has a compound eye composed of approximately 1000 cells (ommatidia), from which nerve fibres emerge in small bundles and come together to form the optic nerve. Hartline and his colleagues recorded the activity produced by one ommatidium while neighbouring ommatidia were stimulated by light. Their results indicated that the ability of the ommatidium to discharge impulses was reduced by illuminating other ommatidia adjacent to it, and the extent of this inhibition depended, amongst other factors, on the intensity and the area of the glaring source.

In a comprehensive series of experiments Holladay (1926) investigated many aspects of disability glare, the most important for which he is remembered being the relationship between the amount of glare (defined in terms of contrast reduction), the position of the glaring source with respect to the observer and the amount of light entering the observer's eye.

His experiments indicated that:

$$\text{Contrast reduction} = \frac{k \times \text{illumination produced by glaring source at the eye}}{(\text{Angle at eye between glare source and object being viewed})^{2 \cdot 4}}$$

The value of k in this formula appears to depend on the age of the observer, as age causes changes in the constituency of the fluids in the eyeball (Christie and Fisher, 1966).

There are two practical implications of Holladay's formula: First the relationship between disability glare and the angle of the glare source to the line of sight implies that the extent of the glare decreases as the angle of the glare source increases. The other important factor concerns the amount of illumination produced by the glare source. As Hopkinson and Collins (1970) point out, the effect will be the same whether the glaring source is a small source of high luminance or a large source of low luminance—provided the illumination at the eye is the same. This means that a dark sky, for example, seen through a large window can cause as much disability glare as a small, more intense, light bulb—even though its brightness may not be sufficiently high to cause any discomfort. Finally, Murrell (1971) points out that, although strictly speaking it is not a case of disability glare, the phototropic effect of the glaring source can also cause performance loss. For example, Hopkinson and Longmore (1959) have demonstrated that the eye makes more frequent, jerky darts towards the brighter area in the visual field. This distracting effect can also be fatiguing.

Discomfort glare

Paradoxically, discomfort glare appears to have been more extensively studied than disability glare, and a number of formulae have been derived which relate various physical parameters of the glare source to levels of 'discomfort'.

The discomfort produced by a glare appears to have a different physiological origin than does disability glare. Thus Hopkinson (1956) has demonstrated a link between the level of discomfort and the activity of the eye musculature which controls the iris. This relationship, however, is not perfect and Hopkinson concluded that discomfort sensations were due only in part to the conflict which arises between the requirements of the areas of the retina stimulated by the glaring source, and those receiving lower levels of illumination for pupil control.

Being a subjective attribute, however, discomfort glare has been studied in the past mainly by asking people for their direct, subjective impressions of different glare 'situations'. In the earliest work, for example, Luckiesh and Holladay (1925) asked their subjects to alter the luminance of a glare source to produce a series of sensations from 'scarcely noticeable', through 'most pleasant' to 'thoroughly uncomfortable' and to 'irritating and painful'. Such work, however, suffers from the obvious disadvantage which can be ascribed to much of the early work dealing with (dis)comfort (whether it be due to glare, vibration, or temperature, etc.) of meaningless subjective categories—for example, how can glare be 'most pleasant'? In addition the categories are unlikely to represent the same to all subjects—due to differences in interpretation, what one observer understands by the term 'uncomfortable' is not necessarily the same as another's.

Nevertheless, work of this nature did at least provide the impetus for a number of further studies, from which a series of 'glare formulae' have been produced (see, for example, Murrell, 1971, for a list and description of these formulae). Unfortunately, however, all have used subjective 'comfort' categories to a great or lesser extent, so all have similar failings. Luckiesh and Guth (1949) did attempt to reduce the possible discrepancies in interpretation between observers by using a single criterion of discomfort called the 'borderline between comfort and discomfort' (or BCD), rather than a number of different criteria. However, their technique was to present the glare source momentarily rather continuously. Their argument for taking this approach was that glare is experienced when the observer looks up from his work for a second or so; continuous discomfort glare is unlikely to be tolerated. Whether or not they are justified in their assertion, this distinction between the techniques should be remembered when considering the literature in more detail. Certainly, as Hopkinson and Collins (1970) point out, it has the effect of increasing the importance of the glare source luminance in the final 'glare formula'.

The most modern glare formula, which has become generally accepted, has been produced by the Illuminating Engineering Society based on empirical work developed by the Building Research Station. This work was carried out over a number of years by Hopkinson and his colleagues (see, for example, Hopkinson, 1940; Hopkinson, 1972; Petherbridge and Hopkinson, 1950). The results of the various investigations produced the following glare formula:

$$\text{glare constant} = \frac{Bs^{1.6} \times w^{0.8}}{Bb \times \theta^2}$$

where
Bs is the luminance of the source
w is the solid angle subtended by the source at the eye (i.e. related to its apparent size),
Bb is the general background luminance,
θ is the angle between the direction of viewing and the direction of the glare source.

Using formulae of this nature, the glare constant obtained can be related to some type of criterion of discomfort. Thus a glare constant of 600 indicates the boundary of 'just intolerable'; 150 the boundary of 'just uncomfortable'; 35 the boundary of 'just acceptable' and 8 would be 'just perceptible'. A more useful criterion, however, might be to express the glare ratings in terms of the probability of producing 'visual comfort'. This has been done by the IES, which McCormick (1976) has adapted to form one, single graph as shown in Figure 11.10.

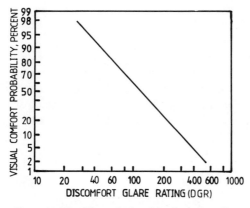

Figure11.10 The relationship between discomfort glare ratings and the probability of retaining visual comfort (from McCormick, 1976, reproduced by permission of the McGraw Hill Book Co.

If more than one glaring source is present in the environment, Hopkinson (1940) showed that the simple arithmetical addition of the glare constants obtained from each source gave a glare value which corresponded closely to the sensation from the complete array of the source. Later (1957), however, he modified his position slightly when he showed that the additive nature was a highly complex phenomenon, the exact function depending, amongst other things, on the luminance of the sources and their position in the field of view.

SUMMARY

As this chapter has demonstrated, small departures from the ideal levels of both illumination and temperature can lead to fairly large reductions in performance and, in the case of temperature, even to death. Both of these environmental parameters, however, are extremely complex in their composition and their effects on the operator, and this chapter has attempted to explain some of these actions.

CHAPTER 12

Ergonomics and Safety

At first sight it might appear strange that the topic of safety is considered as a separate chapter in an ergonomics text which stresses throughout how various aspects of man–environment interactions can lead to unsafe acts. Safety, so the argument continues, was considered for example when discussing the maximum loads the operator should lift (Chapter 3), the design of his displays so that he can perform his tasks accurately (Chapter 5), the design and arrangement of his controls for safe use (Chapters 6 and 7), and so on. What further information can a chapter on safety add?

Whereas it is true that safety and accidents have been examined in detail throughout this book, they were approached primarily from the point of view of the system and the operator's interaction with it (for example, the loads which he can carry; his displays, controls, machines; the noise, temperature; etc.). This chapter examines some of the factors which the operator can introduce into an otherwise 'safe' situation which might possibly make it 'unsafe'. Even if the system has been designed to take account of all the ergonomics principles so far discussed, experience tells us that accidents still occur when a human operator is present. This chapter considers some of the reasons why.

Although the main emphasis is on the characteristics of an operator's behaviour which lead to accidents, the man–machine systems concept is still valid and should not be forgotten. It is still argued that an accident occurs as a result of the environment (including machines) demanding more of the operator than he is able to give, and this is illustrated graphically in Figure 12.1.

In the lower trace the total, fluctuating demands which are made by the environment are shown. For example, if the worker is driving a truck: the environment is likely to demand a high level of coordinating skill to make the truck actually move, to manoeuvre it around corners, to control its speed and to stop it when appropriate, etc. Because he is experienced the operator is capable of doing this, although perhaps his attention wanders at times so his capabilities are reduced (upper trace). At point (a) the environment increases its demands on the driver, perhaps a pedestrian steps into the road and he is forced to swerve. However, his attention has not wandered too much and his abilities are not too reduced, so he is capable of doing this and successfully avoids an accident. At point (b), however, his capabilities have fallen dramatically (he might be looking elsewhere; the sun could be shining in his eyes; he might be fatigued; etc.) and are not sufficient to meet the sudden demands of

248

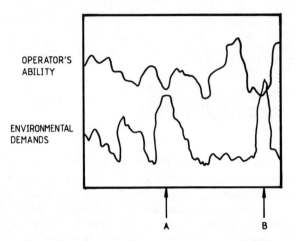

Figure 12.1 A simple model of accident causation

the environment. When this happens an accident occurs—in other words, when the demands of the environment exceed the capacities of the operator.

At this point it is useful to ask the question 'What is an accident?', 'When does an accident become deliberate and hence blamable?' A number of definitions exist but all contain concepts which can be distilled down to three main points (Suchman, 1961); the more of these points which are present, the more likely the event will be called an 'accident'. An accident, then, has

1. a low degree of expectedness,
2. a low degree of avoidability,
3. a low degree of intention to cause the accident.

Thus accidents are unfortunate, unpredictable, unavoidable and unintentional interactions with the environment. This very unpredictability and rarity, however, makes them extremely difficult to investigate directly, so most accident research has been carried out on either accident reports, or on the painstaking method of direct observation of minor 'incidents' or 'near misses' which do not result in reported accidents. Both of these techniques, however, have their drawbacks and can result in biases in interpretation as is discussed in Chapter 14.

THE COST EFFECTIVENESS OF ACCIDENT PREVENTION

In many respects it could be argued that as far as accident prevention is concerned, any consideration of cost-effectiveness is (a) morally wrong, and (b) inherently difficult. How much value does one put on human life, grief, pain, etc? This point is emphasized in Table 12.1, which compares the costs and benefits calculated for incorporating safety features in an automobile petrol

Table 12.1 A cost/benefit analysis produced for the implementation of a new, safe, automobile petrol tank

---

Benefits
Savings: 180 burn deaths, 180 serious burn injuries, 2100 burned vehicles
Unit costs: $200 000 per death, $67 000 per injury, $700 per vehicle
Total benefit: 180 × ($200 000) + 180 × ($67 000) + 2100 × ($700) = $ *49.5 million*

Costs
Sales: 11 million cars, 1.5 million light trucks
Unit costs: $11 per car, $11 per truck
Total cost: 11 000 000 × ($11) + 1 500 000 × ($11) = $*137 million*

---

tank said to have been produced by a major automobile company (reproduced from Biancardi, 1978).

Similar types of analyses have been proposed by other authors. For example Powell *et al.* (1971) initially consider the costs of over 2000 accidents which occurred during their study of four different types of shop floor. They calculated that the number of man-hours lost through accidents was roughly equivalent to the permanent absence of four people. The total wages cost was in the region of £3500 per annum and, assuming the labour content of a factory product to be 20 per cent of the equivalent product, the sales loss was calculated to be £16 500 per annum. As the authors point out, however, 'to a large factory this is peanuts and is no doubt one of the reasons why accidents are not regarded as a serious problem'.

The straightforward loss of productivity from injuries, however, cannot represent all of the cost in any cost/benefit analysis. As Powell *et al.* point out, the burden of an accident is not just a burden on the factory so much as on the community: 'A factory may have a first-aid service included in its on-costs, but the community has a hospital service, a national insurance scheme, and a legal service to pay for, thus relieving the factory of further responsibility for the people it maintains. The factory produces goods and injuries; the community at large pays for both.'

Bearham (1976) takes this point a little further and divides accidents into three types:

1. a lost-time accident—that is, one which causes a loss of time beyond the shift during which the accident occurred due, for example, to hospitalization;
2. a non-lost time accident—that is, one which causes no loss of time beyond the shift during which the accident occurred;
3. a damage accident—resulting in no injury to persons, but involving damage to facilities, equipment or materials.

He further suggests that these three types occur in the ratio of approximately 1:60:400, so that for every lost-time accident occurring in industry, there will be approximately 400 property damage/no injury accidents.

The costs of all three types of accidents, Bearham suggests, can be divided into direct and indirect costs (sometimes referred to as 'known' and 'hidden' costs). Direct costs are the costs incurred through such aspects as the settlement of claims for damages to equipment; compensation for loss of earnings, pain and suffering as a result of injury; legal liability under health and safety acts; and insurance premiums. This type of information is readily available since it is directly paid out. Indirect costs, however, are more difficult to evaluate because they tend to be hidden. However, Bearham's list includes the following items:

1. Safety administration costs: the time of the safety officer and committee dealing with the investigation of the accidents and a proportion of time for any secretarial support.
2. Medical centre costs: the time of the doctors and nurses involved and the medical supplies used in the treatment of injuries. (Bearham does point out that these types of cost would still exist even if accidents were greatly reduced since they promote the wellbeing of employees; they are, however, still costs which result from accidents).
3. Welfare payments: the proportion of the payment made by the company to the employee whilst off work.
4. Ambulance service costs: running and depreciation of the ambulance and the time of the ambulance driver.
5. Cost of time of other employees: including
   (a) the time taken by other employees in helping the injured man,
   (b) the time taken by the foreman in aiding the injured man,
   (c) the time taken by the witnesses in answering questions during the accident investigation, etc.
6. Replacement labour costs: if the man is replaced.
7. Loss of productions costs: due to the unavailability of manpower and machines.
8. Damage to plant and machinery costs: incurred when repairs are needed and replacements are ordered and fitted.
9. Other costs: arising from the investigation of the accident—these include stationery, and secretarial and clerical work performed in actually processing and recording the accident and corresponding with solicitors and insurance companies.

Clearly, when all of these costs are understood and are taken into account, the balance or the cost/benefit equation must tip towards the importance of increasing safety practices. The remainder of this chapter considers some of the reasons (models) why accidents may occur.

MODELS OF ACCIDENT CAUSATION

Behavioural Models

As the name suggests behavioural models imply that an accident occurs due to some deficiencies in the behaviour of the operator in the man–machine system.

Learning theories

The skills which develop at work arise as a result of a complex series of behaviour patterns learned over a long period of time in which feedback, both from the sense organs and in terms of a knowledge of results, plays an extremely important part. One of the basic requirements for learning an action or a skill, therefore, lies in the reinforcement by feedback of the consequences of the response to a particular stimulus. For example, in attempting to knock a nail into a piece of wood a novice carpenter might sometimes hit his fingers (producing negative reinforcement) which would cause him to alter his subsequent behaviour (for example, moving his fingers or altering the hammer swing). Eventually a correct response (hit) to the stimulus (nail) is made, and the resulting (positive) feedback (of a nail properly embedded in the wood) will reinforce the correct behaviour.

This type of learning is very similar to that described when discussing Pavlov's dog who learned to salivate to the sound of a bell, and is known as conditioning. (To distinguish it from the 'passive' behaviour observed in Pavlov's dog—often called 'classical' conditioning—this form of conditioning in which a motor response is made is known as instrumental conditioning.) Two main principles apply in determining the strength of a conditioned response: First, positive reinforcement will tend to make an action more likely to be carried out, and second, the more frequently the action is reinforced (either positively or negatively) the greater will be the learning effect.

Since skilled behaviour is composed of a collection of skilled responses, it is possible to understand how inappropriate (dangerous) behaviour could be incorporated into the repertoire. First, in many respects safe behaviour is negatively reinforcing. It is often time-consuming, it may involve the use of safety clothing and it sometimes attracts unwelcome comments from fellow workers. (Although this attitude is gradually being eroded, in many industries it is still considered 'unmanly' to take safety precautions.) Unsafe behaviour, on the other hand, is often quicker (one talks of 'cutting corners'), more comfortable and is often more socially acceptable. For example Winsemius (1965) describes the operation of a machine used to punch the index letters on the sides of, for example, address books.

In order to start the machine a pedal was pushed down. As long as it was down the machine ran; as soon as the foot was lifted from the pedal the machine stopped almost instantaneously....
The machines were operated by female workers. During their learning period they stopped the machine at every page so that the operator fixed the speed of work. Experienced workers however started the machine after having inserted a booklet and did not stop it again until the whole booklet had been punched.... Experienced workers thus moved their fingers in the same rhythm as the machine.... As the booklet approached completion, the index finger came nearer and nearer to the edge of the punching die, and was indeed very close to it during the punching of the last two or three leaves.
(Reprinted from *Ergonomics*, with permission of the publishers.)

Wheareas the experienced operators were able to maintain this efficient, although unsafe, cycle without halting the machine, sometimes the pages stuck and they would then attempt to unstick them, again without stopping the machine. On most occasions they would be successful in this attempt but the rare accident occurred when they were unable to remove their finger in time before the die fell again on the next cycle.

In this case, therefore, the unsafe behaviour (unjamming a stuck page without stopping the machine) was easier and less time-consuming than the safe operation. Thus it was positively reinforced, whereas the safe behaviour led to fatigue, lost time, and no obvious rewards. Only when the press descended on to the operator's fingers was the negative reinforcement stronger than that resulting from the safe behaviour.

The second principle, the number of times that the event is reinforced, is also important in the application of learning theory to safety. Thus accidents are extremely rare occurrences, even when the behaviour is unsafe. For example, Winsemius (1965) calculated that accidents using the alphabet punch occurred with a frequency of less than 1 in 6 000 000 operations. The stronger negative reinforcement of a punched finger, therefore, is unlikely to be experienced.

Taken together these two aspects (unsafe behaviour is positively reinforcing and its negative effects are extremely infrequent) suggest that training alone is unlikely to reduce unsafe behaviour. Unless the negatively reinforcing aspects have been sufficiently incorporated into the training repertoire, almost by a change in attitudes, the unsafe behaviour is likely to become positively reinforced as soon as the training period ends (as is often the case when learner drivers have recently passed their driving test).

Memory lapses

Many of the tasks commonly carried out at work involve large memory components. Operators have to remember sequences of operations, the meanings of different stimuli and the responses to make, etc. However, just as our capacity to process information is limited, so is our ability to retain it. The more that is needed to be remembered but which is not in the normal repertoire of behaviour, the more likely are particular aspects to be forgotten. Forgetting (the loss, permanent or temporary, of the ability to recall or to recognize something learned earlier), therefore, is a clear candidate for causing accidents.

The reasons why we forget are complex and are still not fully understood. In some cases it may be the result of the repression of the action—a person blocks the recall of the operation. Repression, however, a Freudian concept, is extremely difficult to investigate scientifically. More likely answers lie in the decay of the memory trace over time, the interaction of the memory trace with other traces, or errors in the storage and/or selection of the material to be remembered.

It is common experience that our ability to retain information decays from the time when the information was stored in memory. Originally this was thought to be due to a physiological decay in the brain which loses the information (much like a leak in a cylinder). However, more recent evidence suggests that the laying down of the memory trace may be impaired by the presence of previously learned material (proactive interference) or that its retention is impaired by material which was learned subsequently (retroactive interference). For example, an operator's inability to remember a correct sequence of control actions could be due either to the fact that he had previously learned other, different, sequences which interfere with his memory of the new ones, or that he subsequently has to learn and carry out other tasks, again, interfering with the memory trace.

In most cases, however, a complete memory breakdown of, say, an appropri-

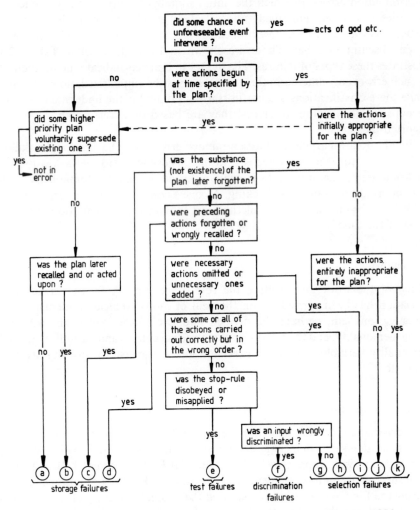

Figure 12.2 Pathways to an absentminded act (from Reason, 1976, reproduced by permission of *New Society*)

ate control sequence, is unlikely to lead to accidents since performance will stop. Accidents are more likely to occur when an incorrect memory trace is recalled—in other words, when the wrong control sequence is carried out without the operator's awareness.

The causes of such 'absentmindedness' have not been studied in such depth as have the causes of forgetting. However, Reason (1976) asked people to keep a diary of their 'non-planned actions' over a continuous 2 week period. From his analysis of the 35 completed diaries (23 women and 12 men), he divided absentminded acts into one of four types: storage errors, in which the original memory trace was incorrectly stored; test errors, in which the operator incorrectly checks the progress and outcomes of a sequence of actions; discrimination errors, in which the initial stimulus to carry out a sequence of activities is wrongly identified; and selection errors, in which the incorrect response is selected. From these four categories, Reason produced 11 'pathways' to an absentminded act. These are shown in Figure 12.2, while Table 12.1 illustrates these types of failures. The figures in brackets indicate the percentage of such errors encountered in the diaries. It must be emphasized, however, that while these classifications are useful starting-points for the understanding and perhaps the rectification of errors, they are based on a very small sample of reported incidents.

Finally, Reason argues that absentminded errors are a hazard for the skilled rather than for the unskilled operator. 'They seem to be a feature of well-practised or habitual tasks which are carried out fairly automatically, with only intermittent conscious checking. As one's conscious mind is rarely still, it is likely to be engaged with something other than the present routine action most of the time.'

Personality

Much of the work investigating the role of personality factors in accident causation has been carried out in the field of motor vehicle accidents. These investigations can, perhaps, be divided into two further areas: the accident prone personality, and the relationship between accidents and various measured personality variables such as intraversion and extraversion, curiosity, aggression, etc.

Accident proneness

The idea that an individual might have some personality trait which predisposes him to accidents was first suggested by three statisticians, Greenwood, Woods, and Yule in 1919. Supported by the Industrial Fatigue Research Board, they published an account of accidents sustained by workers in a munitions factory during the First World War, and showed that a small minority of workers had more accidents than they would have done if chance factors alone were operating.

Table 12.2 Categories of absentmindedness (from Reason, 1976)

Presumed type of failure	Category	Examples and percentages
Storage	A	undetectable errors—both the original intention and the failure to execute it are forgotten. (0)
Storage	B	forgetting items in the plan—part of the plan is temporarily lost but later recalled—e.g. forgetting to post a letter during a shopping expedition. (22.4)
Storage	C	forgetting the plan—where the existence of the plan is not forgotten, indeed its actions are usually underway, but its substance cannot be recalled—e.g. 'Went over to open drawer, but forgot what I wanted.' (6.5)
Storage	D	forgetting preceding actions—preceding actions are either forgotten or incorrectly recalled. Often results in 'losing one's place' in the plan—e.g. boiling a second kettle of water after the tea has already been made. Another common variant is forgetting where articles were put down. (12.5)
Test	E	verification errors—where a planned sequence is terminated too early, or is allowed to continue beyond its intended conclusion. (6.0)
Discrimination	F	classification errors—input wrongly identified often leading to actions appropriate for the erroneous classification but not for the plan in operation—e.g. taking out one's own front door key on approaching friend's house, removing contact lenses instead of combing hair on looking into mirror normally used for adjusting lenses. (10.9)
Selection	G	branching errors—where two different outcomes have initial actions in common and where the actions proceed toward the unplanned (for outcome)—e.g. On passing through back porch on the way to his car, one subject put on the gardening jacket and wellington boots that are kept there (14.3)
Selection	H	misordering errors—the correct actions are carried out, but in the wrong order—e.g. while filling a bucket with water, subject replaced lid before turning off the tap (6.0)
Selection	I	insertions and omissions—when unwanted actions are added to a sequence—e.g. turning on a light when leaving a room in broad daylight; or when necessary actions are omitted—e.g. filling electric kettle, switching on socket but faling to insert plug (12.5)
Selection	J	corrected errors—where a sequence of actions initially deviates from the plan, but is later corrected—e.g. starting to look for a jar of coffee in a cupboard and then realizing that it is always kept in the pantry. (1.2)
Selection	K	total errors—where all the actions carried out during the period specified by the plan are inappropriate—e.g. using a box of matches to light an electric fire. (3.2)

The next major study in the area was carried out by Newbold in 1926. She described a similar analysis on nearly 9000 workers in factories manufacturing products of all sorts from motorcars to optical instruments, chocolates and cardboard boxes, and her data substantiated many of the findings of Greenwood, Woods, and Yule.

Both sets of authors, however, were careful to emphasize that the statistical analysis of accident records is an analysis only of *what* happened—not of *why* or *how* it happened. In addition, they recorded only the occurrence of an accident—not those of near accidents. Merely because an 'accident' was not sufficiently serious for it to be reported does not mean that the individual or the situation is not prone to causing accidents. Furthermore, Newbold pointed out that in her particular sample there were definite indications that youth, inexperience, and possibly even poor health, were associated with accident occurrence, and that it was not possible to rule out increased risk of exposure to dangerous situations as an important factor in the individual differences in 'proneness' to accidents.

The concept of accident proneness as a personal idiosyncrasy predisposing the individual to relatively high accident rate was first suggested by Farmer and Chambers in 1939. They examined the accident records of a large group of drivers and found, once again, that a few had a disproportionately large number of accidents. However, in their description of 'accident proneness', they went a stage further than their statistical data should have allowed, to argue that we can talk about an individual's personal proneness (regardless of the environmental circumstances) rather than the individual's and the environment's tendency to *create* a situation in which an accident is likely to occur.

If, as Farmer and Chambers suggest, accident proneness as a trait is 'no longer a theory but an established fact', it should be possible, using appropriate personality tests, to differentiate between the personalities of 'accident-free' people and 'accident repeaters'. This a number of investigators have tried to do, but without much success. Many traits have been established—aggressiveness, inconsiderateness, hostility, timerity, etc.—but no single personality dimension has been found. For example Tillman and Hobbs (1949) investigated the lifestyle of accident repeaters and showed that they had marked aggressive, antisocial tendencies, more frequent appearances in court in both youth and adult life for non-traffic offences, a higher incidence of venereal disease, and a greater rate of attendance at social and welfare agencies. In their results, however, there is no indication of any particular group of traits consistently emerging to produce an 'accident-prone' individual.

Such difficulties have led to a reformulation of the concept of an accident-prone personality. Reason (1974) for example suggests that if the trait of accident-proneness does exist, it is one which is not a permanent, unchangeable state of the individual, but rather an attribute influenced by both personal and environmental factors which vary from one period to another. He argues that 'examination of accident-repeaters over a lengthy period indicates that they are members of a club which is continuously changing its membership. New people are added, while longstanding members cease to qualify. It is possible that in some people accident proneness is a passing phase, while in others it is more enduring.' Thus age, experience, hazard exposure, and a multitude of other factors, may contribute to cause an individual at some times to be more liable to have an accident than at others. As Reason summarizes 'the whole business [of accident proneness] is a good deal more complicated than was first imagined . . . it is reasonable to assume that each person has a range of behaviour, any part of which may be safe or unsafe depending on the type of hazards to which he is exposed'. Accident proneness, therefore, if it exists at all, is likely to be a fault of the current man–machine system with all of its complexities, rather than only of the man.

Other personality traits

Because it is unlikely that a single dimension of 'accident proneness' exists, it is not necessarily the case that personality traits or dimensions do not predispose an individual towards interacting with his environment in a dangerous manner. Using the Eysenck Personality Inventory, for example, Shaw and Sichel (1971) compared the personality profiles of a group of South African bus drivers with their accident records (categorized as 'good', 'fair', 'poor' and 'bad'). All but one of the 'good' drivers had personalities which were measured as being towards the 'stable introvert' type (thoughtful, peaceful, controlled, calm, etc.), whilst the 'bad' and 'poor' drivers tended towards the 'unstable extravert' ends of the dimensions (outgoing, anxious, touchy, active, etc.). Within this classification, however, their results indicated that the 'introversion-extraversion' personality dimension correlated more highly with their accident criterion ($r = 0.61$). than did the 'stability' dimension ($r = 0.47$). Similar results have been found by other workers. For example, Fine (1963) found that drivers scoring highly on the extraversion scale had more recorded accidents and driving violations than those who scored as introverts or intermediate between the two. Similarly, Mackay *et al.* (1969) have demonstrated that a sample of accident drivers produced a mean extraversion score higher than the norms for the general population.

Aggression is another trait which is likely to lead to a person causing or being involved in an accident. For example, Wrogg (1961) has pointed to the evidence of a marked increase in emotionality, particularly just before an accident occurs. Similarly, Suchman (1965) has explained the greater severity rate of accidents among male drivers in terms of the role of the male in American society as an aggressive risk-taker. Unfortunately, however, very little statistical evidence has been advanced to link aggressiveness *per se* with accidents.

Risk-taking

Although there appears to be no single personality trait of 'accident proneness', an individual's predisposition to create an accident will be related to his willingness to 'cut corners' and to take risks.

Risk is the subjective probability level which an operator ascribes to an event occurring. However, the level of risk is related to the degree of danger and hazard observed by the operator. In this respect 'danger' represents the presence of a situation which could inflict injury or damage if an error is made. 'Hazard', then, is the objective (that is, measured) probability that a man will err in the presence of danger, and risk is the subjective probability (that is, the individual's estimate) that he will err in the presence of known danger. Risk-taking, therefore, is the extent to which an individual will perform an action which he has previously judged to be dangerous and have some degree of hazard.

The relationship between risk-taking and the extent to which the operator feels able to perform the task has been demonstrated by a number of experimenters. For example, Cohen, Dearnaley, and Hansel (1958) compared the anticipated abilities (risk) of bus drivers to drive through a progressively narrow gap with their actual performance. Their results indicated that a group of 15 experienced drivers took less risk (in other words, were more realistic in their assessment of their abilities) than did a group of inexperienced drivers. Furthermore, the one experienced driving instructor whom they tested took no risk at all. Skill, therefore, is related to risk-taking, and a component of skilled behaviour is the ability to be able to predict accurately the outcome of one's actions.

These suggestions lead to a possible means of reducing the degree of risk-taking behaviour. First, the requirement for more appropriate training in avoiding the consequences of risky behaviour cannot be denied. However, designing the machine system so that 'short cuts' cannot be taken, and ensuring that the operator is aware that he will be unable to perform the risky act, should reduce the incidence of risk-taking behaviour. Guards and mechanical or electronic 'interfering' devices can help in this respect.

Age and experience

Age is one of the most frequently considered factors in accident research, but it is a factor confounded by the effects of experience, the task and, indeed, by the effects of the accidents themselves resulting in a tendency to leave the employment. For example it is safe to assume that employees who have a high accident rate will tend to drop out either due to injury, separation or leaving voluntarily. Any simple fall in accident rate with age, therefore, might reflect this trend.

Despite these confounding variables, the overall general trend which can be extracted from the available data indicates that during the teens and early twenties the number of accidents is high; it then drops sharply, levelling out in the mid-twenties. When such factors as employee turnover were taken into account, some authors have then found a slight rise in accident rate towards ages in the mid-fifties.

The effects of experience and training, however, cannot be excluded from these age effects—younger workers are less likely to be experienced and thus more likely to make an error. However, even in studies which match different age groups for their level of experience, it appears that younger workers are still more likely to have accidents. For example, Van Zelst (1954) analysed the accident records of two groups of workers (over 500 people in each group), with both groups matched for level of experience. Throughout the 18 months of the survey period, the younger group (mean age 28.7 years) had consistently more accidents than the older group (mean age 41.1 years).

Many reasons have been advanced why younger workers are more likely to have accidents, although few are based on any empirical evidence. In reviewing the literature, for example, Hale and Hale (1972) suggest that factors such as inattention, indiscipline, impulsiveness, recklessness, misjudgement, overestimation of capacity, pride, and lack of family responsibilities can be blamed. Although it is possible that all of these suggestions are correct in particular circumstances, as the authors point out they are based on very slim evidence and mainly on the subjective impressions of individual investigators.

Whereas younger workers labour under the handicap of the traits of impulsiveness, etc., often ascribed to youth, the problems of older workers lie in their reduced and failing capacities which come with age. For example Murrell (1962a) has suggested that tasks which have heavy perceptual demands, particularly when these are accompanied by speed, are not well tolerated by older people. This was supported by earlier studies which demonstrated that fewer older men were employed on jobs involving severe demands on attention to fine detail, or sustained care and attention. When older men are employed on such tasks, however, the evidence appears to suggest that their accident rates rise.

Life stress

Life stress is a concept which is becoming widely recognized as an influence on both health and behaviour. Significant stressful life experiences (such as a family bereavement, divorce, taking out a house mortgage, etc.) have been shown to play a role in precipitating episodes of serious illnesses and, lately, even in the susceptibility to the common cold (Totman and Kiff, 1979).

The effects of life experiences on a worker's predisposition to accidents, however, have not been studied in any great detail. Verhaegen et al. (1976), however, quote a study which apparently demonstrated that the subjects they studied who were responsible for an accident differed by their heavier 'psychological burden' from subjects who were mere victims of an accident. 'Heavier psychological burden' was defined in terms of worries about children, housing, money, a marital partner, etc. However, when the authors attempted to replicate this study, they obtained conflicting results. In one industrial plant that they investigated, no differences were obtained between the stress experiences of accident causers and accident victims. In another plant, however, they obtained differences in the expected direction for worries about children's health and satisfaction with the home, but not for the subject's own or spouse's health, the educational progress of their children, or the repayment of mortgages.

Despite these conflicting conclusions, some other authors have found a significant relationship between accidents and life stresses. For example, Whitlock, Stoll, and Rekhdahl (1977) compared the life experiences of 71 17–65-year-old orthopaedic patients who had sustained accidents, with a further 71 control subjects drawn from routine surgery wards who were matched for age, sex and marital status. Their results demonstrated that the accidentally injured patients had experienced more changes in their lives over the past six months. Unfortunately, however, a number of the accident victims had been taking various drugs for complaints such as hypertension, depression, diabetes, etc. The possibility that the life stresses had helped to cause these complaints, and therefore that the drugs had contributed to the accidents, cannot be ruled out.

In summary, it is difficult to decide on the importance of life stressors in accident causation. As has been demonstrated throughout it is certainly the case that distraction at an important point in the carrying out of the task can lead to an accident, and personal worries can be distracting. Whether life stresses *per se* (rather than specific worries) can make an individual liable to an accident, however, cannot be decided until more evidence is available.

Accidents as withdrawal behaviour

In 1953 Hill and Trist advanced the theory that a worker may be motivated to have an accident in order to be able to take time off work. Their withdrawal hypothesis was supported by comparing the uncertified absence rates of 200 men who had remained free from accidents with those of 89 men who had sustained one or more accidents. From these figures their results indicated that those sustaining accidents had had significantly more other absences than those who remained free of accidents. Unfortunately, however, since the authors only reported the number of accidents, it is not possible to determine whether the people who had had high accident rates had more accidents or merely reported more. Only the former would directly tend to support the theory. Hill and Trist's survey can also be criticized on the grounds that they did not match the jobs of their high and low accident groups. As Hale and Hale (1972) point out, it is plausible to suggest that those with higher accident rates came from dangerous jobs, which also tend to be heavier jobs making greater physical demands on people. The workers might therefore feel incapable of doing the job when slightly 'under the weather' and so take more days off work, uncertified. This view is lent some support by the fact that Castle (1956) found no relation between accidents and uncertified absence in a photographic processing works where the work was much lighter. Powell *et al.* (1971), however, did show such a relationship in both men and women in a machine shop (heavier work), but not in an assembly shop.

It would appear, therefore, that there is very little evidence to support the hypothesis that workers either consciously or unconsciously cause an accident in order to escape from work.

Physiological Models

Physiological models suggest that accidents are caused as a result of the worker's body not being able to cope with the task requirements. In this respect, disabilities such as deafness, blindness (or at least poor eyesight), a lack of muscular strength, etc., might each be a cause of an accident occurring. Such factors, however, have been discussed elsewhere. This section discusses two factors which might detrimentally influence the body's processing capacities—physiological rhythms and drugs (particularly alcohol).

Physiological rhythms

Our bodies contain a number of systems which regulate functions in a rhythmical fashion. Perhaps the most obvious is the heart which beats consistently throughout life at between about 60 and 80 beats per minute. Less obvious are the rhythmical contractions of the muscles in the intenstine, and the fairly high frequency voltage changes (about 13 Hz depending on one's alertness) generated by the brain (as measured by electroencephalography, EEG). Over slightly longer periods, but still within 24 hours, we have rhythms of waking and sleeping and of temperature regulation. These daily variations are known as circadian rhythms (from the Latin *circa* (around), *dies* (day)). Over even longer periods, up to a month, other bodily functions also occur, for example the changes which occur in the lining of women's uterus' culminating each month in menstruation. In addition, there are thought by some to be cycles of emotionality, and intellectual and physical performance. These longer-term variations are commonly known as biological rhythms, or biorhythms.

Circadian rhythms and accidents

Daily variations in human performance and efficiency have long been recognized. Indeed Rutenfranz and Colquhoun (1979) have traced these observations back to Kraepelin in 1893:

With continued work the performance usually rises until noon. Soon after the meal it has decreased very significantly. Simultaneously a predominance of external associations and lower stability of imaginative combinations can be shown. These phenomena are no indication of work fatigue, since they disappear after 2–3 hours, even when work is continued. (After the noon meal) the performance again increases slowly. Sooner or later, however, without fail, fatigue gains the upper hand.

Rutenfranz and Colquhoun add to these the observation that much greater overall variation can be expected when the night hours are included.

Rhythmical variations in performance on a number of tasks have been shown to occur under laboratory conditions. For example, Folkard and Monk (1978) demonstrated that the amount of information able to be remembered is influenced by the time of day at which it was presented, although there was no such influence in the ability to retrieve information from memory. Using a number of tasks (reaction time, calculations, card sorting, letter cancellation, etc.), Blake (1971) demonstrated increases in performance between 08.00 and 10.30 hours with a 'post-lunch' dip at about 14.00 hours. However, on one task (digit span—that is, the number of digits which can be remembered) the post-lunch performace dip did not occur until late evening. Rutenfranz and Colquhoun (1979) interpret these findings to imply that just as *physiological* processes show different circadian rhythms, so do different *psychological* functions.

Industrial studies relating circadian rhythms to performance, however, are fewer, a fact which Folkard and Monk (1979) ascribe to factors such as union or management opposition and problems in finding appropriate measures. However there appears to be agreement between the studies which have been carried out, primarily in relation to shiftwork. For example, Hildebrandt, Rohmert, and Rutenfranz (1974) demonstrated a typical 24-hour circadian pattern of unintended brakings by train drivers (in other words, when the train came to a stop because the driver failed to respond to a signal in the cab). Other, similar, conclusions from different situations have been reviewed by Folkard and Monk (1979).

The relationship between such rhythms and accidents, and the implications for shiftwork, have been discussed by Colquhoun (1975b). Investigating sailors on a 4-hour watchkeeping schedule, he demonstrated a circadian variation in alertness (detecting a very faint auditory or visual signal) and in reaction time, implying that the potential for accident causation also varies rhythmically. Similar results were obtained using 8-hour shifts, without any 'weekend break'. Furthermore, Folkard, Monk, and Chobban (1978) demonstrated a highly significant rhythmical variation over the day in the frequency of 'minor accidents' occurring to patients during their stay in hospital. The changes in the rhythm coincided well with the timings of the nurses' early-day, late-day and night shifts. Although the authors point out that the relationship between the two functions could be spurious (the tendency for accident peaks to occur in early morning or late night could be attributed to a patient's need to urinate immediately before going to sleep or on awakening) they also suggest that a causal relationship might exist.

All these data, therefore, do suggest that the potential for accidents could follow some sort of circadian rhythm.

Biorhythms and biological rhythms

Whereas circadian rhythms have been demonstrated, and occur over a 24-hour period, the evidence for biorhythms is not so secure. If they exist they occur over periods of up to a month. The biorhythm theory suggests that there are cycles of 23, 28 and 33-day duration that govern physical, emotional and intellectual performance respectively. Each cycle is described by a sinusoidal curve having a positive and a negative phase, the cycle starting on a positive half-cycle at the moment of birth. (Why the theory should consider that the rhythm should begin at birth rather than from the time of conception, however, is not fully clear). The positive phase corresponds to periods when performance is best, the negative phase to periods of poorer performance, and the crossover points are termed 'critical'. These critical periods are said to represent times of poorest performance and of greatest accident susceptibility. The length of this critical period is usually considered to be 24-hours. According to the theory accidents would be more likely to occur on critical days; in other words, on days when more than one cycle is in the critical phase, or days when one cycle is in a critical phase and the others are in a negative phase. In addition, days when all cycles are in a negative phase are said to be critical because of the change from 'positive' to 'negative' behaviour—'just as the most 'dangerous' time for a light bulb to burn out is when power surges through it as it is switched on' (Gittelson, 1978).

Controversy rages in both the scientific and popular press as to the existence of such cycles—whether they exist at all and, if so, whether they are fixed from birth—and it is not the purpose of this chapter to fuel the arguments. However, the theory does propose testable hypotheses with respect to accidents, and a few investigators have attempted to correlate recorded accident data with that which would have been predicted according to the theory. After reviewing these studies, however, and after carrying out a study of their own on over 4000 pilot accidents, Wolcott et al. (1977a,b) concluded that there is no statistically significant correlation between biorhythms and accident occurrence.

Support for these conclusions have also been given by Khalil and Kurucz (1977). They investigated the relationship between the timing of 63 aircraft accidents in which 'pilot error' was said to be to blame, and the position of each pilot's biorhythm. Again their analyses indicated that 'biorhythm' had no significant influence on accident likelihood.

Alcohol

A great deal of work has been conducted on the effects of alcohol on judgement and on the performance of skilled tasks. In general, the experimental work has shown that alcohol has a deleterious effect on performance because of its effects on vision, perceptual motor functions, judgement, reasoning and memory. Jellinek and McFarland (1940), for example, reporting on experiments of alcohol effects up to 1940, decided that the speed of reaction is 'lengthened' by alcohol, despite faulty techniques in almost every experiment. Similar conclusions were reached by Carpenter (1959) using alcohol levels as low as 40 mg/kg of body weight.

Many of the studies, however, have been related to the task of driving although similar conclusions have been reached. (In these cases the normal way of expressing the extent of intoxication has been to measure the volume of alcohol present in 100 ml of blood, that is, to produce a percentage of blood alcohol level). Bjerver and Goldberg (1950), for example, used a special track designed to measure the ability to operate a car within close limits. Such manoeuvres as parallel parking, driving out of a garage, and turning around in a narrow roadway were required. Their results showed that the time for skilled drivers to perform tests correctly was significantly lengthened when the subjects were at a blood level of about 40 mg/100 ml of blood.

Because of studies such as these which have consistently shown a positive relationship between alcohol level and performance decrement, most countries have fixed by law a blood level beyond which it is an offence to drive. These levels vary from 50 mg/100 ml of blood to 150 mg/100 ml.

The role of alcohol in industrial accidents, however, has been less extensively studied. Some studies have certainly indicated, as did the Bjerver and Goldberg study, that the accident rate rises with the proportion of drinkers in the workforce. However, these are cases of alcohol consumed whilst at work—the question might well be posed as to the accident likelihood of the alcoholic worker who drinks heavily but not at work. As Trice and Roman (1972) point out, it is logical to expect a deviant drinker to have more accidents than other workers; he would be less steady, and his coordination, timing, motor responses and sense of danger would be impaired. In addition, hangovers on the job could also cause accidents.

Unfortunately, the research data which have emerged over the years make generalizations about the relationships between alcohol and industrial accidents difficult. For example 'Observer' and Maxwell (1959) ('Observer' was a pseudonym) concluded that problem drinkers do have more accidents than other workers. They recorded three times as many accidents caused by problem drinkers than by members of their control group. Furthermore, in another study Brenner (1967), who considered the death rate due to fatal accidents in a sample of 1343 patients in four alcoholism rehabilitation clinics, concluded that the alcoholics were seven times as likely to become the victims of fatal accidents than were his non-alcoholic controls. Of the 35 fatal accidents, however, only one was reported as occurring whilst at work.

A direct comparison of alcoholism and accident rate, however, has many pitfalls. The first concerns the problem of accident statistics as a source of data; they give no indication of the situation leading to the accident, nor of the accident severity. Second, the evidence appears to suggest that the risk of accidents from alcohol is related to the extent to which the alcoholic is able to cope with his 'disability'. When breaking down their accident data for different age groups, for example, 'Observer' and Maxwell considered that while problem drinkers do account for more accidents than other workers of the same age during the early problem-drinking years, by the time that the problem has developed fully problem drinkers are generally no more prone to on-the-job accidents than other employees.

These findings have been explained, by both 'Observer' and Maxwell (1956) and Trice and Roman, (1972) as being due to the improved coping behaviour of the committed alcoholic and his colleagues. Such factors include the routine nature of the job, extra caution, protection from accidents by fellow workers, absenteeism when incapacitated (for example, from hangovers), and assignment by supervisors to less hazardous work. Apparently these factors—or at least some of them—do not play the same role during the early problem-drinking period in protecting the individual from on the job accidents.

SUMMARY

Although one of the main themes of this book has been to illustrate ways in which the environment can be designed to promote safety, this chapter deals specifically with human accident behaviour. It has taken as its thesis the suggestion that accidents occur when the environmental demands surpass the operator's capabilities. Various behavioural and physiological models were proposed to account for why the operator's capacities may be reduced at the important point in time.

CHAPTER 13

Inspection and Maintenance

In many respects a chapter dealing with inspection and maintenance in a general ergonomics text could be considered as somewhat out of place. Ergonomics is an applied discipline; an ergonomist's role is to apply his knowledge and experience concerning man's interaction with his environment to ensure that the environment fits the man. Throughout this book, this theme has been stressed and various routes to this goal have been illustrated. On the other hand, both inspection and maintenance are areas of ergonomics application, in the same way that the ergonomics of space flight, of agriculture or of medicine, are different areas of interest to different ergonomists. Why, therefore, should inspection and maintenance be singled out for special consideration?

The responses to this question are two-fold. First although it is true that inspection and maintenance are specific areas of interest, both exemplify the need to consider the total working system. They both provide ideal examples of how ergonomics can be implemented to help to solve specific problems. Second, both do need to be considered in their own right when designing a system, just as much as the design of individual displays, controls or workplaces. It is a fact of life that components within a system occasionally break down, and only if the maintainability of the system had been considered in the original design can the fault be rectified quickly and easily.

The distinction between inspection and maintenance is, in many ways, a distinction without a difference. It is hard to draw the dividing line between the two processes, since both involve the detection and elimination of a fault. However, for the purposes of the present discussion inspection will be conceived as being the process by which a faulty component is detected, and then rejected. Maintenance, however, involves the rectification of the fault, usually while within the system itself.

INSPECTION

With the increasing efficiency and output of machines, the task of an industrial inspector has become more complex over the years. Faster, more productive, more accurate machinery is likely to produce goods at a pace which causes him more difficulties in detecting product defects. In addition the consequences of missing a defect have also increased in importance. At one level a lack of vigilance or bad inspection performance could cause dissatisfaction among customers, evidenced by the return of faulty goods and perhaps a loss of custom. At another level, however, poor inspection may raise production costs by causing unnecessary machine stoppages, by interruptions in the production flow, or by wasted material which had escaped the inspector's surveillance.

It is possible, of course, to design machines to carry out the inspector's job with almost 100 per cent efficiency but, at present, this is feasible only if one or two specific types of fault are possible in any one type of product. As the list of functions best carried out by man and machine, discussed in Chapter 1, illustrate man is normally more efficient in complex, variable situations, particularly when the perception of fine form, depth or texture is required. It would be difficult to find a machine that can match human powers of examining for numbers of different faults at once over a wide range of products, or that can assess the 'finish' of a surface or the uniformity of colour, or can take note of a rare, perhaps unspecified, fault. Man is as yet unsurpassed where discrimination is needed between a large number of faults, where classification and diagnosis are needed, and where eliminating the fault involves liaison with other men or machines.

A Simple Model of Industrial Inspection

Thomas (1968) suggests a very simple model to explain the extremely complex process of inspection. It is based on the reports of inspectors who suggest that their task is to keep the idea of what constitutes a 'good' item in mind, and then to look for faults in the object being inspected. In other words they appear to be comparing the appearance of the 'external' object with an internal representation of the acceptable item. If there is any mismatch between the two then the decision is made to reject the item. This can explain, therefore, why if an inspector is asked to identify the fault which has caused him to reject the item, sometimes he will need to re-examine the item. He will have learned not to waste time in the additional scanning activity needed to classify the nature of the mismatch when all that he needs to do is to identify its existence by comparing his model with his object.

Thomas also points out, however, that an inspector's job often requires him to recognize the nature of the faults; he may be instructed to send defective items for appropriate remedial treatment, or he may have to record on a chart the frequency with which various faults occur. To do this the inspector's model must contain sets of alternatives, similar to subroutines in a computer programme, so that he can continue his scanning beyond the stage of identifying a rejectable mismatch, to match the object's appearance with the detailed characteristics of specific 'fault subsection' of his model.

Finally Thomas's model recognizes one very important consequence of repetitive inspection—that of a drift in inspection standards. Since the task involves matching an external stimulus to an 'intenal' representation, it is quite possible that the internal representation may undergo small, subtle changes, perhaps because of past experience, social pressures, or a host of other factors.

These changes will cause a corresponding change in the criterion which the inspector uses in deciding to accept or reject an item.

The ergonomist's task, therefore, is to help the inspector to obtain the best possible stimulus of the object which he is to perceive, and also to assist him to develop the most appropriate internal model. Both of these aspects can be carried out by implementing many of the ergonomics principles discussed elsewhere in this book. This, however, is an extremely complex task, as will become evident.

Factors Affecting Inspection Accuracy

Megaw (1978) suggests four groups of factors which could influence inspection accuracy.

Into the first category fall all the variables which the inspector brings to the task: his visual ability and performance; his age and experience at inspection; his personality and intelligence; etc. There is little that the ergonomist can do to affect these variables alone although, having recognized their effects, other aspects of the work can be altered to either compensate for or accentuate them.

Some of these other aspects fall into Megaw's second category—the physical and environmental factors pertaining to the environment in which the inspection task is carried out. Thus aspects such as the lighting, the workplace layout, the background noise and the presence or absence of visual aids, are important.

The other factors over which the ergonomist may have some control appear in Megaw's third category: the organizational factors, and the characteristics of the inspector's overall job. These include the number of inspectors available; the type of training and feedback received; rest pauses and shiftwork; social aspects of the job; etc.

Finally, Megaw collects together factors which pertain specifically to the inspector's task—whether the objects to be viewed are moving or stationary; the probability that any one piece is defective; the complexity of the object; the density of the items; etc.

These factors are shown in Table 13.1, and some are discussed in more detail below.

Inspector Variables

Visual acuity

Since, in the majority of cases, inspection is solely a visually based task, it is reasonable to expect that inspection performance will degrade as the inspector's visual ability decreases. Furthermore, it is not unreasonable to expect the reverse—that inspection performance increases with increasing visual ability.

For some time the criterion of visual ability for an inspection task has been some measure of visual acuity. As was discussed earlier (see Chapters 2 and 11), acuity is a measure of the eye's ability to be able to perceive fine detail,

Table 13.1 Factors which could affect inspection performance (from Megaw, 1978)

Subject factors
Visual acuity
Eye movement scanning strategies
Age
Experience
Personality
Sex
Intelligence

Physical and Environmental factors
Lighting
Workplace
Aids
   Magnification
   Overlays
   Viewing screen
   Closed circuit TV
Background noise
Background music

Organizational factors
Numbers of inspectors
Briefing/instructions
Feedback
Feedforward
Training
Selection
Standards
Time-on-task
Shift hours
Rest pauses
Social factors
Motivation
Incentives
Job rotation

Task factors
Inspection time
Paced versus unpaced tasks
Viewing area
Fault probability
Fault conspicuity
Product complexity
Density of items

and is normally calculated in terms of the minimum angle needed to be subtended at the eye by the test object. As an indication of the relationship between acuity and performance, McCormick (1950) measured the visual acuity of over 5000 different employees doing different jobs (the sample included sewing machinists and close assembly workers in addition to visual inspectors). The criterion of job performance varied from plant to plant and from job to job, and were based on such factors as production data or earnings, or were the results of employer ratings. For this reason each employee was only identified as being 'high' or 'low' on some criterion of job performance.

The relationship obtained by McCormick between visual acuity and performance is shown in Figure 13.1. It would appear from the graph, for example, that approximately 55 per cent of the employees having fairly good near acuity (minimum visual angle of about 1 min of arc) were from the high criterion group, whereas all of the employees having poor acuity (for example, 10 min of arc) were from the low criterion group.

Whereas McCormick's data suggest a trend of reduced performance with much reduced acuity the converse is by no means the case—about 45 per cent of the employees with good acuity fell into the low criterion category. As Megaw concludes 'the results demonstrate that the possession of good acuity does not ensure that person as a high criterion inspector, merely that the chances are relatively high'.

It is not altogether surprising that a perfect relationship between acuity and performance was not obtained. Being a laboratory-based measure, the acuity task does not test the operator's performance within his work system; most acuity tasks present stimuli to the subject, while most of the objects in most inspection tasks are moving. Most acuity tests are carried out under 'ideal' environmental conditions; this is not necessarily so at work; and most acuity

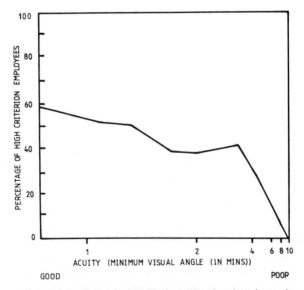

Figure 13.1 The relationship between visual acuity and the percentage of high criterion employees (from McCormick, 1950)

tasks carry no reward or punishment for the testee. Again, this is not necessarily the case at work.

Given these drawbacks it is unlikely that a simple relationship can, or will, be found. Nevertheless some attempts at more realism have been made. For example Nelson and Barany (1969) have developed a test of dynamic visual acuity where the test material and conditions incorporate many of the features of the inspection task for which the inspectors are being selected. However as Magaw points out, such tests are difficult and expensive to develop if they are to be in any way predictive.

Another approach has been taken by some workers who feel that the size of the inspector's visual field (essentially a test of peripheral acuity) should be related to inspection performance. Johnston (1965), for example, investigated the relationship between the size of visual fields of observers and the time required to locate targets on static displays. Her data suggested that people with large visual fields can find targets more rapidly than those with small fields.

Other studies have indicated that an important difference between best and poorest performance is the length of time which the observer takes between eye fixations during the search task. (Our eyes scan the visual field as a series of 'jumps', or saccades, rather than as a smooth scan.) For example, Boynton, Elworth, and Palmer (1958) found that their best subject averaged nearly twice the length of sweep between fixations as did their 'worst' subject, so they concluded that longer sweeps are an aid to efficient search. This is presumably because longer sweeps between fixations result in fewer fixations of the eyeball; thus it takes a shorter time to scan the material so that more scans can be made in the time available.

It still remains the case, however, that all these are results obtained from subjects in the laboratory and have not been validated in the case of industrial tasks.

Age and experience

It has already been noted, when considering safety and accidents, how difficult it is to isolate age as a single variable. Increasing age is often associated with failing health, which can detrimentally affect performance, and increased experience, which can improve performance. It is not surprising, therefore, that few studies are available which have investigated the effects of ageing on inspectors.

Jamieson (1966) recorded the ages of his inspectors when carrying out two types of task: either inspecting switches for mechanical faults as they left the assembly line (which involved a minimum of 190 discriminations per switch); or checking telephone exchange racks for faulty soldered joints and connections. (Unfortunately no indication was given of the experience of his subjects.) In both types of task, inspection performance increased with increasing age, although the relationship was significant only in the telephone rack task. These results were confirmed with a further group of workers. Unfortunately, however, since it is likely that the sight of many of the older workers would have begun to fail, the study did not answer the question of whether experience and other variables such as conscientiousness outweighted the negative factors of ageing.

Physical and Environmental Factors

Lighting

Because inspection is primarily a visual task, it should be obvious that the inspector's main environmental needs are for adequate illumination. This implies the provision of enough light to be able to view the objects without undue fatigue, but not so much light that glare occurs. For these reasons all of the factors discussed in the previous chapter which can affect visual performance need to be considered.

However, two additional aspects must be emphasized when considering the illumination of the workplace for inspection tasks: first, the type of lighting, and second, its position.

Regarding the type of lighting, Lion (Lion, 1964; Lion, Richardson, and Browne, 1968) performed a series experiments to compare inspection and manipulation performance using two types of illumination: fluorescent and tungsten lights. In the first experiment she asked her subjects to carry out a variety of tasks (grading ballbearings, threading needles, reading numbers, and measuring the lengths of rods) using either an 80 watt 'warm' fluorescent tube or a 100 watt clear tungsten bulb. Both produced an equal level of illumination of 14 lumens/sq foot at the subject's work surface. Her subjects took a significantly shorter time to complete the tasks using a fluorescent rather than a tungsten bulb (in other words, more units were performed in the time allotted), although the number of errors made in each case were not significantly affected. Since all the tasks used stimuli which had a high degree of reflectivity, Lion argued that the performance difference was due to the filament (tungsten) light producing a concentrated light source and thus more glare. This was not enough to produce reduced performance in terms of increased errors but it was enough to slow down the rate of work.

In her second series of experiments, Lion and her colleagues employed tasks more akin to those encountered in an inspection task, in other words, searching for defects in small objects moving on a conveyor belt. The defective objects were either buttons with off-centred holes or black plastic discs with a broken white design etched on to their surface. The items were equally spaced 2 in. (5 cm) apart and travelled at a rate of 7 in./sec (17.5 cm/sec). Again tungsten (four 200 watt bulbs) and fluorescent (four 80 watt bulbs) room lighting was compared, which produced equivalent illumination levels of 30 lumens/sq foot at the working level.

In this experiment the results were slightly less conclusive. Lion's subjects overlooked significantly fewer faulty broken designs on the disc-sorting task under fluorescent than tungsten lighting, but there were no significant differences on the button-sorting task. The authors explain the discrepancy by pointing to the type of task involved in each case. The 'faulty design' task required the operator to perceive a break in a line which is essentially simply an acuity task. The button-sorting task, however, required a more 'global' picture of the object to be sought, since the criterion of rejection was whether or not the four button holes were central—that is, equidistant from the button circumference. The authors suggest, therefore, that it is only the visual acuity task which is at a disadvantage under point-source illumination, possibly because of the glaring effects. However, more data would be required before the assertion could be substantiated.

With respect to the position of the light source, Fox (1977) describes an interesting series of experiments designed to investigate problems experienced in the quality control of coins produced at the Royal Mint. In one investigation he compared the error detection rate when a bank of five 40 watt fluorescent lights were placed immediately above the workplace in a horizontal arrangement, and when the bank was angled on its supports. This second condition purposefully produced specular glare from the coin surfaces. The interesting aspect, however, was that the quality of the glare changed with the absence, presence, and even type of coin defect. Thus the difficulty of the inspector's task was reduced to one of detecting the type (intensity) of glare rather than the presence of a misshapen or damaged coin (in other words, contrast between the faulty part of the item and its background increased). Although the fault detection rate did not differ using the two arrangements, Fox argues that the angled position allowed a more 'standard' performance from the inspectors. Unfortunately, he provides no details of the angle used, and made no attempt to carry out a controlled study of this question. Nevertheless, the use of glare to benefit the inspector is an interesting concept, although it should be pointed out that the prolonged use is likely to cause visual fatigue.

Visual aids

Although the level and type of illumination are extremely important aspects of the visual environment, as was pointed out in Chapter 1 the inspector's task can often be helped by the use of visual aids such as magnifiers or visual overlays.

Magnification is an obvious aid for the inspector—particularly if the task includes significant acuity components. Increasing the apparent size of the object will also increase the size of the defect to be detected. However, the term 'magnification' should not be limited solely to discussing 'size'. As Spencer (1968) points out, the term can be used in its broadest sense to denote an increase of any type which enhances the perceived magnitude of the object or event being examined. Thus, an inspection task which involves detecting small colour differences may be aided (that is, the differences 'magnified') by using different coloured lights; a specific signal in an auditory discrimination task can be 'magnified' (that is, the tone made more apparent) by modulation; etc.

If the apparent size of an object is magnified by a magnifying glass, however, Spencer recommends the use of a jig or holder for the object being inspected. This has two advantages: First, because most optical magnifiers have a restricted depth of field, the jig will immediately place the object at the appropriate distance from the magnifying glass, ensuring a standard appearance of all the inspected objects; second, although the natural hand tremor does not interfere with the normal perception of objects, behind a magnifying glass the movement caused by tremor is increased and this might affect visual acuity performance, in the way described when discussing the effects of vibration.

A note of caution must be sounded, however, against the 'unrestricted' use of linear, optical magnification. From the scanty data available it would appear that, as with most encounters with the physical world, there is an optimum level of magnification, and levels both above and below this tend to result in reduced performance. One of the first investigators to demonstrate this effect were Nayyar and Simon who, in 1963, asked subjects to perform a task which consisted of picking up, transporting, and assembling metal dots 0.01 in. (0.25 mm) in diameter. Their subjects performed the task under three magnifications: $20 \times$, $30 \times$, and $40 \times$, and the results indicated that the subjects took a shorter time to pick up the dots using the $30 \times$ magnification than either of the other two. In a subsequent study Simon (1964) demonstrated that this optimum magnification was strongly related to the size of the object to be manipulated. Again comparing $20 \times$ and $30 \times$ magnification, he demonstrated the advantage of the $30 \times$ magnification for a 0.01 in. (0.25 mm) dot but, when a larger dot was used; (0.015 in.; 0.3 mm) the relative advantages of the two magnification sizes were reversed.

These results suggest, therefore, that the important consideration is not so much the magnification size but the size of the image on the retina. This contention is supported by Smith and Adam's (1971) results when they demonstrated that no matter what size object was used, the optimum magnification was that producing an image which subtends an angle of about 8–9 mins of arc at the eye. This factor remained constant even under different illumination levels.

Visual aids in the form of overlays have also been shown to be beneficial to inspection performance. Using such aids it is often possible to help the inspector to spot defects by making them stand out from the background. The inspection task thus becomes one of looking for things that are highlighted, as opposed to looking at each and every individual item. For example Teel, Springer, and Sadler (1968) describe one such 'mask' which was used to overlay photographic negatives used in the microelectronic industry. Using these masks, defects in the placement of components in the original negatives appeared as misplaced lines or circles. Clear and significant increases in the detection of faulty negatives were demonstrated which, the authors calculated, would save the company approximately $ 10 000 per year.

Noise

The efects of noise on performance have already been discussed in Chapter 10, and will not be discussed further here. However, it should be remembered that under some conditions the 'correct' type of noise might increase performance. Thus Poulton (1977) has suggested that noise may act as an arouser when dealing with boring, repetitive tasks, while Fox (1971) has demonstrated significant improvements in performance with background music.

Organizational Factors

In addition to the factors which the physical environment brings to the inspection task (in particular the quality of the illumination and the noise), the social and organizational environments should not be forgotten. These include the way in which the inspector interacts with his colleagues and his organization, and the arrangement of his job as a whole.

Social factors

The inspection model described earlier relies heavily on the subjective judgement made by the inspector as to whether or not the object in his field of view matches the ideal model in his 'mind'. However, it is well known that judgements of this type can be greatly influenced by social pressures on the observer to conform to the group norm. As a famous example Asch (1952) asked subjects to judge the length of a line. When subjected to various types of group pressures, individual's estimates could be observed to conform strongly to the group's estimates, even when, by manipulating 'stooges' in the group, the group decision was wildly in error.

As an example of how such pressures to conform may influence an inspector's judgement Belbin (1957) describes how, in a knitwear factory, inspectors reported very few 'seconds'; instead they classified nearly all defective work as 'mendable'. The point of this observation is that while the knitters lost pay for defective work, they lost more for 'seconds' than for 'menders'. It would appear, therefore, that lacking clear standards, the inspectors could not resist pressure from the knitters to class most defects as 'menders' rather than 'seconds'.

In addition to pressures from the workforce, an inspector often has to face similar types of coercion from his production management. As a part of the production team, the inspector's task includes the maintenance of a certain production flow. As Thomas and Seaborne (1961) point out, when the production department is in difficulties there will be considerable pressure to maintain a certain output by accepting some components which, in more favourable circumstances, he might have rejected. On the other hand, when output is high his criterion for acceptance could be made more stringent without an appreciable reduction in the productive flow.

McKenzie (1957) summarizes some of these aspects by pointing out that, particularly when objects are individually made, 'in the factory inspection is always, if implicitly, of people; inspection decisions about a man's work directly reflect on him'. Again, therefore, the implication is made that there are a number of social pressures placed on the inspector not to reject an item. These pressures may arise from the workplace (increased rejection might result in reduced bonuses; increased rejection implies reduced work quality) or even from the production management (increased rejection means reduced output schedules and higher costs).

Feedback

The beneficial role of adequate feedback to performance has been stressed throughout this book, and its effects on inspection performance are no different. Since inspection involves the comparison of the external object with an 'internal' model it is essential that the inspector is given every opportunity to 'update' his model in the light of experience. Two separate roles can be identified for feedback: first, the maintenance of motivation, and second, the provision of information regarding efficiency.

The value of feedback as a motivator is difficult to assess in any controlled way. However, Megaw considers that there is a 'general consensus of opinion' that feedback helps inspectors to appreciate the importance of their work and thus become more involved in their task.

As with most tasks, giving the inspector details of his performance standards is likely to increase his accuracy. If he realizes that he is rejecting too many good items, missing too many faulty items, or performing well, he can adjust his 'internal model'—his criterion of rejection— accordingly. That feedback is effective in this role was demonstrated by Drury and Addison (1973) on the inspection of complex glass items. For a controlled period the output of their inspectors was checked by a second group of inspectors who informed the first group immediately they had made an error (that is, missing a faulty item or rejecting a good one). The introduction of this undelayed feedback resulted in superior performance such that the probability of one of the first group's inspectors missing a fault was halved after the change in procedure. However, the authors rightly caution that the improvement may not have occurred solely as a result of improved feedback since the feedback change also involved other procedural changes, including changes in the method of supervision.

Rest pauses

Inspection is normally a light, repetitive task, the speed of which is governed by the speed of the industrial process that produces the objects to be inspected. It is also a task with a high vigilance component. Since most people require a short break after an activity has been continued for some time, even if highly motivated, it would not be unreasonable to suggest that the same principle holds for industrial inspectors.

Murrell (1962b) has proposed the concept of an 'actile period', which he defines as 'a period during which there is a state of preparedness to respond optimally to stimulation either discretely or continuously'—in other words, the period during which a worker can maintain concentration on the task in hand.

He further proposes that this period has finite duration depending on the individual involved and the task undertaken. At the end of the actile period, performance will start to deteriorate and, if activity is continued, the deterioration will become serious and output will begin to fall. In addition to the projected fall in output, a further important consideration of an actile period concept is that, if no break is forthcoming, the operator will himself impose such a break of a few seconds at a time. Thus Branton and Oborne (1979) have inferred the presence of 'mini-sleeps' among anaesthetists who had to monitor long, boring operations.

The results of a number of studies are now available which have demonstrated the need for rest pauses throughout the working day (see, for example, Murrell, 1971). However, Colquhoun (1958) has demonstrated the need for such pauses at more frequent intervals during inspection work. In a laboratory study in which his subjects were required to spot faulty discs on a moving drum, his results indicated that when the work was performed uninterrupted for a period of an hour the efficiency, although initially high, declined after about 30 minutes. If a 5 minute rest pause was inserted at this point, however, the performance was maintained at a high level throughout the hour.

Task Factors

So far, consideration has been given to the aspects of the inspection task which the inspector himself brings, and those which are added by the physical, social and organizational environments. The final aspect of inspection which needs considering is the factors supplied by the task itself. In this respect, important considerations are the pace of the work, its complexity, and the likelihood of a fault being present.

Speed of working

Drury (1973) has pointed out that the rate at which an inspector can work is likely to be one of the most vital pieces of information required when designing industrial inspection systems. For a fixed production rate, knowing the time required for the inspection of an item determines the manning level and hence, to a large extent, the design of the physical facilities. Poor quality control is likely to occur if too slow or too fast a pace is imposed or too few or too many inspectors are installed.

In his investigation of the quality control of coin production, Fox (1977) varied the speed at which the coins passed the inspector's field of view, either in pairs or in triplets. Whereas no significant difference in inspection accuracy was obtained between the inspection speeds when only two coins appeared at any one time (inspection time/coin varied between 0.05 sec and 0.125 sec), when three coins were presented at a time a significant increase from 70.8 to 82.7 per cent correct detection occurred (inspection time/coin ranged from 0.033–0.83 sec).

The increase in detection efficiency with inspection time, however, is not linear. In fact most of the studies reviewed by Drury (1973) demonstrate an exponential increase in fault detecting as the time available for inspection is increased. The curve flattens off below perfect detection, with the critical inspection time unfortunately depending largely on the complexity of the task and on the conspicuousness of the fault. No general guidelines can be given for appropriate times, therefore. In addition to this trend, however, there appears to be a very slight, although generally non-significant, increase in the probability of making the other type of inspection error (for instance, rejecting a good item rather than a bad one) with time. Although this trend is very slight, and in some studies completely absent, it does imply again that a slight change may occur in the inspector's criterion of a 'good' item.

Product complexity

The importance of product complexity in affecting the optimum time per item available for inspection has already been considered. However, even when inspectors are allowed long periods for inspection, the complexity of the items can have a detrimental effect on performance. Thus Harris (1966) demonstrated an almost linear reduction in detection performance with increasing item complexity (complexity was defined in two ways—either according to a rated complexity index, or in terms of the number of components—circuit boards, transistors, wires, etc.—in the equipment). Since the detection decrease could not be overcome simply by increasing inspector's time, Harris suggests using visual aids with complex equipment. As has already been seen, these can be useful in such circumstances.

The probability of a defect occurring

Since most of human behaviour is shaped by the probability of an event occurring or not, it is not unreasonable for this also to be the case for industrial inspectors. Thus it has already been discussed how unsafe behaviour may arise as a result of faulty learning and the positive reinforcement of unsafe acts, or how perceptual 'set' from past experience can cause an operator to misperceive his displays, and the same is true for vigilance performance. In a vigilance test, for example, Jenkins (1958) demonstrated that as the number of signals to be detected was increased, the number of correct detections also increased. This view was later altered slightly, however, to suggest that it was not the frequency of 'wanted' signals *per se* which was the important criterion, but the ratio of wanted to the total number of signals (that is, the probability of a wanted signal occurring) (see, for example, Jerison, 1966).

These assertions were tested in a factory environment by Fox and Haslegrave (1969) who varied the probability of a defect occurring in batches of screws and bolts. Their results indicated that detection efficiency increased almost linearly with an increase in the fault probability. However, a corresponding increase in false detections (that is, the rejection of a good item) also occurred. Again, therefore, it appears that the inspectors were altering their criterion of rejection rather than simply becoming more efficient detectors. Increasing the probability of a defective item occurring, therefore, simply causes more items to be rejected, rather than more items to be rejected *correctly*.

MAINTENANCE

From the discussion so far, it is clear that inspection is essentially a perceptual problem—the operator's task is to perceive faulty aspects of an object and to make judgements on the basis of these perceptions. Any aspect of the task, operator, environment, or organization which affects his perceptual ability is also likely to affect his inspection performance. Similarly, any way in which the ergonomist can increase the operator's perceptual acuity should help to increase inspection performance.

If this very crude analogy is extended along a similar theme, it could be argued that maintenance performance contains significant proportions of motor and control components. Thus, although the maintenance operator needs, perhaps, to scan his 'faulty' equipment to isolate the fault, his subsequent problems are likely to centre around the accessibility of the component, the controllability of his equipment and/or his own ability to manipulate, lift or pull out the relevant machinery. In essence, therefore, this crude distinction could be explained in terms of the aspects of the body sensory nervous structure discussed in Chapter 2 (inspection) and its motor control structure considered in Chapter 3 (maintenance).

It is a fact of life that no machine or component is likely to survive the life of the total system. Unless we accept the concept of a totally 'throwaway' society in which any machine which breaks down is discarded for a new model, the continual maintenance and replacement of faulty parts will be necessary. However, although great strides have been made in reducing the failure rate of various components (particularly in the electronics industry), such advances have not kept pace with system complexity or costs. As the system becomes more complex, the maintenance of parts of it is likely to become more difficult and costly. For example it has been estimated that the US Department of Defense allocates one-quarter of its total budget to maintenance costs (Smith, Westland, and Crawford 1970). There is no reason why this proportion should alter significantly in any large organization which uses large and complex man–machine systems. Maintenance, therefore, costs money, and it is in the interests of all who operate a system to reduce the costs by considering (or having others consider) at an early stage in the product or system design the ease with which it can be maintained.

The remainder of this chapter discusses some of the ways that this ideal state of affairs can be approached but, before doing so, it is well to consider the conditions under which equipment often has to be maintained.

First, extremes of environmental conditions: When having to locate faulty components or otherwise maintain systems under field conditions, any of the environmental considerations discussed in Chapters 10 and 11 are likely to be far from optimum. In particular, the illumination levels in the interiors of machines are often very low, usually requiring the maintenance engineer also to have to cope with some type of torch. At the other end of the illumination scale, bright sunlight or even the position of unguarded lights in the machine mean that maintenance often has to be carried out under conditions of sometimes high disability glare. Other environmental conditions, for example temperature and weather, can also vary depending on the site of the equipment when it breaks down. A long-distance lorry driver, for example, may have to carry out repairs to his vehicle on an isolated, cold, windswept road in the highlands of Scotland on one day, and in the middle of a burning desert a few days later.

Second, the workplace of the maintenance engineer is often very cramped, and this point becomes more salient with the continuing trend towards miniaturization. This problem applies not only to the engineers' difficulty of being able to climb into appropriate parts of the equipment as necessary, but to difficulties in being able to reach, grasp and manipulate components within the machine itself.

Finally, it should be remembered that maintenance is often carried out under conditions of stress, and this can take many forms. Since a machine which is out of operation is costing money, the maintenance engineer is often under time stress from both the operator and from 'management' to rectify the fault. He is also possibly faced with danger from his immediate work environment. In combat this danger is clear and takes the form of enemy action, but other examples may be adduced from civilian maintenance, for example, the presence of high-voltage electricity or perhaps toxic chemicals. Finally, stress can also take the form of some fear of failure to rectify the fault. When the machine breaks down it is the maintenance engineer's profession which is called upon to make an unworkable machine workable. 'Professional pride' and worry about his inability to carry out the task as the job progresses may itself create a stressful environment.

Clearly, therefore, it is important to consider maintainability in the design of equipment. The length of time for which a machine is out of operation is wasted time for that machine. The more difficult the maintenance task, the longer the engineer is likely to take, and this increases his own cost and reduces the machine's value.

Designing for Maintainability

Although it is so important to consider the ease of maintenance at an early stage in the design process, in many respects the techniques for doing so have evolved through commonsense rather than through scientific investigation. Unfortunately it can be argued that even today designing for maintainability is more of an art than a science, and that good maintenance design is more a state of mind than an established procedure.

However, various suggestions can be made to help the designer to produce equipment which is easily maintainable, and all tend to follow very closely the various suggestions made throughout this book to aid the operator in his task—for example, considerations of display and control design and arrangement; appropriate workplace and workspace design; the effect of various environmental parameters, etc. Many of these suggestions are described and exemplified well by Morgan *et al.* (1963) but it is useful to reiterate the importance of some of these aspects with respect to maintenance design.

Anthropometric considerations

As was discussed above, with the increasing trend towards miniaturization the problem of the maintenance worker's accessibility to the machine or to the component in question becomes of paramount importance. This applies equally well to large pieces of machinery into which the operator might need to climb or behind which he may need to crouch, as to small machines (currently electronic machines) into which the operator may need to be able to manipulate his fingers or hand. At both extremes, and in all cases in between, efficiency is likely to be reduced if insufficient consideration is given to the anthropometric dimensions of the appropriate body or limb, plus its associated (protective) clothing.

Communication: maintenance manuals, labelling and symbols

Since he is likely to have to maintain a number of different machines it is not unreasonable for the maintenance operator to require detailed information about the particular type of machine with which he has to work—its performance characteristics, its mode of construction, possible faults, details and values of its components, etc. This he is likely to obtain from the maintenance manual. As Sinaiko and Brislin (1973) demonstrated (p. 76), however a poorly designed manual is likely to lead to reduced performance in terms of increased errors and maintenance time. Thus much of the information provided in Chapter 4 (on man—man communication) is important in this context.

Being provided with a good maintenance manual, however, is only half of the battle for the maintenance operator. Having located a possible fault using, perhaps, a diagram, he then needs to transfer the appropriate image from the diagram to the machine with which he is working to locate the relevant component. The use of labels, numbers, or symbols can greatly expedite matters in this respect.

Information displays

In addition to providing means of communication between the operator and the machine, displays can also be used to tell the maintenance man whether the system, or parts of it, are performing normally. In this respect, the important principles for display design were considered in detail in Chapter 5.

Controls

Again, most of the controls likely to be used by maintenance men will also be used by the normal operator, and the important principles in control design were considered in both Chapter 3 and 6.

One type of 'control' which is likely to be used more by the maintenance operator, however, is the connector or fastener. In the strict sense of the word, of course, connectors are not controls, since they perform no controlling function between man and his machine. However, they have to be operated in much the same ways as controls (using the limbs), and their design involves many of the same principles (for example ensuring that the connector size is appropriate to fit the limb (fingers) operating it; ensuring that the forces required to connect two connectors together are within the abilities of the operator, etc.).

Each of the principles suggested for the appropriate design and location of connectors and fasteners follow two main themes:

1. Ease of connection and disconnection. For easy maintenance, plug-in electrical connectors are likely to be better than screw terminals, and screw terminals better than solder connections. Similarly U-shaped lugs at the end of cables which can be connected without removing the securing screw are likely to be easier to maintain than O-shaped lugs with a hole through which the screw is inserted. For connectors which fasten equipment together, wing nuts are easier to remove than bolts, while a bolt which can be unscrewed using either a spanner or a screwdriver (has both hexagonal sides and a screwdriver slot) is likely to be more easily operated than a bolt which has only one such feature.
2. Ensuring that the connectors are not easily connected either the 'wrong way round' or to the wrong socket. Various techniques are available to ensure that the appropriate connections are made. These include using sets of connectors and sockets of different sizes, colour coding, cutting and shaping cables so that a connector cannot reach a wrong socket, and using lugs or other aligning pins inside the cannector and socket. If lugs are employed, however, it is sensible to ensure that they are arranged unsymmetrically so the plug cannot be inserted 180 degrees from the correct position.

Arranging the workspace and workplace

Whereas the normal operator has to contend with a number of groups of men and machines in his environment, the nature of the maintenance operator's problems with respect to his workspace and workplace are much the same, although scaled down in size since they concern only the particular machine on which he is working. Thus his problems involve access to the appropriate part of the machine (primarily an anthropometric problem) and the layout of components inside the machine. Each of these problems, therefore has been considered in Chapters 8 and 9.

Finally, with regard to the layout of components within the machine, McKendry, Corso, and Grant (1960) have demonstrated that an appropriate arrangement can considerably reduce the time required to locate a fault. In their experiment the components in either a simple radio receiver or a complex radar simulator were arranged according to one of four principles:

1. Component packaging, in which all similar components, for example, all tubes or all transistors, were grouped together in one place on the equipment. Using this procedure, once the area of a fault is determined, for cheap components at least, a whole plug-in section could be replaced.
2. Circuit packaging, in which components relating to the various circuits (for example amplification, tuning, trimming, etc.) were placed together.
3. Logical flow, which represents a combination of two approaches: first, the use of modules and subassemblies so that only a single, simple input and output check is necessary to isolate trouble to that unit, and second, the modules are arranged in a clear sequence representing the 'flow' through the circuitry, for example tuning—amplification—speaker.
4. The 'standard' packaging method, which did not employ any grouping or modules nor did it attempt to suggest a logical flow in components. The parts of the equipment on top of the chassis were appropriately spaced to distribute weight and heat.

In all conditions the subject's task was to isolate a fault to a given component. The results demonstrated that the subjects found the faulty component significantly faster when using the logical flow method than the standard arrangement. The 'circuit' and 'component' arrangements produced intermediate times. Clearly, therefore, simply varying the arrangement of the components to be maintained can have significant effects on maintenance time.

SUMMARY

Both of the topics discussed in this chapter, inspection and maintenance, provide fitting summaries for many of the topics discussed elsewhere in this book. In their own right, both represent problems in the design of man—machine systems. However, the solution of these problems involves considering all aspects of the man—machine system—communication, displays, controls, workspace and workplace design, and the physical and social environment—and ensuring that they interact to the benefit of the operator.

CHAPTER 14

Ergonomics Investigations: Some Problems and Techniques

The purpose of this final chapter is not to present the reader with a list of different techniques and methods available for investigating the multitude of problems and questions which are likely to occur when considering the interactions between man and his environment. These are covered well enough by authors such as Chapanis (1959) or Oppenheim (1966). Rather, its aim is to illustrate and emphasize the pitfalls and problems of which the unsuspecting investigator may be unaware—particularly when investigating the behaviour of *human* operators at work.

When faced with a problem, there are normally two avenues available to the intelligent investigator. First, he can attempt to discover how other people have solved similar, though possibly not precisely similar, problems; he can read accounts of their investigations; and he can try to apply their results to the particular problem. In essence this is what this book helps him to do. The other, not necessarily mutually exclusive, approach is for our investigator to solve the problem by independent study. This has the major advantage that the investigation concerns the problems experienced in a specific workplace, and that the results and conclusions obtained will be applicable to that particular problem. If only one message has been extracted from the book, it should have been that ergonomics is a discipline which attempts to fit together the man and his environment within a system, and that the results and conclusions from one system are not always applicable to another.

By deciding to investigate a particular man–machine relationship in the system, however, the potential investigator immediately encounters a multitude of further problems—all of which are caused by the human part of the system. There are a number of ways in which man's complexity can bias or distort the outcome of experiments in which he takes part, and there are a number of ways in which he can make the investigator's task harder or even impossible.

Perhaps the first bias, about which any aspiring psychologist is taught, is that of respondent bias (sometimes called volunteer error).

For obvious reasons, any experiment normally has to be carried out on a sample of subjects which has been selected from the population being studied. For example, if the investigation is concerned with the speed of response of women workers on a shop floor, it would be sensible only to use those particular workers in the experiment, and not to draw subjects from any other population. However, it may not be feasible to measure the speed of response of *all* women workers in the factory, so a sensible solution might be to select randomly a dozen or so individuals as being a representative sample. But how does the investigator go about his selection?

Since it is the normal expectation in a democratic society to be asked to participate, rather than to be commanded, the random sample is likely to be drawn from those individuals who were actually asked and who agreed to participate. Unfortunately, however, although the observance of democratic principles meets the ethical requirements for research, it leaves open the question of whether the 'representative' sample was truly representative or whether it could have been biased: the sample contains only those subjects who agreed to be studied (the volunteers). But why did they volunteer? Are they any different from those who refused, or were not asked (personality, fears, age, intellectual level, etc.)? Why were some not asked? The answers given to any of these questions could raise serious doubts regarding the impartiality of the sample selected.

Although some degree of volunteer bias normally has to be accepted, other biases can also easily be introduced into the investigation. These can occur simply because the investigation is being carried out in the first place; because of the interest which is being shown towards the 'man' who is in the system.

Perhaps the most appropriate example of this bias is illustrated by a series of studies conducted at the Hawthorne plant of the Western Electric Company, under the direction of Elton Mayo, in the 1920s. The studies began by examining ways in which changes in the worker's physical environment would affect production, the first aspect to be investigated being the illumination level. The results obtained, however, proved to be rather surprising, in that they indicated that illumination changes resulted in productivity *increases*—whatever the direction of the changes. Indeed in one experiment the illumination level was reduced considerably, almost to moonlight levels, yet productivity still increased.

These paradoxical results have since been explained by suggesting that the individuals increased their output because they were in an experimental situation. Perhaps they acted in this way because they felt that the experimenters were taking an interest in them. Whatever the reason, this 'Hawthorne Effect'—that subjects may improve performance due solely to the experimental situation—has now become a recognized pitfall in psychological investigations.

If these biases were constant and always acted in one direction (for example, always caused the subject to perform better or worse), the problems that they cause would not be too great. Unfortunately, they are neither constant nor consistent. For a number of reasons, the same pressures which produce the 'Hawthorne Effect' can act in the reverse direction—to cause the subject to perform at a slower, less accurate rate than normal.

One reason for this effect may be due to attitudes which the workforce, the unions and/or the management may have towards the investigation. As was discussed in Chapter 1, ergonomics has many features in common with work study and time and motion study. Unfortunately, however, both of these disciplines enjoy considerable notoriety which could interfere with the investigation. If the interference is extreme enough to be symptomatized by a complete withdrawal of cooperation and work then this is bad enough. From the point of view of the investigator, however, a worse situation would be if the subjects are apparently cooperative but are, nevertheless, antagonistic to the aims of the study. In these cases the investigator has no idea of the effect which these antagonistic attitudes may have on the subject's behaviour—particularly if the subject appears to cooperate.

Another negative 'Hawthorne Effect' can arise in studies carried out in industrial premises, when measures of worker's productivity are recorded—particularly if the operators feel that the results of the investigation might affect future productivity and payment norms being set (again the memory of time and motion study is difficult to erase). In this case productivity may be depressed purposefully during the period of the investigation so that, once the norms have been set and after the investigation has ended, 'normal' production will appear as increased productivity.

The message to be gained concerning each of these biases, therefore, suggests that as far as possible the confidence of the subjects (whether they are subjects in the laboratory or operators at work) should be sought. For many reasons it may not be possible to explain all the reasons for the experiment (for example, telling subjects that an experiment will investigate the effects of different noise levels on behaviour may cause them to react differently to the noise than they otherwise might have done). However, in many cases it should be possible to allay fears, to discuss the aims of the investigation, and to ensure the subjects that their results are likely to be of benefit (if, indeed, they are).

At this stage it is well to consider where the study can be carried out—in the laboratory or, in situ, in the field. Both environments have their advantages and their disadvantages.

If the research is to be carried out in the laboratory, precise questions can be asked and controlled stimuli can be presented to the subject, ensuring that all subjects receive the same stimulus. In the field, however, the investigator is not so well acquainted with the nature of the stimulus which the operator is receiving. Different operators might be receiving (either subjectively or physically) different stimuli due to factors which are outside the investigator's control, such as variations in seating positions or even the operator's states of health at the time.

In contrast, however, field investigations have one major advantage over laboratory studies. This concerns the validity of the results which are obtained finally, and have to be related back to the working situation. In this respect it is well to notice that in the preceding paragraph 'subjects' were discussed as receiving stimuli in the laboratory, whereas 'operators' were said to receive the stimulus in the field. This simple semantic distinction highlights the advantage of the field over the laboratory investigation—namely that if the results of the investigation are to benefit any particular section of the community, then members of that group should be employed in the investigation.

The advantage of field studies, therefore, relates to the fact that a laboratory experiment cannot take account of all extraneous variables which may influence the operator at work. For example, a laboratory subject is not having to earn his living at that particular point in time; he may not fully understand the various consequences of his actions; he is not likely to be at ease—in fact he may enter the laboratory fearing the worst; and he is not being subjected to all the extraneous stimuli that occur in normal work, for example the noise, temperature and even pressures from management and colleagues.

It is possible, of course, to combine the two research environments of 'laboratory' and 'field' to produce a third arrangement. In this case subjects' responses are measured under field conditions—that is, personnel are 'employed' specifically to carry out an experiment (they then become 'subjects') but the investigation is carried under field conditions. This field (subject) approach has been used quite extensively in the past (see Oborne, 1978b) and has much to recommend it. At least the experiment is carried out in the appropriate environment with all the relevant extraneous stimuli. However, it still suffers from the drawback that 'subjects' rather than 'operators' are used.

Before discussing some of the techniques suitable for an ergonomics investigation it is well to pose one further problem—whether it is possible to investigate behaviour at all, without affecting the behaviour being investigated.

This question is perhaps better understood by nuclear physicists since the proposition of Heisenberg's 'Uncertainty Principle' in the 1930s. Heisenberg was concerned with measuring the behaviour of atoms, but concluded that it is not possible to measure their behaviour without affecting them: 'Every subsequent observation...will alter the momentum (of the atom) by an unknown and undeterminable amount such that after carrying out the experiment our knowledge of the electronic motion is restricted by the uncertainty relation'. The same is also true, unfortunately, for human behaviour. As soon as the investigator enters the system and inserts his measuring equipment (whether the equipment is in the form of 'hardware' such as meters or probes, or 'software' such as questions or observation) he is likely to affect the subsequent behaviour of that system in an uncertain manner. Unfortunately, however, this cannot be totally overcome, although it can be reduced by using unobtrusive techniques such as observation.

OBSERVATION AND OTHER UNOBTRUSIVE MEASURES

Whatever, technique is chosen to investigate the performance of the human operator in the system, Heisenberg's Uncertainty Principle still operates. However some techniques produce less interference than others and, providing it is carried out properly, perhaps the least intrusive in this respect is simply to observe behaviour or its concomitants.

Direct Observation

Directly observing what the operator does when he interacts with his environment is possibly one of the most commonly used techniques. At a basic level it is simple, requires little equipment and often produces data which can be easily interpreted. However direct observation can be misused, producing misleading results—particularly if the observer intrudes into the system; than any of the biases and hindrances discussed earlier can occur.

The most important principle of direct observation, therefore, is that the observer (and any equipment which he uses) should be unobtrusive. This is not to say that he should be invisible—merely that he should not interfere in any way with the system which he is observing. In many cases, of course, an invisible observer (for example, one who is behind a screen) is also likely to be unobtrusive. However, situations can arise in which the absence of an observer who otherwise would have been expected could lead to mistrust or worry. Knowing (or feeling) that he is being observed through a one-way screen, for example, could upset the subject more than if the observer were visible.

The second important principle of direct observation is that a permanent record of the behaviour should be made for future, detailed analysis. This is normally done in one of two ways: either by recording all of the activities using cinefilm or videotape, or by sampling the operator's activities and recording aspects of his behaviour on either precoded paper sheets or directly into a computer.

The advantages of using cinefilms or videofilms should be obvious: all the operator's activities are recorded and the film can subsequently be played back many times during the analysis. Providing that the camera has been positioned appropriately, therefore, the investigator is ensured of obtaining a complete, permanent record of the operator's behaviour. Furthermore, when analysing the record he can play the film back at a slower speed than normal (again ensuring that no activity, however slight, is missed), or at a faster-than-normal speed.

The ability to play the record back at a faster speed than that recorded has two advantages. Firstly it speeds up the analysing time. By the time it has been stopped, started and return a few times, a 2-hour record, for example, is likely to take at least three or four times as long (6–8 hours) to analyse at normal speed. Playing it back at a faster rate should reduce this time. Second, fast playback allows the observer to 'see' motions which might otherwise be too slow to observe at normal speeds. For example Branton and Grayson (1967) used cinefilm to record the sitting behaviour of passengers in two types of train seats throughout a 5-hour journey. When the films were played back at a faster speed than normal, however, it became apparent that one type of seat caused the sitter to gradually slip forward and out of the seat, requiring him to adjust his position constantly. This 'slipping' action, however, was too slow to be observed in 'real' time. As a further example, Branton and Oborne (1979) recorded the behaviour of anaesthetists during various surgical operations. When the videofilm was speeded up (20 × normal speed) a head-'nodding' behaviour was observed which the authors suggest could be related to the onset and continuation of fatigue and monotony.

The disadvantage of cine or video records has already been discussed—namely the mass of data which is obtained and which needs to be analysed. For example,

Lovesey (1975) used a cinefilm to record pilot activity during different types of helicopter flying. Just over 2 hours of flying was recorded, which needed to be analysed on a frame-by-frame basis. Since the film speed was eight frames per second, Lovesey needed to analyse the head and hand positions of his pilots in each of 63 370 frames.

Analysing the behaviour as it is carried out obviates this problem. This is the one advantage of sampling behaviour and recording the operator's activities, say, every minute on a precoded sheet or directly into a computer. Unless he is fully conversant with the activities likely to have to be recorded, however, and unless the sampling time periods are sufficiently long, this method can place the observer under considerable time stress. As an example of this approach, Christensen (1950) recorded the activities of navigators and radar operators during 15-hour flights. The observer used a metronome to produce a tone once a minute and recorded his subject's activities when he heard the tone.

Indirect Observation

No matter how well the observer is integrated in the system, Webb *et al.* (1966) argue that he still has the potential to bias the results in many of the ways discussed above. For this reason, they suggest techniques by which the activities of the subject(s) can be observed indirectly.

Physical traces

If an activity is carried out frequently, over time the equipment, or the part of it which is operated, is likely to become well used. To the observant, this is useful data. As an example, Webb *et al.* cite the wear on library books; the less popular books remain in a newer condition than do the books which are borrowed frequently, and the page corners and edges of the popular parts of the book are likely to be dirtier and more dogeared than the less popular parts. As a further example, the wear and tear on various parts of a floor can indicate the extent to which various machines are used.

The same type of approach can be used to determine the nature of the activity which is carried out. For example dial settings which an operator leaves on his machine, say at night, may indicate the way in which he has used the machine, as may the positions in which moveable machines and components are left after use.

Statistical records

In many respects, by using statistical records the investigator is merely observing what has happened in the past rather than what is happening at present. As was discussed in Chapter 12, records are most frequently used to study trends in accidents, their types, places and personnel involvement. They can be very useful in locating areas of potential danger—either in the man or in the environment. As long as a large enough pool of accidents is sampled useful statistical analyses can be computed and, of course, since the investigator is never at the scene of the accident, he cannot be intrusive.

Statistical records, however, have many limitations in their value. As Chapanis (1959) points out, accidents or near accidents are infrequent events and it may take years to collect enough instances to form any sort of reasonable classification of the kinds of troubles that can occur. Second, it is virtually impossible to compare or contrast different accident situations because of the difficulty of assessing the risk of exposure to an accident. This point was emphasized, for example, when considering the concept of accident proneness — 'accident-prone' individuals may simply be in riskier situations. Third, the statistics tell us simply that an event (accident) had occurred. It is difficult, if not impossible, to reconstruct what happened or why. For these reasons, statistical evidence is only useful as 'back-up' information or for helping to generate hypotheses which can later be tested by more controlled observation or experimentation.

SUMMARY

This chapter has not attempted to equip the reader with the practical or statistical skills required to carry out an ergonomic investigation. The techniques available are so numerous and so diverse that to be able to do so one would need to produce a separate book. Instead this chapter has provided the reader with a background to the problems involved with human experimentation, the interpersonal problems which can occur but about which the investigator may be unaware. It has also suggested simple, unobtrusive measures which can be used. Armed with an understanding of these problems, the potential ergonomist will be equipped to investigate further the multitude of ways in which people and their environments interact.

References

Agate, J. N. (1949). An outbreak of cases of Raynaud's phenomenon of occupational origin. *British Journal of Industrial Medicine*, **6**, 144–163.

Allen, G. R. (1975). *Ride quality and International Standard 2631*. Paper given to Ride Quality Symposium, Williamsburg, August. NASA TM X-3295.

Allgeier, A. R. and Byrne, D. (1973). Attraction towards the opposite sex as a determinant of physical proximity. *Journal of Social Psychology*, **90**, 213–219.

American Society of Heating and Refrigeration Engineers (1965). *ASHRAE Guide and Data Book: Fundamentals and Equipment*. (ASHRAE).

American Standards Association (1960). *American Standards Acoustical Terminology S1-1960*. (NY)

Andersson, G. B. J. (1980). The load on the lumbar spine in sitting postures. In D. J. Oborne and J. A. Levis (eds.) *Human Factors in Transport Research, Volume II* (London: Academic Press).

Andersson, G. B. J., Ortengren, R., Nachemson, A. L., Elfstrom, G. and Broman, H. (1975). The sitting posture: An electromyographic and discometric study. *Orthopedic Clinics of North America*, **6**, 105–120.

Andrews, I. and Manoy, R. (1972). Anthropometric survey of British Rail footplate staff. *Applied Ergonomics*. **3**, 132–135.

Anon. (1966) *Lifting in Industry*. Pamphlet prepared by the posture subcommittee (London: Charted Society of Physiotherapy).

Argyle, M. and Cook, M. (1976). *Gaze and Mutual Gaze*. (Cambridge: Cambridge University Press).

Asch, S. E. (1952) *Social Psychology* (N. J.: Prentice Hall)

Azer, N. Z., McNall, P. E. and Leurig, H. C. (1972). Effects of heat stress on performance. *Ergonomics*, **15**, 681–691.

Baddeley, A. D. (1966). Time estimation at reduced body temperature. *American Journal of Psychology*, **79**, 475–479.

Baddeley, A. D., Cuccaro, W. J., Egstrom, G. H., Weltman, G. and Willis, M. A. (1975). Cognitive efficiency of divers working in cold water. *Human Factors*. **17**, 446–454.

Barkala, D. (1961). The estimation of body measurements of British population in relation to seat design. *Ergonomics*, **4**, 123–132.

Barnard, P. and Marcel, A. (1977). A preliminary investigation of factors influencing the interpretation of pictorial instructions for the use of apparatus. *Proceedings of the 8th International Symposium on Human Factors in Telecommunications*, Cambridge, September, pp. 379–392.

Barnes, R. M. (1963). *Motion and Time Study* (London: John Wiley & Co).

Bartlett, F. C. (1950). Programme for experiments on thinking. *Quarterly Journal of Experimental Psychology*, **2**, 145–152.

Bauer, D. and Cavonius, C. R. (1980). Improving the legibility of visual display units through contrast reversal. In E. Grandjean and E. Vigliani (eds.) *Ergonomic Aspects of Visual Display Terminals*. (London: Taylor and Francis).

Baughn, W. L. (1966). Noise control—percent of population protected. *International Audiology*, **5**, 331–338.

291

Bearham, J. (1976). *The Cost of Accidents Within the Port Industry*. (London: Manpower Development Division, National Ports Council).

Beevis, D. and Slade, I. M. (1970). Ergonomics—costs and benefits. *Applied Ergonomics*, 1, 79–84.

Belbin, R. M. (1957). New fields for quality control. *British Management Review*, 15, 79–89.

Bell, C. R., Crowder, M. J. and Walters, J. D. (1971). Durations of safe exposure for men at work in high temperature environments. *Ergonomics*, 14, 733–757.

Benn, R. T. and Wood, P. H. N. (1975). Pain in the back: An attempt to estimate the size of the problem. *Rheumatology and Rehabilitation*, 14, 121–128.

Benson, A. J., Huddleston, J. H. F. and Rolfe, J. M. (1965). A psychophysiological study of compensatory tracking on a digital display. *Human Factors*, 7, 457–472.

Berger, C. (1944). Stroke-width, form and horizontal spacing of numerals as determinants of the threshold of recognition. *Journal of Applied Psychology*, 28, 208–231.

Berry, P. C. (1961). The effect of coloured illumination upon perceived temperature. *Journal of Applied Psychology*, 45, 248–250.

Biancardi, M. (1978). The cost/benefit factor in safety decisions. *Professional Safety*, **November**, 17–22.

Bilger, R. C. and Hirsh, I. J. (1956). Masking of tones by bands of noise. *Journal of the Acoustical Society of America*, 28, 623–630.

Bjerver, K. and Goldberg, L. (1950). Effects of alcohol ingestion on driving ability. *Quarterly Journal on the Study of Alcohol*, 11, 1–30.

Blake, M. J. F. (1971). Temperament and time of day. In W. P. Colquhoun (ed.) *Biological Rhythms and Human Performance*. (London: Academic Press).

Bonney, M. C. and Williams, R. W. (1977). CAPABLE: A computer program to layout controls and panels. *Ergonomics*, 20, 297–316.

Booher, H. R. (1975). Relative comprehensibility of pictorial information and printed words in proceduralized instructions. *Human Factors*, 17, 266–277.

Bowen, H. M. (1968). Diver performance and the effects of cold. *Human Factors*, 10, 445–464.

Boynton, R. M., Elworth, C. and Palmer, R. M. (1958). *Laboratory studies pertaining to visual air reconnaissance*. USAF, WADC TR 55–304.

Bradley, J. V. (1967). Tactual coding of cylindrical knobs. *Human Factors*, 9, 483–496.

Bradley, J. V. (1969a). Glove characteristics influencing control manipulability. *Human Factors*, 11, 21–36.

Bradley, J. V. (1969b). Optimum knob diameter. *Human Factors*, 11, 353–360.

Bradley, J. V. (1969c). Desirable dimensions for concentric controls. *Human Factors*, 11, 213–226.

Bradley, J. V. (1969d). Optimum knob crowding. *Human Factors*, 11, 227–238.

Branton, P. (1966). *The Comfort of Easy Chairs*. Furniture Industry Research Association Report No 22.

Branton, P. (1969). Behaviour, body mechanics and discomfort. *Ergonomics*, 12, 316–327.

Branton, P. (1972). Ergonomic research contributions to the design of the passenger environment. Paper presented to Institute of Mechanical Engineers Symposium on Passenger Comfort, London.

Branton, P. and Grayson, G. (1967). An evaluation of train seats by an observation of sitting behaviour. *Ergonomics*, 10, 35–51.

Branton, P. and Oborne, D. J. (1979). A behavioural study of anaesthetists at work. In D. J. Oborne, M. M. Gruneberg and J. R. Eiser (ed.) *Research in Psychology and Medicine. Volume I*. (London: Academic Press).

Brebner, J. and Sandow, B. (1976). The effect of scale side on population stereotype. *Ergonomics*, 19, 571–580.

Brenner, B. (1967). Alcoholism and fatal accidents. *Quarterly Journal on Studies of Alcohol*, **28**, 517–528.

British Standards Institute (1964). *Recommendations for the design of scales and indexes. Part I Instruments of bold presentation and for rapid reading.* BS 3693.

British Standards Institute (1969) *Recommendations for the design of scales and indexes. Part II Indicating instruments to be read to 0.33–1.25% resolution.* BS 3693.

Broadbent, D. E. (1954). Some effects of noise on visual performance. *Quarterly Journal of Experimental Psychology*, **6**, 1–5.

Broadbent, D. E. (1977). Language and ergonomics. *Applied Ergonomics*, **8**, 15–18.

Brookes, M. J. (1972). Office landscape: Does it work? *Applied Ergonomics*, **3**, 224–236.

Brookes, M. J. and Kaplan, A. (1972). The office environment: Space planning and affective behaviour. *Human Factors*, **14**, 373–391.

Brown, I. D. (1965). A comparison of two subsidiary tasks used to measure fatigue in car drivers. *Ergonomics*, **8**, 467–473.

Brown, I. D. and Poulton, E. C. (1961). Measuring the spare 'mental capacity' of car drivers by a subsidiary task. *Ergonomics*, **5**, 35–40.

Brumaghim, S. H. (1967). *Subjective Reaction to Dual Frequency Vibration.* Boeing Co (Wichita) Report D8-7562.

Buckhout, R. (1964). Effects of whole-body vibration on human performance. *Human Factors*, **6**, 157–163.

Buckler, A. T. (1977). *A Review on the Legibility of Alphanumerics on Electronic Displays.* May 1977. AD-A040-625.

Burrows, A. A. (1965). Control feel and the dependent variable. *Human Factors*. **7**, 413–421.

Burt, C. (1959). *A Psychological Study of Typography.* (Cambridge: Cambridge University Press).

Byrne, D., Baskett, G. D. and Hodges, L. (1971). Behavioural indicators of interpersonal attraction. *Journal of Applied Psychology*, **1**, 137–149.

Caldwell, L. S. (1963). Relative muscle loading and endurance. *Journal of Engineering Psychology*, **2**, 155–161.

Caldwell, L. S. (1964). Measurement of static muscle endurance. *Journal of Engineering Psychology*, **3**, 16–22.

Carlson, B. R. (1969). Level of maximum isometric strength and relative load isometric endurance. *Ergonomics*, **12**, 429–435.

Carlsoo, S. (1972). *How Man Moves.* (London Heinemann).

Carpenter, J. A. (1959). The effect of caffeine and alcohol on simple visual reaction time. *Journal of Comparative and Physiological Psychology*, **52**, 491–496.

Cashen, V. M. and Leicht, K. L. (1970). Role of the isolation effect in a formal eductional setting. *Journal of Eductional Psychology*, **61**, 484–486.

Castle, P. F. C. (1956). Accidents, absence, and withdrawal from the work situation. *Human Relations*, **9**, 223–233.

Cavanaugh, W. J., Farrell, W. R., Hirtle, P. W. and Watters, B. G. (1962). Speech privacy in buildings. *Journal of the Acoustical Society of America*, **34**, 475–492.

Chambers, J. B. and Stockbridge, H. C. W. (1970). Comparison of indicator components and push button recommendations. *Ergonomics*, **13**, 401–420.

Chaney, F. B. and Teel, K. S. (1967). Improving inspector performance through training and visual aids. *Journal of Applied Psychology*, **51**, 311–315.

Chaney, R. E. (1964). *Subjective Reation to Whole-Body Vibration.* Boeing Co (Wichita) Report D3-6474.

Chaney, R. E. (1965). *Whole-Body Vibration of Standing Subjects.* Boeing Co (Wichita) Report D3-6779.

Chapanis, A. (1951) Studies of manual rotary positioning movements II: The accuracy of estimating the position of an indicator knob. *Journal of Psychology*, **31**, 65.

294

Chapanis, A. (1959). *Research Techniques in Human Engineering.* (Baltimore: Johns Hopkins University Press).

Chapanis, A. (1960). Human engineering. In C. D. Flagle, W. H Higgins and R. N. Roy (eds.) *Operations Research and Systems Engineering.* (Baltimore: The Johns Hopkins University Press).

Chapanis, A. (1965a). On the allocation of functions between men and machines. *Occupational Psychology,* **39,** 1–11.

Chapanis, A. (1965b). Words, words, words. *Human Factors,* **7,** 1–17.

Chapanis, A. (1974). National and cultural variables in ergonomics. *Ergonomics,* **17,** 153–175.

Chapanis, A. (1976). Engineering psychology. In M. D. Dunnette (ed.) *Handbook of Industrial Psychology.* (Chicago: Rand McNally Corp).

Chapanis, A. and Lindenbaum, L. E. (1959). A reaction-time study of four control-display linkages. *Human Factors,* **1,** 1–7.

Chapanis, A. and Lockhead, G. R. (1965). A test of the effectiveness of sensor lines showing linkages between displays and controls. *Human Factors,* **7,** 219–229.

Chapanis, A. and Mankin, D. A. (1967). Tests of ten control-display linkages. *Human Factors,* **9,** 119–126.

Chorley, R. A. (1973). Human factors. *Aviation Review,* (**March**), 12–15.

Christ, R. E. (1975). Review and analysis of colour coding research for visual displays. *Human Factors,* **17,** 542–570.

Christensen, J. M. (1950). A sampling technique for use in activity analysis. *Personnel Psychology,* **3,** 361–368.

Christie, A. W. and Fisher, A. J. (1966). The effect of glare from street lighting lanterns on the vision of drivers of different ages. *Transactions of the Illuminating Engineering Society,* **31,** 93–108.

Clark, R. E. (1961). The limiting hand skin temperature for unaffected manual performance in the cold. *Journal of Applied Psychology,* **45,** 193–194.

Clark, R. E. and Cohen, A. (1960). Manual performance as a function of rate of change in hand skin temperature. *Journal of Applied Physiology,* **15,** 496–498.

Clarke, R. S. J., Hellon, R. F. and Lind, A. R. (1958). The duration of sustained contractions of the human forearm at different muscle temperatures. *Journal of Physiology,* **143,** 454–473.

Cohen, E. and Follert, R. L. (1970). Accuracy of interpolation between scale gradations. *Human Factors,* **12,** 481–483.

Cohen, J., Dearnaley, E. J. and Hansel, C. E. M. (1958). The risk taken in driving under the influence of alcohol. *British Medical Journal,* **1,** 1438–1442.

Coles, A. R. A., Garinther, G. R., Hodge, D. C. and Rice, C. G. (1968). Hazardous exposure to impulse noise. *Journal of the Acoustical Society of America,* **43,** 336–343.

Colquhoun, W. P. (1958). The effect of a short rest pause on inspection accuracy. *Ergonomics,* **2,** 367–372.

Colquhoun, W. P. (1975a). Evaluation of auditory, visual and dual mode displays for prolonged sonar monitoring in repeated sessions. *Human Factors,* **17,** 425–437.

Colquhoun, W. P. (1975b). *Accidents, Injuries and Shiftwork.* Paper presented to NIOSH Symposium on 'Shiftwork and Health'; Cincinnati.

Connell, S. C. (1948). *Psychological Factors in Check Reading Single Instruments.* USAF Air Material Command Memo Report No MCREXD-694-17A.

Conrad, R. (1962). The design of information. *Occupational Psychology,* **36,** 159–162.

Conrad, R. and Hull, A. J. (1968). The preferred layout for numerical data-entry keysets. *Ergonomics,* **11,** 165–173.

Corlett, E. N., Morcombe, V. J. and Chanda, B. (1970). Shielding factory noise by work-in-progress storage. *Applied Ergonomics,* **1,** 73–78.

Corlett, E. N., Hutcheson, C., DeLugan, M. A. and Rogozenski, J. (1972). Ramps or stairs: The choice using physiological and biomechanical criteria. *Applied Ergonomics*, **3**, 195–201.

Corlett, E. N. and Parsons, A. T. (1978). Measurement of changes: What criteria do we adopt? *International Journal of Management Sciences*, **6**, 399–406.

Corso, J. F. (1963). Age and sex differences in pure-tone thresholds. *Acta Otolaryngollogica*, **77**, 385–405.

Crawford, B. H. and Stiles, W. S. (1937). The effect of a glaring light source on extra-foveal vision. *Proceedings of the Royal Society (Series B)*, **122**, 255–280.

Crook, M. A. and Langdon, F. J. (1974). The effects of aircraft noise in schools around London airport. *Journal of Sound and Vibration*, **12**, 221–232.

Crouse, J. H. and Idstein, P. (1972). Effects of encoding cues on prose learning. *Journal of Educational Psychology*, **61**, 484–486.

Cushman, W. H. (1980). Selection of filters for dark adaptation goggles in the photographic industry. *Applied Ergonomics*, **11**, 93–99.

Damon, A. and Stoudt, H. W. (1963). The functional anthropometry of old men. *Human Factors*, **5**, 485–491.

Damon, A., Stoudt, H. W. and McFarland, R. A. (1971). *The Human Body in Equipment Design*. (Mass.: Harvard University Press).

Dashevsky, S. G. (1964). Check reading accuracy as a function of pointer alignment, patterning and viewing angle. *Journal of Applied Psychology*, **48**, 344–347.

Davies, B. T. (1972). Moving loads manually. *Applied Ergonomics*, **3**, 190–194.

Davis, P. R. and Stubbs, D. A. (1977a). Safe levels of manual forces for young males (1). *Applied Ergonomics*, **8**, 141–150.

Davis, P. R. and Stubbs, D. A. (1977b). Safe levels of manual forces for young males (2). *Applied Ergonomics*, **8**, 219–228.

Davis, P. R. and Stubbs, D. A. (1978). Safe levels of manual forces for young males (3). *Applied Ergonomics*, **9**, 33–37.

Davis, P. R. and Troup, J. D. G. (1964). Pressures in the trunk cavities when pulling, pushing and lifting. *Ergonomics*, **7**, 465–474.

Davis, P. R. Ridd, J. E., and Stubbs, D. A. (1980). Acceptable loading levels for British workers. Paper presented at the Annual Conference of the Ergonomics Society, Nottingham.

Deatherage, B. H. and Evans, T. R. (1969). Binaural masking: Backward, forward and simultaneous effects. *Journal of the Acoustical Society of America*, **46**, 362–371.

de Jong, J. R. (1967). The contribution of ergonomics to work study. *Ergonomics*, **10**, 579–588.

Dempsey, C. A. (1963). The design of body support and restraint systems. In E. Bennett, J. Degan and J. Spiegel (eds.) *Human Factors in Technology*. (New York: McGraw-Hill).

Dempster, W. T. (1955). *Space Requirements of the Seated Operator: Geometrical, Kinematic, and Mechanical Aspects of the Body With Special Reference to the Limbs.* WADC Technical Report 55–159. (WPAFB, Ohio).

Dennis, J. P. (1965). The effect of whole-body vibration on visual performance task. *Ergonomics*, **8**, 193–205.

Dewar, M. E. (1977). Body movements in climbing a ladder. *Ergonomics*, **20**, 67–86.

Dickinson, J. (1974). *Proporioceptive Control of Human Movement*. (London: Lepus Books).

Diebschlag, W. and Muller-Limroth, W. (1980).Physiological requirements on car seats: Some results of experimental studies. In D. J. Oborne and J. A. Levis (eds.) *Human Factors in Transport Research, Volume II*. (London: Academic Press).

Diffrient, N., Tilley, A. R. and Bardagy, J. C. (1974). *Humanscale 1/2/3*. (Mass.: MIT Press).

296

Dooling, D. J. and Lachman, R. (1971). Effects of comprehension on the retention of prose. *Journal of Experimental Psychology*, **88**, 216–222.

Drillis, R. and Conti, R. (1966). *Body Segment Parameters*. Report No 1166-03 (Office of Vocational Rehabilitation, Dept Health, Education and Welfare) (New York: New York School of Engineering and Science).

Drury, C. G. (1973). The effect of speed of working on industrial inspection accuracy. *Applied Ergonomics*, **4**, 2–7.

Drury, C. G. and Addison, J. L. (1973). An industrial study of the effects of feedback and fault density on inspection performance. *Ergonomics*, **16**, 159–169.

Dwyer, F. M. (1967). Adapting visual illustrations for effective learning. *Harvard Educational Review*, **37**, 250–263.

Easterby, R. S. (1970). The perception of symbols for machine displays. *Ergonomics*, **13**, 149–158.

Eckstrand, G. A. and Morgan, R. L. (1956). *The Influence of Training on the Tactual Discriminability of Knob Shapes*. WADC Technical Report 56-8.

Edholm, O. G. and Murrell, K. F. H. (1973). *The Ergonomics Society: A History 1949–1970*. (London: Ergonomics Research Society).

Ellis, N. C. and Hill, S. E. (1978). A comparative study of seven segment numerics. *Human Factors*, **20**, 655–660.

Engel, F. L. (1980). Information selection from visual display units. In E. Grandjean and E. Vigliani (eds.) *Ergonomic Aspects of Visual Display Terminals* (London: Taylor and Francis).

Evans, G. W. and Howard, R. B. (1973). Personal space. *Psychological Bulletin*, **80**, 334–344.

Fanger, P. O. (1970). *Thermal Comfort*. (New York: McGraw-Hill).

Fanger, P. O., Breum, N. O. and Jerking, E. (1977). Can colour and noise influence man's thermal comfort? *Ergonomics*, **20**, 11–18.

Farmer, E. and Chambers, E. G. (1939). *A Study of Accident Proneness Amongst Motor Drivers*. Industrial Health Research Board Report No 84.

Ferguson, D. and Duncan, J. (1974). Keyboard design and operating posture. *Ergonomics*, **13**, 731–744.

Fine, B. J. (1963). Introversion, extraversion and motor driver behaviour. *Perceptual and Motor Skills*, **16**, 95.

Fisher, B. A. (1978). *Perspectives on Human Communication*. (London: Collier Macmillan).

Fitch, J. M., Templer, J. and Corcoran, P. (1974). The dimensions of stairs. *Scientific American*, **231**, 82–90.

Fitts, P. M. (1962). Functions of man in complex systems. *Aerospace Engineering*, **21**, 34–39.

Fitts, P. M. and Jones, R. E. (1947a). Psychological aspects of instrument display. I Analysis of 270 'pilot error' experiences in reading and interpreting aircraft instruments. Aeromedical Laboratory Report AMRL-TSEAA-694-12A, July. In W. Sinaiko (ed.) *Selected Papers in the Design and Use of Control Systems*. 1961. (New York: Dover).

Fitts, P. M. and Jones, R. E. (1947b). Analysis of factors contributing to 460 'pilot error' experiences in operating aircraft controls. Aeromedical Laboratory Report TSEAA-694-12. July. In W. Sinaikoio (ed.) *Selected Papers in the Design and Use of Control Systems*. 1961. (New York: Dover).

Fitts, P. M. and Seeger, C. M. (1953). S-R compatibility: Spatial characteristics of stimulus and response codes. *Journal of Applied Psychology*, **46**, 199–210.

Fleishman, E. A. (1966). Human abilities and the acquisition of skill. In E. A. Bilodeau (ed.) *Acquisition of Skill* (New York: Academic Press).

Flesch, R. (1948). A new readability yardstick. *Journal of Applied Psychology*, **32**, 221–233.

Fletcher, H. (1940). Auditory patterns. *Review of Modern Physics*, **12**, 47–65.

Fletcher, H. and Munson, W. A. (1937). Relation between loudness and masking. *Journal of the Acoustical Society of America*, **9**, 1–10.

Floyd, W. F. and Ward, J. S. (1969). Anthropometric and physiological considerations in school, office and factory seating. *Ergonomics*, **12**, 132–139.

Folkard, S. and Monk, T. H. (1978). Time of day effects in immediate and delayed memory. In M. M. Gruneberg, P. E. Morris and R. N. Sykes (eds.) *Practical Aspects of Memory*. (London: Academic Press).

Folkard, S. and Monk, T. H. (1979). Shiftwork and performance. *Human Factors*, **21**, 483–492.

Folkard, S., Monk, T. H. and Chobban, M. C. (1978). Short and long term adjustment of circadian rhythms in 'permanent' night nurses. *Ergonomics*, **21**, 785–799.

Foster, J. and Coles, P. (1977). An experimental study of typographic cueing in printed text. *Ergonomics*, **20**, 57–66.

Fowler, R. L., Williams, W. E., Fowler, M. G. and Young, D. D. (1968) *An Investigation of the Relationship between Operator Performance and Operator Panel Layout for Continuous Tasks*. USAF AMRL-TR-68-170

Fowler, R. L. and Barker, A. S. (1974). Effectiveness of highlighting for retention of text material. *Journal of Applied Psychology*, **63**, 309–313.

Fox, J. G. (1971). Background music and industrial productivity—a review. *Applied Ergonomics*, **2**, 70–73.

Fox, J. G. (1977). Quality control of coins. In J. S. Weiner and H. G. Maule (eds.) *Human Factors in Work, Design and Production*. (London: Taylor and Francis).

Fox, J. G. and Embrey, E. D. (1972). Music—an aid to productivity. *Applied Ergonomics*, **3**, 202–205.

Fox, J. G. and Haslegrave, C. M. (1969). Industrial inspection efficiency and the probability of a defect occurring. *Ergonomics*, **12**, 713–721.

Fox, W. F. (1967). Human performance in the cold. *Human Factors*, **9**, 203–220.

Fried, M. L. and DeFazio, V. J. (1974). Territoriality and boundary conflicts in the subway. *Psychiatry*, **37**, 47–59.

Gagge, A. P., Stolwijk, J. A. J. and Nishi, Y. (1971). An effective temperature scale based on a simple model of human physiological regulatory response. *ASHRAE Transactions*, **77**, Part I, 247–262.

Gane, C. P., Horabin, I. S. and Lewis, B. N. (1966). The simplification and avoidance of instruction. *Industrial Training International*, **1**, 160–166.

Garrett, J. W. (1971). The adult human hand: Some anthropometric and biomechanical considerations. *Human Factors*, **13**, 117–131.

Gibson, J. J. (1950). The perception of visual surfaces. *American Journal of Psychology*, **63**, 367–384.

Gilbert, M. and Hopkinson, R. G. (1949). The illumination of the Snellen chart. *British Journal of Ophthalmology*, **33**, 305–310.

Gittelson, B. (1978). *Biorhythm: A Personal Science (2nd Edition)*. (New York: Arco Publishing Co).

Gladstones, W. H. (1969). Some effects of commercial background music on data preparation operators. *Occupational Psychology*, **43**, 213–222.

Glass, D. C. and Singer, J. E. (1972). *Urban Stress*. (London: Academic Press).

Gorrill, R. B. and Snyder, F. W. (1957). *Preliminary Study of Aircrew Tolerance to Low Frequency Vertical Vibration*. Boeing Co. (Wichita) Report D3-1189.

Gould, J. D. (1968). Visual factors in the design of computer controlled CRT displays. *Human Factors*, **10**, 359–376.

Graham, N. E. (1954) The human response to variations in the design of a visual indicator. In W. Floyd and A. Welford (eds.) *Human Factors in Equipment Design*. (London: H. K. Lewis).

Grandjean, E. (1973). *Ergonomics in the Home*. (London: Taylor and Francis).

298

Greene, J. M. (1970). The semantic function of negatives and passives. *British Journal of Psychology*, **61**, 17–22.

Greene, J. M. (1972). *Psycholinguistics: Chomsky and Psychology*. (Harmondsworth. Middlesex: Penguin).

Greenwood, M., Woods, H. M. and Yule, G. U. (1919). A report on the incidence of industrial accidents upon individuals with special reference to multiple accidents. Report 4. Industrial Health Research Board. In W. Haddon, E. A. Suchmann and D. Flein (eds.) *Accident Research* (1964) (New York: Harper).

Grether, W. F. (1949). Instrument reading. I. The design of long-scale indicators for speed and accuracy of quantitative readings. *Journal of Applied Psychology*, **33**, 363–372.

Grether, W. F. (1971). Vibration and human performance. *Human Factors*, **13**, 203–216.

Grieve, D. W. (1979a). The postural stability diagram (PSD): Personal constraints on the static exertion of force. *Ergonomics*, **22**, 1155–1164.

Grieve, D. W. (1979b). Environmental constraints on the static exertion of force: PSD analysis in task-design. *Ergonomics*, **22**, 1165–1175.

Grieve, D. W. and Pheasant, S. T. (1981) Biomechanics. In *The Body at Work–Biological Ergonomics* (ed. W. T. Singleton) (London: Cambridge University Press).

Griffin, M. J. (1976). Vibration and visual acuity. In W. Tempest (ed.) *Infrasound and Low Frequency Vibration*. (London: Academic Press).

Griffin, M. J. and Lewis, C. H. (1978). A review of the effects of vibration on visual acuity and continuous manual control. Part I: Visual acuity. *Journal of Sound and Vibration*, **56**, 383–413.

Griffiths, I. D. and Langdon, F. J. (1968). Subjective response to road traffic noise. *Journal of Sound and Vibration*, **8**, 16–32.

Guillemin, V. and Wechsberg, P. (1953). Physiological effects of long term repetitive exposure to mechanical vibration. *Journal of Aviation Medicine*, **24**, 208–221.

Guillien, J. and Rebiffé, R. (1980). Anthropometric models of a population of bus drivers. In D. J. Oborne and J. A. Levis (eds.) *Human Factors in Transport Research, Vol I*. (London: Academic Press).

Haines, R. F. and Gilliland, K. (1973). Response time in the full visual field. *Journal of Applied Psychology*, **58**, 289–295.

Hale, A. R. and Hale, M. (1972). *A Review of the Industrial Accident Research Literature*. Committee on Safety and Health at Work Paper. (London: HMSO).

Hall, E. T. (1976) The anthropology of space: An organising model. In H. M. Proshansky, W. H. Ittleson and L. G. Rivlin (eds.) *Environmental Psychology (2nd Ed)*. (N. Y.: Holt, Rineholt and Winston, Inc)

Hanes, R. M. (1970) *Human Sensitivity to Whole-body Vibration in Urban Transport Systems: A Literature Review*. Johns Hopkins University Transportation Programs Report APL/JHU-TPR 004.

Hardyck, C. and Petrinovich, L. F. (1977). Left-handedness. *Psychological Bulletin*, **84**, 385–404.

Harris, C. S., Sommer, H. C. and Johnson, D. L. (1976). Review of the effects of infrasound on man. *Aviation, Space and Environmental Medicine*, **47**, 430–434.

Harris, D. (1966). Effect of equipment complexity on inspection performance. *Journal of Applied Psychology*, **50**, 236–237.

Hartley, J. and Burnhill, P. (1976). *Textbook Design: A Practical Guide*. (Paris: UNESCO).

Hartline, H. K., Wagner, H. G. and Ratliffe, F. (1956). Inhibition in the eye of LIMULUS. *Journal of General Physiology*, **39**, 651–673.

Haslegrave, C. M. (1979). An anthropometric survey of British drivers. *Ergonomics*, **22**, 145–153.

Helberg, W. and Sperling, E. (1941). Critical appraisal of the riding properties of railway vehicles. *Org. Fortschr. Eisenbahnwesens*, **96**, 12. (Translated from the German by the Research Department Translation Service, British Railways Board Translation 743).

Helmholtz, H. Von (1889) *Popular Scientific Lectures.* (London: Longmans).

Hemingway, J. C. and Erickson, R. A. (1969). Relative effects of raster scan lines and image subtense on symbol legibility on television. *Human Factors*, **11**, 331–338.

Hepburn, H. A. (1958). Portable ladders. I: The quarter-length rule. *British Journal of Industrial Safety*, **4**, 155–158.

Hertzberg, H. T. E. and Burke, F. E. (1971). Foot forces exerted at various aircraft brake-pedal angles. *Human Factors*, **13**, 445–456.

Hettinger, T. (1961). *Physiology of Strength.* (Springfield, Ill.: Charles C. Thomas).

Hildebrandt, G., Rohmert, W. and Rutenfranz, J. (1974). Twelve and twenty four hour rhythms in error frequency of locomotive drivers and the influence of tiredness. *International Journal of Chronobiology*, **2**, 175–180.

Hill, J. M. M. and Trist, E. L. (1953). A consideration of industrial accidents as a means of withdrawal from the work situation. *Human Relations*, **6**, 357–380.

Hitt, W. D. (1961). An evaluation of five different abstract coding methods. *Human Factory*, **3**, 120–130.

Hochberg, J. (1972). Perception. I Colour and shape. In J. W. Kling and L. Riggs, (eds.) *Woodworth and Schlosberg's Experimental Psychology.* (London: Methuen).

Hofmann, M. A. and Heimstra, N. W. (1972). Tracking performance with visual, auditory or electrocutaneous displays. *Human Factors*, **14**, 131–138.

Holladay, L. L. (1926). The fundamentals of glare and visibility. *Journal of the Optical Society of America*, **12**, 271–319.

Hopkinson, R. G. (1940). Discomfort glare in lighted streets. *Transactions of the Illuminating Engineering Society*, **5**, 1–30.

Hopkinson, R. G. (1956). Glare discomfort and pupil diameter. *Journal of the Optical Society of America*, **46**, 694–656.

Hopkinson, R. G. (1957). Evaluation of glare. *Illumination Engineering (NY)*, **52**, 305–316.

Hopkinson, R. G. (1972). Glare from daylight in buildings. *Applied Ergonomics*, **3**, 206–215.

Hopkinson, R. G. and Collins, J. B. (1970) *The Ergonomics of Lighting.* (London: McDonald Technical and Scientific).

Hopkinson, R. G. and Longmore, J. (1959). Attention and distraction in the lighting of workplaces. *Ergonomics*, **2**, 321–333.

Hopkinson, R. G., Waldram, J. M. and Stevens, W. R. (1941). Brightness and contrast in illuminating engineering. *Transactions of the Illuminating Engineering Society (London)*, **6**, 37–47.

Houghton, F. C. and Yaglou, C. P. (1923). Determining lines of equal comfort. *ASHRAE Transactions*, **29**, 163–176.

Huddleston, J. H. F. (1970). Tracking performance on a visual display apparently vibrating at one to ten Hertz. *Journal of Applied Psychology*, **54**, 401–408.

Huddleston, J. H. F. (1974). A comparison of two 7 × 9 matrix alphanumeric designs for TV displays. *Applied Ergonomics*, **5**, 81–83.

Hunt, D. P. (1953). *The Coding of Aircraft Controls.* USAF, WADC, Technical Report 53-221.

Illuminating Engineering Society (1973). *IES Code for Interior Lighting.* (IES).

International Organisation for Standardisation (1974) *Guide for the Evaluation of Human Exposure to Whole-body Vibration.* ISO 2631.

Jacklin, H. M. and Liddell, G. J. (1933). *Riding Comfort Analysis.* Purdue University Reasearch Bulletin No 44.

Jamieson, G. H. (1966). Inspection in the telecommunications industry: A field study of age and other performance variables. *Ergonomics*, **9**, 297–303.

Jellinek, E. M. and McFarland, R. A. (1940). Analysis of psychological experiments on the effects of alcohol. *Quarterly Journal on the Study of Alcohol*, **1**, 272–371.

Jenkins, H. M. (1958). The effect of signal rate on performance in visual monitoring. *American Journal of Psychology*, **71**, 647–661.

Jenkins, W. O. (1947). The tactual discrimination of shapes for coding aircraft-type controls. In P. M. Fitts (ed.) *Psychological Research on Equipment Design*. USAF Research Report No 19.

Jenkins, W. L. and Connor, M. B. (1949). Some design factors in making settings on a linear scale. *Journal of Applied Psychology*, **33**, 395–409.

Jerison, H. J. (1959). Effects of noise on human performance. *Journal of Applied Psychology*, **43**, 96–101.

Jerison, H. J. (1966). Remarks on Colquhoun's 'effect of unwanted signals on performance in a vigilance task'. *Ergonomics*, **9**, 413–416.

Johnsgard, K. W. (1953). Check reading as a function of pointer symmetry and uniform alignment. *Journal of Applied Psychology*, **37**, 407–411.

Johnson, D. L., Nixon, C. W. and Stephenson, M. R. (1976). Long-duration exposure to intermittent noises. *Aviation, Space and Environmental Medicine*, **47**, 987–990.

Johnson, S. L. and Roscoe, S. N. (1972). What moves, the airplane or the world? *Human Factors*, **14**, 107–129.

Johnston, D. M. (1965). Search performance as a function of peripheral activity. *Human Factors*, **7**, 527–535.

Jones, A. J. and Saunders, D. J. (1972). Equal comfort contours for whole-body, vertical, pulsed sinusoidal vibration. *Journal of Sound and Vibration*, **23**, 1–4.

Jones, J. C. (1969). Methods and results of seating research. *Ergonomics*, **12**, 171–181.

Jones, M. R. (1962). Colour coding. *Human Factors*, **4**, 355–365.

Jones, S. (1968). *Design of Instruction. Training Information Paper 1; Department of Employment and Productivity*. (London: HMSO).

Jorgensen, K. and Poulsen, E. (1974). Physiological problems in repetitive lifting with special reference to tolerance limits to the maximum lifting frequency. *Ergonomics*, **17**, 31–39.

Kamman, R. (1975). The comprehensibility of printed instructions and the flowchart alternative. *Human Factors*, **17**, 183–191.

Keegan, J. J. and Radke, A. O. (1964). Designing vehicle seats for greater comfort. *SAE Journal*, **September**, **72**, 50–55.

Kemsley, W. F. F. (1950). Weight and height of a population in 1943. *Annals of Eugenics*, **15**, 161–183.

Kennedy, K. W. (1975). International anthropometric variability and its effects on aircraft cockpit design. In A. Chapanis (ed.) *Ethnic Variables in Human Factors Engineering*. (Baltimore: Johns Hopkins University Press).

Khalil, T. M. and Kurucz, C. N. (1977). The influence of 'biorhythm' on accident occurrence and performance. *Ergonomics*, **20**, 389–398.

Kimura, D. and Durnford, M. (1974). Normal studies on the function of the right hemisphere in vision. In S. J. Dimond and J. G. Beaumont (eds.) *Hemisphere Function in the Human Brain*. (London: Elek Science).

Kimura, D. and Vanderwolf, C. H. (1970). The relation between hand preference and the performance of individual finger movements by left and right hands. *Brain*, **93**, 767–774.

Klare, G. R. (1963). *The Measurement of Readability*. (Des Moines, Iowa, Iowa State University Press).

Klemmer, E. T. (1971). Keyboard entry. *Applied Ergonomics*, **2**, 2–6.

Korn, T. S. (1954). Effect of psychological feedback on conversational noise reduction in rooms. *Journal of the Acoustical Society of America*, **26**, 793–794.

Kroemer, K. H. E. (1970). Human strength: Terminology, measurement, and interpretation of data. *Human Factors*, **12**, 297–313.

Kroemer, K. H. E. (1971). Foot operation of controls. *Ergonomics*, **14**, 333–361.
Kroemer, K. H. E. (1974). Horizontal push and pull forces. *Applied Ergonomics*, **5**, 94–102.
Kroemer, K. H. E. and Robinette, J. C. (1968). *Ergonomics in the Design of Office Furniture: A Review of European Literature*. AMRL-TR-68-80.
Kryter, K. D. (1970). *The Effects of Noise on Man*. (New York: Academic Press).
Kryter, K. D. and Pearsons, K. S. (1963). Some effects of spectral content and duration on perceived noise level. *Journal of the Acoustical Society of America*, **35**, 866–883.
Kryter, K. D., Ward, W. D., Miller. J. D. and Eldredge, D. H. E. (1966). Hazardous exposure to intermittent and steady-state noise. *Journal of the Acoustical Society of America*, **39**, 451–464.
Kurke, M. I. (1956). Evaluation of a display incorporating quantitative and check-reading characteristics. *Journal of Applied Psychology*, **40**, 233–236.
Lee, R. A. and King, A. I. (1971). Visual vibration response. *Journal of Applied Psychology*, **30**, 281–286.
Leventhall, H. G. and Kyriakides, K. (1976). Environmental infrasound: Its occurrence and measurement. In W. Tempest (ed.) *Infrasound and Low Frequency Vibration*. (London: Academic Press).
Lewin, T. (1969). Anthropometric studies on Swedish industrial workers when standing and sitting. *Ergonomics*, **12**, 883–902.
Lewis, C. H. and Griffin, M. J. (1978). A review of the effects of vibration on visual acuity and continuous manual control. Part II: Continuous manual control. *Journal of Sound and Vibration*, **56**, 415–457.
Licklider, J. C. (1948). The influence of interaural phase relations upon the masking of speech by white noise. *Journal of the Acoustical Society of America*, **20**, 150–159.
Liebman, M. (1970). The effects of sex and race norms on personal space. *Environmental Behaviour*, **2**, 208–246.
Lion, J. S. (1964). The performance of manipulative and inspection tasks under tungsten and fluorescent lighting. *Ergonomics*, **7**, 51–61.
Lion, J. S., Richardson, E. and Browne, R. C. (1968). A study of the performance of industrial inspectors under two kinds of lighting. *Ergonomics*, **11**, 23–34.
Little, K. B. (1965). Personal space. *Journal of Experimental Social Psychology*, **1**, 237–247.
Lockhart, J. M. (1968). Extreme body cooling and psychomotor performance. *Ergonomics*, **11**, 249–260.
Lockhart, J. M. and Keiss, H. O. (1971). Auxiliary heating of the hands during cold exposure and manual performance. *Human Factors*, **13**, 457–465.
Loeb, M. (1965) *A Further Investigation of the Influence of Whole-body Vibration and Noise on Tremor and Visual Acuity*. AMRL Rpt No 165 (6-95-20-001).
Loveless, N. E. (1962). Direction-of-motion stereotypes: A review. *Ergonomics*, **5**, 357–383.
Lovesey, E. J. (1975). The helicopter—some ergonomic factors. *Applied Ergonomics*, **6**, 139–146.
Luckiesh, M. and Guth, S. K. (1949). Brightness in the visual field at borderline between comfort and discomfort (BCD). *Illuminating Engineering*, **44**, 650–670.
Lukiesh, M. and Holladay, L. L. (1925). Glare and visibility. *Transactions of the Illuminating Engineering Society*, **20**, 221–252.
Lundberg, U. (1976). Urban commuting: Crowdedness and catecholamine-excretion. *Journal of Human Stress*, **2**, 26–32.
Mackay, G. M., DeFoneka, C. P., Blair, I. and Clayton, A. B. (1969). *Causes and Effects of Road Accidents*. Department of Transportation, University of Birmingham.
Mackworth, N. H. (1950). Researches on the measurement of human performance. MRC special report series 268. (London: HMSO). In W. Sinaiko (ed.) *Human Factors in the Design and Use of Control Systems*. (1961) (New York: Dover).

Maddox, M. E., Burnette, J. T. and Gutmann, J. C. (1977). Font comparisons for 5 × 7 dot matrix characters. *Human Factors*, **19**, 89–93.

Magid, E. B., Coermann, R. R., Lowry, R. D. and Bosley, W. J. (1962). *Physiological and Mechanical Response of the Human to Longitudinal Whole-body Vibration as Determined by Subjective Response.* Biomedical Research Laboratory's Technical Document. MRL-TDR-62-66.

Magora, A. (1972). Investigation of the relation between low back pain and occupation. Three physical requirements: sitting, standing and weight lifting. *Industrial Medicine*, **41**, 5–9.

Mahoney, E. R. (1974). Compensatory reactions to spatial immediacy. *Sociometry*, **37**, 423–431.

Mandal, A. C. (1976). Work chair with tilting seat. *Ergonomics*, **19**, 157–164.

Mandel, M. J. and Lowry, R. D. (1962). *One-Minute Tolerance in Man to Vertical Sinusoidal Vibration in the Sitting Posture.* USAF AMRL Report AMRL-TDR-62-121.

Martin, A. (1972). A new keyboard layout. *Applied Ergonomics*, **3**, 48–51.

Matsumoto, S. (1977). New multicolour liquid crystal display. *Toshiba Review*, **March**, 1–4.

McBride, G., King, M. G. and James, J. W. (1965). Social proximity effects on GSR in adult humans. *Journal of Psychology*, **61**, 153–157.

McClelland, I. and Ward, J. S. (1976). Ergonomics in relation to sanitary ware design. *Ergonomics*, **4**, 465–478.

McCormick, E. J. (1950). An analysis of visual requirements in industry. *Journal of Applied Psychology*, **34**, 54–61.

McCormick, E. J. (1976). *Human Factors in Engineering and Design.* (New York: McGraw-Hill).

McKendry, J. M., Corso, J. F. and Grant, G. (1960). The design and evolution of maintainable packaging methods for electronic equipment. *Ergonomics*, 255–272.

McKenzie, R. M. (1957). On the accuracy of inspectors. *Ergonomics*, **1**, 258–272.

McLaughlin, G. H. (1966). Comparing styles of presenting technical information. *Ergonomics*, **9**, 257–259.

McNall, P. E., Ryan, P. W., Rohles, F. H., Nevins, R. G. and Springer, W. E. (1968). Metabolic rates at four activity levels and their relationship to thermal comfort. *ASHRAE Transactions*, **74**, Part I, IV.3.1 to IV.3.20.

Megaw, E. D. (1978). Some factors affecting inspection accuracy. Paper presented to Symposium on Ergonomics and Visual Inspection. Birmingham.

Mehrabian, A. and Diamond, S. G. (1971). Effects of furniture arrangements, props, and personality on social interaction. *Journal of Personality and Social Psychology*, **20**, 18–30.

Meisels, M. and Canter, F. M. (1970). Personal space and personality characteristics: A non confirmation. *Psychological Reports*, **27**, 287–290.

Middlemist, R. D., Knowles, E. S. and Matter, C. F. (1976). Personal space invasions in the lavatory: Suggestive evidence for arousal. *Journal of Personality and Social Psychology*, **33**, 541–546.

Miller, G. A. (1956). The magical number seven plus or minus two: Some limits on our capacity to process information. *Psychological Review*, **63**, 81–97.

Miller, G. A. and Licklider, J. C. R. (1950). The intelligibility of interrupted speech. *Journal of the Acoustical Society of America*, **22**, 167–173.

Miller, G. R. (1972). *An Introduction to Speech Communication, 2nd Ed.* (Indianapolis: The Bobbs-Merrill Co. Inc).

Milroy, R. and Poulton, E. C. (1978). Labelling graphs for improved reading speed. *Ergonomics*, **21**, 55–61.

Moore, T. G. (1974). Tactile and kinaesthetic aspects of push-buttons. *Applied Ergonomics*, **5**, 66–71.

Moore, T. G. (1975). Industrial push-buttons. *Applied Ergonomics*, **6**, 33–38.

Moore, T. G. (1976). Controls and tactile displays. In K. F. Kraiss and J. Moraal (eds.) *Introduction to Human Engineering* (Koln: TUV Rhineland).

Moores, B. (1972). Ergonomics—or work study. *Applied Ergonomics*, **3**, 147–154.

Morgan, C. T. (1965). *Physiological Psychology*. (New York: McGraw-Hill).

Morgan, C. T., Cook, J. S., Chapanis, A. and Lund, M. (1963). *Human Engineering Guide to Equipment Design*. (New York: McGraw-Hill).

Morgensen, M. F. and English, H. B. (1926). The apparent warmth of colours. *American Journal of Psychology*, **37**, 427–428.

Morris, D. (1977). *Manwatching: A Field Guide to Human Behaviour*. (London: Jonathan Cape).

Morton, R. and Provins, K. A. (1960). Finger numbness after acute local exposure to cold. *Journal of Applied Physiology*, **15**, 149–154.

Mortimer, R. G. (1974). Foot brake pedal force capability of drivers. *Ergonomics*, **17**, 509–513.

Muller, E. A. (1965). Physiological methods of increasing human physical work capacity. *Ergonomics*, **8**, 409–424.

Mundel, M. E. (1950). *Motion and Time Study*. (New York: Prentice-Hall).

Murrell, K. F. H. (1958). The relationship between dial size, reading distance and reading accuracy. *Ergonomics*, **1**, 182–190.

Murrell, K. F. H. (1962a). Industrial aspects of ageing. *Ergonomics*, **5**, 147–153.

Murrell, K. F. H. (1962b). Operator variability and its industrial consequence. *International Journal of Production Research*, **1**, 39.

Murrell, K. F. H. (1967). Why ergonomics? *Occupational Psychology*, **44**, 17–24.

Murrell, K. F. H. (1969). Beyond the panel. *Ergonomics*, **12**, 691–700.

Murrell, K. F. H. (1971). *Ergonomics: Man in His Working Environment*. (London: Chapman Hall).

Murrell, K. F. H. and Kingston, P. M. (1966). Experimental comparison of scalar and digital micrometers. *Ergonomics*, **9**, 39–47.

Nason, W. E. and Bennett, C. A. (1973). Dials *v.* counters: Effects of precision on quantitative reading. *Ergonomics*, **16**, 749–758.

Nayyar, R. M. and Simon, J. R. (1963). Effects of magnification on a subminiature assembly operation. *Journal of Applied Psychology*, **47**, 190–195.

Nelson, J. D. and Barany, J. W. (1969). A dynamic visual recognition test for paced industrial inspection. *AIIE Transactions*, **1**, 327–332.

Nemecek, J. and Grandjean, E. (1973). Noise in landscaped offices. *Applied Ergonomics*, **4**, 19–22.

Nevins, R. G. and Gagge, A. P. (1972). The new ASHRAE comfort chart. *ASHRAE Journal*, **14**, 41–43.

Nevins, R. G., Rohles, F. H., Springer, W. and Feyerherm, A. M. (1966). A temperature-humidity chart for thermal comfort of seated persons. *ASHRAE Transactions*, **72**, 283–291.

Newbold, E. M. (1926). *A Contribution to the Study of the Human Factor in the Causation of Accidents*. Report 34 Industrial Health Research Board.

Nixon, J. C. and Glorig, A. (1961). Noise-induced permanent threshold shift at 2000 cps and 4000 cps. *Journal of the Acoustical Society of America*, **33**, 904–908.

Oborne, D. J. (1976). A critical assessment of studies relating whole-body vibration to passenger comfort. *Ergonomics*, **19**, 751–774.

Oborne, D. J. (1978a). Vibration and passenger comfort: Can data from subjects be used to predict passenger comfort? *Applied Ergonomics*, **9**, 155–161.

Oborne, D. J. (1978b) Techniques for the assessment of passenger comfort. *Applied Ergonomics*, **9**, 45–49.

Oborne, D. J. and Humphreys, D. A. (1976). Individual variability in human response to whole-body vibration. *Ergonomics*, **19**, 719–726.

'Observer' and Maxwell, M. A. (1959). A study of absenteeism, accidents and sickness payments in problem drinkers in one industry. *Quarterly Journal of Studies on Alcohol*, **20**, 302–312.

Oppenheim, A. N. (1966). *Questionnaire Design and Attitude Measurement*. (London: Heineman).

Parks, D. L. and Snyder, F. W. (1961). *Human Reaction to Low Frequency Vibration*. Boeing Co (Seattle) Report D3-3512-1.

Paterson, D. G. and Tinker, M. A. (1946). Readability of newspaper headlines printed in capitals and lower case. *Journal of Applied Psychology*, **30**, 161–168.

Patterson, M. L. and Sechrest, L. B. (1970). Interpersonal distance and impression formation. *Journal of Personality*, **38**, 161–166.

Patterson, M. L., Mullens, S. and Romano, J. (1971). Compensatory reactions to spatial intrusion. *Sociometry*, **34**, 144–121.

Peacock, L. J. (1956). *A Field Study of Rifle Aiming Steadiness and Serial Reaction Performance as Affected by Thermal Stress and Activity*. US Army Medical Research Laboratory Report 231.

Peizer, E. and Wright, D. W. (1974). Human locomotion. In: Institute of Mechanical Engineers (ed.) *Human Locomotion Engineering* (London).

Pepler, R. D. (1958). Warmth and performance: An investigation in the tropics. *Ergonomics*, **2**, 63–88.

Perry, D. K. (1952). Speed and accuracy of reading Arabic and Roman numerals. *Journal of Applied Psychology*, **36**, 346–347.

Peters, G. A. and Adams, B. B. (1959). These three criteria for readable panel markings. *Product Engineering*, **May 25th**, 30, 55–57.

Petherbridge, P. and Hopkinson, R. G. (1950). Discomfort glare and the lighting of buildings. *Transactions of the Illuminating Engineering Society*, **15**, 39–79.

Pheasant, S. and O'Neill, D. (1975). Performance in gripping and turning—a study in hand/handle effectiveness. *Applied Ergonomics*, **6**, 205–208.

Plath, D. W. (1970). The readability of segmented and conventional numerals. *Human Factors*, **12**, 493–497.

Pollack, I. and Ficks, L. (1954). Information of elementary multi-dimensional auditory displays. *Journal of the Acoustical Society of America*, **26**, 155–158.

Pook, G. K. (1969). Colour coding effects in compatible and noncompatible display-control arrangements. *Journal of Applied Psychology*, **53**, 301–303.

Pottier, M., Dubreuil, A. and Mond, H. (1969). The effects of sitting posture on the volume of the foot. *Ergonomics*, **12**, 753–758.

Poulsen, E. and Jorgensen, K. (1971). Back muscle strength, lifting, and stooped working postures. *Applied Ergonomics*, **2**, 133–137.

Poulton, E. C. (1967). Searching for newspaper headlines printed in capitals or lower-case letters. *Journal of Applied Psychology*, **51**, 417–425

Poulton, E. C. (1969a). How efficient is print? *New Society*, **5th June**, 869–871.

Poulton, E. C. (1969b). Skimming lists of food ingredients printed in different sizes. *Journal of Applied Psychology*, **53**, 55–58.

Poulton, E. C. (1975). Colours for sizes: A recommended ergonomic colour code. *Applied Ergonomics*, **6**, 231–235.

Poulton, E. C. (1976). Continuous noise interferes with work by masking auditory feedback and inner speech. *Applied Ergonomics*, **7**, 79–84.

Poulton, E. C. (1977). Continuous intense noise masks auditory feedback and inner speech. *Psychological Bulletin*, **84**, 977–1001.

Poulton, E. C. (1978). Increased vigilance with vertical vibration at 5 Hz: An alerting mechanism. *Applied Ergonomics*, **9**, 73–76.

Poulton, E. C., Warren, T. R. and Bond, J. (1970). Ergonomics in journal design. *Applied Ergonomics*, **4**, 207–209.

Powell, P. I., Hale, M., Martin, J. and Simon, M. (1971). *2000 Accidents*. (London: NIIP).

Radl, G. W. (1980). Experimental investigation for optimal presentation-mode and colours on the CRT screen. In E. Grandjean and E. Vigliani (eds.) *Ergonomic Aspects of Visual Display Terminals*. (London: Taylor and Francis).

Rasch, P. J. and Pierson, W. P. (1963). Some relationships of isometric strength, isotonic strength, and anthropometric measures. *Ergonomics*, **6**, 211–215.

Reason, J. T. (1974). *Man in Motion: The Psychology of Travel*. (London: Weidenfeld and Nicolson).

Reason, J. T. (1976). Absent minds. *New Society*, **4th November**, 244–245.

Reason, J. T. (1978). Motion sickness—some theoretical and practical considerations. *Applied Ergonomics*, **9**, 163–167.

Reason, J. T. and Brand, J. J. (1975). *Motion Sickness*. (London: Academic Press).

Redgrove, J. (1979). Fitting the job to the woman: A critical review. *Applied Ergonomics*, **10**, 215–223.

Roberts, D. F. (1960). Functional anthropometry of elderly women. *Ergonomics*, **3**, 321–327.

Roberts, D. F. (1975). Population differences in dimensions, their genetic basis and their relevance to practical problems of design. In A. Chapanis (ed.) *Ethnic Variables in Human Factors Engineering*. (Baltimore: Johns Hopkins University Press).

Rodger, A. and Cavanagh, P. (1962). Training occupational psychologists. *Occupational Psychology*, **36**, 82–88.

Roebuck, J. A., Kroemer, K. H. E. and Thomson, W. G. (1975). *Engineering Anthropometry Methods*. (New York: John Wiley and Co.)

Rohles, F. H. (1967). Environmental psychology. *Psychology Today*, **June**, 54–63.

Rohles, F. H. (1969). Preference for the thermal environment by the elderly. *Human Factors*, **11**, 37–41.

Rolf, J. M. (1969a). Human factors and the display of height information. *Applied Ergonomics*, **1**, 16–24.

Rolfe, J. M. (1969b). *Some Investigations Into The Effectiveness of Numerical Displays For The Presentation of Dynamic Information*. IAM Technical Report R470.

Rolfe, J. M. and Allnutt, M. F. (1967). Putting the man in the picture. *New Scientist*, **16th February**, 401–406.

Ronnholm, N. (1962). Physiological studies on the optimum rhythm of lifting work. *Ergonomics*, **5**, 51–52.

Rosegger, R. and Rosegger, S. (1960). Health effects of tractor driving. *Journal of Agricultural Engineering Research*, **5**, 241–275.

Ross, S., Katchmar, L. T. and Bell, H. (1955). Multiple-dial check reading: Pointer symmetry compared with uniform alignment. *Journal of Applied Psychology*, **39**, 215–218.

Royal Society for the Prevention of Accidents. (1977). *Factory Accidents*.

Russek, A. S. (1955). Medical and economic factors relating to the compenstable back injury. *Archives of Physical Medicine and Rehabilitation*, **36**, 316–323.

Rutenfranz, J. and Colquhoun, W. P. (1979). Circadian rhythms in human performance. *Scandinavian Journal of Work, Environment and Health*, **5**, 167–177.

Savinar, J. (1975). The effect of ceiling height on personal space. *Man–Environment Systems*, **5**, 321–324.

Sell, R. G. (1977). Ergonomics as applied to crane cabs. In J. S. Weiner and H. G. Maule (eds.) *Human Factors in Work, Design and Production*. (London: Taylor and Francis).

Shackel, B. (1959). A note on panel layout for numbers of identical items. *Ergonomics*, **2**, 247–253.

Shackel, B. (1962). Ergonomics in the design of a large digital computer console. *Ergonomics*, **5**, 229–241.

Sharp, E. D. and Hornseth, J. P. (1965). *The Effect of Control Location Upon Performance*

Time For Knob, Toggle Switch and Push Button. AMRL Technical Report TR-65-41.

Shaw, L. and Sichel, H. S. (1971) *Accident Proneness: Research in the Occurrence, Causation, and Prevention of Road Accidents.* (Oxford: Pergamon).

Shibolet, S., Lancaster, M. C. and Danon, Y. (1976). Heat stroke: A review. *Journal of Aviation, Space and Environmental Medicine,* **47,** 280–301.

Shinar, D. and Acton, M. B. (1978). Control–display relationships on the four-burner range: Population stereotypes versus standards. *Human Factors,* **20,** 13–17.

Shoenberger, R. W. (1967). Effects of vibration on complex psychomotor performance. *Aerospace Medicine,* **38,** 1264–1269.

Shoenberger R. W. (1974). An investigation of human information processing during whole-body vibration. *Aerospace Medicine,* **45,** 143–153.

Seigel, A. I. and Brown, F. R. (1958). An experimental study of control console design. *Ergonomics,* **1,** 251–257.

Simon, C. W. and Roscoe, S. N. (1956). *Altimeter Studies Part II. A Comparison of Integrated Versus Separated Displays.* Hughes Aircraft Company, Culver City, California, Technical Memo No 435.

Simon, J. R. (1964). Magnification as a variable in subminiature work. *Journal of Applied Psychology,* **48,** 20–24.

Simon, J. R. and Rudell, A. P. (1967). Auditory S-R compatibility: The effect of an irrelevant cue on information processing. *Journal of Applied Psychology,* **51,** 300–304.

Sinaiko, H. W. and Brislin, R. W. (1973). Evaluating language translations: Experiments on 3 assessment methods. *Journal of Applied Psychology,* **57,** 328–334.

Singleton, W. T. (1967). Ergonomics in systems design. *Ergonomics,* **10,** 541–548.

Smith, G. L. and Adams, S. K. (1971). Magnification and microminiature inspection. *Human Factors,* **13,** 247–254.

Smith, R. L., Westland, R. A., and Crawford, B. M. (1970). The status of maintainability models: A crtical review *Human Factors,* **12,** 271–283

Smith, S. L. and Thomas, D. W. (1964). Colour versus shape coding in information displays. *Jouranl of Applied Psychology,* **48,** 137–146.

Snook, S. H. (1978). The design of manual handling tasks. *Ergonomics,* **21,** 963–985.

Sommer, R. (1968). Intimacy ratings in five countries. *International Journal of Psychology,* **3,** 109–114.

Sommer, R. (1969). *Personal Space: The Behavioural Basis of Design.* (New York: Prentice-Hall Inc.).

Spencer, J. (1963). Pointers for general purpose indicators. *Ergonomics,* **6,** 35–49.

Spencer, J. (1968). Inspection and human factors. In *Human Factors in the Management of Industrial Inspection.* Papers presented to the Industrial Section of the Ergonomics Research Society and the Institute of Mechanical Engineering.

Spoor, A. (1967). Presbycusis values in relation to noise-induced hearing loss. *International Audiology,* **6,** 48–57.

Stevens, S. S. (1972). Stability of human performance under intense noise. *Journal of Sound and Vibration,* **21,** 35–56.

Stiles, W. S. (1929). The scattering theory of glare. *Proceedings of the Royal Society (Series B),* **105,** 131–146.

Stockbridge, H. C. W. and Lee, M. (1973). The psycho-social consequences of aircraft noise. *Applied Ergonomics,* **4,** 44–45.

Stoudt, H., Damon, A., McFarland, R. A. and Roberts, J. (1965). *Weight, Height and Selected Body Dimensions of Adults US 1960–1962.* Report No 8; National Center for Health Statistics, Public Health Service Publication No 1000, Washington DC.

Suchman, E. A. (1961). A conceptual analysis of the accident phenomenon. *Social Problems,* **8,** 241.

Suchman, E. A. (1965). Cultural and social factors in accident occurrence and control. *Journal of Occupational Medicine,* **7,** 487.

Szlichcinski, K. P. (1979a). Diagrams and illustrations as aids to problem solving. *Instructional Science*, **8**, 253–274.

Szlichcinski, K. P. (1979b). Telling people how things work. *Applied Ergonomics*, **10**, 2–8.

Taylor, F. V. and Garvey, W. D. (1966). The limitations of a 'Procrustean' approach to the optimization of man-machine systems. *Ergonomics*, **9**, 187–194.

Teel, K. S. (1971). Is human factors engineering worth the investment? *Human Factors*, **13**, 17–21.

Teel, K. S., Springer, R. M. and Sadler, E. E. (1968). Assembly and inspection of microelectronic systems. *Human Factors*, **10**, 217–224.

Teichener, W. H. and Kobrick, J. L. (1955). Effects of prolonged exposure to low temperature on visual-motor performance. *Journal of Experimental Psychology*, **49**, 122–126.

Thomas, J. C. (1965). Use of piezoaccelerometer in studying eye dynamics. *Journal of the Optical Society of America*, **55**, 534–537.

Thomas, L. F. (1968). Setting subjective and objective standards and making judgements. In *Human Factors in the Management of Industrial Inspection*. Papers presented to the Industrial Section of the Ergonomics Research Society and the Institute of Mechanical Engineering.

Thomas, L. F. and Seaborne, A. E. M. (1961). The socio-technical context of industrial inspection. *Journal of Occupational Psychology*, **35**, 36–43.

Tichauer, E. R. (1971). A pilot study of the biomechanics of lifting in simulated industrial work situations. *Journal of Safety Research*, **3**, 98–115.

Tichauer, E. R. (1975). *Occupational Biomechanics*. Rehabilitation Monograph No 51 (New York: Institute of Rehabilitation Medicine).

Tillman, W. A. and Hobbs, G. E. (1949). The accident-prone automobile driver. *American Journal of Psychiatry* **106**, 321–331.

Timbal, J., Loncle, M. and Boutelier, C. (1976). Mathematical model of man's tolerance to cold using morphological factors. *Aviation, Space and Environmental Medicine*, **47**, 958–964.

Timmers, H., Van Nes, F. L. and Blommaert, F. J. J. (1980). Visual word recognition as a function of contrast. In E. Grandjean and E. Vigliani (eds.) *Ergonomic Aspects of Visual Display Terminals* (London: Taylor and Francis).

Tinker, M. A. (1960). Legibility of mathematical tables. *Journal of Applied Psychology*, **44**, 83–87.

Totman, R. G. and Kiff, J. (1979). Life stress and the susceptibility to colds. In D. J. Oborne, M. M. Gruneberg, and J. R. Eiser (eds.) *Research in Psychology and Medicine. Volume I*. (London: Academic Press).

Trice, H. M. and Roman, P. M. (1972). *Spirits and Demons at Works: Alcohol and Other Drugs on the Job*. (New York: New York State School of Industrial and Labour Relations, Cornell University).

Van, Nes, F. L. and Bouma, H. (1979). Legibility of segmented numerals. Paper presented to the 30th Annual Conference of the Ergonomics Society, Oxford.

Van Nes, F. L. and Bouma, H. (1980). On the legibility of segmented numerals. *Human Factors*, **22**, 463–474.

Van Zelst, R. H. (1954). The effects of age and experience upon accident rate. *Journal of Applied Psychology*, **38**, 313–317.

Vartabedian, A. G. (1971). Legibility of symbols on CRT displays. *Applied Ergonomics*, **2**, 130–132.

Vartabedian, A. G. (1973). Developing a graphic set for cathode ray tube display using a 7×9 dot pattern. *Applied Ergonomics*, **4**, 11–16.

Vernon, M. D. (1953). Presenting information on diagrams. *Audio-Visual Communication Review*, **1**, 147–158.

Vernon, H. M. and Warner, C. G. (1932). The influence of the humidity of the air on capacity for work at high temperatures. *Journal of Hygiene (Cambridge)*, **32**, 431.

Verhaegen, P., Vanhalst, B., Derycke, H. and VanHoecke, M. (1976). The value of some psychological theories of industrial accidents. *Journal of Occupational Psychology*, **1**, 39–45.

Ward, J. S. and Beadling, W. (1970). Optimum dimensions for domestic staircases. *Architects Journal*, **151**, 513–520.

Ward, J. and Fleming, P. (1964). Change in body weight and body composition in African mine recruits. *Ergonomics*, **7**, 83–90.

Ward, J. S. and Kirk, N. S. (1967). Anthropometry of elderly women. *Ergonomics*, **10**, 17–24.

Warner, H. D. and Mace, K. C. (1974). Effects of platform fashion shoes on brake response time. *Applied Ergonomics*, **5**, 143–146.

Warrick, M. J. (1947). Direction of movement in the use of control knobs to position visual indicators. In P. M. Fitts (ed.) *Psychological Research on Equipment Design*. (US Army Air Force, Aviation Program Research Department Report No 19).

Webb, E. J., Campbell, D. T., Schwartz, R. D., and Sechrest, L. (1966). *Unobtrusive Measure: Nonreactive Research in the Social Sciences*. (Chicago: Rand McNally).

Wegel, R. L. and Lane C. E. (1924). The auditory masking of a pure tone by another, and its probable relation to the dynamics of the inner ear. *Physics Review*, **23**, 266–285.

Weiner, J. S. and Hutchinson, J. C. D. (1945). Hot humid environment: Its effect on the performance of a motor co-ordination task. *British Journal of Industrial Medicine*, **2**, 154–157.

Welford, R. T. (1976). *Skilled Performance: Perceptual and Motor Skills*. (Illinois Scott, Foresman and Co.).

White, R. M. (1975). Anthropometric measurements of selected populations of the world. In A. Chapanis (ed.) *Ethnic Variables in Human Factors Engineering*. (Baltimore: Johns Hopkins University Press).

White, W. J., Warrick, M. J. and Grether, W. F. (1953). Instrument Reading II: Check reading of instrument groups. *Journal of Applied Psychology*, **37**, 302–307.

Whitlock, F. A., Stoll, J. R. and Rekhdahl, R. J. (1977). Crisis, life events and accidents. *Australia and New Zealand Journal of Psychiatry*, **11**, 127–132.

Whitney, R. J. (1958). The strength of the lifting action in man. *Ergonomics*, **1**, 101–128.

Wilkinson, R. T. and Gray, R. (1975). *Effects of Duration of Vertical Vibration Beyond Proposed ISO 'Fatigue Decreased Proficiency' Time, on the Performance of Various Tasks*. AGARD-CP-145.

Wilson, A. (1963). *Noise—Final Report*. (London: HMSO).

Willis, F. N. (1966). Initial speaking distance as a function of the speakers' relationship. *Psychonomic Science*, **5**, 221–222.

Wing, J. F. (1965). Upper tolerance limits for unimpaired mental performance. *Aerospace Medicine*, **36**, 960–964.

Winsemius, W. (1965). Some ergonomic aspects of safety. *Ergonomics*, **8**, 151–162.

Wohlwill, J. F., Nasar, J. L., DeJoy, D. M. and Foruzami, H. H. (1976). Behavioural effects of a noisy environment: Task involvement versus passive exposure. *Journal of Applied Psychology*, **61**, 67–74.

Wolcott, J. H., McMeekin, R. R., Burgin, R. E. and Yanowitch, R. E. (1977a). Correlation of occurrence of aircraft accidents with biorhythmic criticality and cycle phase in US Air Force, US Army and civilian aviation pilots. *Aviation, Space and Environmental Medicine*, **48**, 976–983.

Wolcott, J. H., McMeekin, R., Burgin, R. E. and Yanowitch, R. E. (1977b). Correlation of general aviation accidents with the biorhythm theory. *Human Factors*, **19**, 283–293.

Wright, P. (1975). Forms of complaint. *New Behaviour*, **7 August**, 206–209.

Wright, P. (1977a). Decision making as a factor in the ease of using numerical tables. *Ergonomics*, **20**, 91–96.

Wright, P. (1977b). Presenting technical information: A survey of research findings. *Instructional Science*, **6**, 93–134.

Wright, P. (1978). Feeding the information eaters: Suggestions for integrating pure and applied research on language comprehension. *Instructional Science*, **7**, 249–312.

Wright, P. and Barnard, P. (1975a). Effects of 'more than' and 'less than' decisions on the use of numerical tables. *Journal of Applied Psychology*. **60**, 606–611.

Wright, P. and Barnard, P. (1975b). 'Just fill in this form'—a review for designers. *Applied Ergonomics*, **6**, 213–220.

Wright, P. and Fox, K. (1970). Presenting information in tables. *Applied Ergonomics*, **1**, 234–242.

Wright, P. and Fox, K. (1972). Explicit and implicit tabulation formats. *Ergonomics*, **15**, 175–187.

Wright, P. and Reid, F. (1973). Written information: Some alternatives to prose for expressing the outcomes of complex contingencies. *Journal of Applied Psychology*, **57**, 160–166.

Wrogg, S. G. (1961). The role of emotions in industrial accidents. *Archives. of Environmental Health*, **3**, 519.

Wyburn, G. M., Pickford, R. W. and Hirst, R. J. (1964). *Human Senses and Perception*. (London: Oliver and Boyd).

Wyndham, C. H. (1966). A survey of the causal factors in heat stroke and of their prevention in the gold mining industry. *Journal of the South African Industry of Mining and Metallurgy*, **1**, 245–258.

Wyndham, C. H., Strydom, N. B., Benade, A. J. S. and van Rensburg, A. J. (1970). Tolerance times of high wet bulb temperatures by acclimatized and unacclimatized men. *Environmental Research*, **3**, 339–352.

Yaglou, C. P. (1947). A method for improving the effective temperature index. *ASHVE Transactions*, **53**, 307–326.

Zeff, C. (1965). Comparison of conventional and digital time displays. *Ergonomics*, **8**, 339–345.

Ziegenruecker, G. H. and Magid, E. B. (1959). *Short time tolerance to sinusoidal vibration*, Wright Air Development Center Technical Report No 59–391.

Author Index

Subject Index